MW00911914

THE NEW NORMAL

THE NEW NORMAL

The Canadian Prairies in a Changing Climate

EDITED BY

David Sauchyn, Harry Diaz and Suren Kulshreshtha

University of Regina

Printed and bound in Canada at Friesens.
The text of this book is printed on 100% post-consumer recycled paper with earth friendly vegetable-based inks.

Cover and text design: Duncan Campbell, CPRC.
Editor for the Press: Brian Mlazgar, CPRC.

COVER PHOTOS: FRONT COVER: "High angle view of snow melting in a field" by Harri Tahvanainen, Gorilla Creative/Getty Images. BACK COVER: "Patch of grass in thawing snow" by Image Source/Getty Images.

Library and Archives Canada Cataloguing in Publication

The new normal : the Canadian prairies in a changing climate / edited by David Sauchyn, Harry Diaz, and Suren Kulshreshtha.

(University of Regina publications, 1480-0004 ; 27)
Includes bibliographical references and index.
ISBN 978-0-88977-231-1

1. Climatic changes—Economic aspects—Prairie Provinces.
2. Climatic changes—Social aspects—Prairie Provinces.
3. Climatic changes—Environmental aspects—Prairie Provinces.
I. Sauchyn, David J. II. Diaz, Harry P. III. Kulshreshtha, Surendra N. (Surendra Nath), 1941- IV. University of Regina. Canadian Plains Research Center V. Series: University of Regina publications ; 27.

QC903.2.C3N49 2010 577.2'209712 C2010-906148-9

10 9 8 7 6 5 4 3 2 1

Canadian Plains Research Center
University of Regina
Regina, Saskatchewan, Canada, S4S 0A2
TEL: (306) 585-4758 FAX: (306) 585-4699
E-MAIL: canadian.plains@uregina.ca WEB: www.cprcpress.ca

We acknowledge the financial support of the Government of Canada through the Canada Book Fund for our publishing activities.

Canadian Heritage Patrimoine canadien

CONTENTS

3. STUDIES IN IMPACTS AND ADAPTATIONS

FOREWORD

As I write this, it appears that our political leaders will do nothing in the foreseeable future to substantially reduce the rate of climate warming. The much-anticipated climate summit in Copenhagen, Denmark yielded almost nothing, and prospects for international agreement on controlling greenhouse gases at the G8 and G20 summits in June of 2010 look very dim, despite a budget for the "summits" of over $1 billion. Politicians show no leadership, continuing to squabble over which country must lead the necessary reduction in global greenhouse gas emissions, literally fiddling while the planet burns.

Public understanding of the urgency of the climate warming problem is equally poor. Many consider a few leaked emails taken out of context to be evidence that climate warming is a scientific hoax, despite considerable evidence that is visible to even lay people with rudimentary computer skills. In contrast to the public view that climate change has been exaggerated, many of the papers that have crossed my desk in the past few years indicate that climate warming is causing Earth to change much more rapidly than the Intergovernmental Panel on Climate Change (IPCC) predicted. One need not be a scientist to see some of the most obvious under-predictions: sea ice and glaciers are disappearing more rapidly than predicted, as anyone familiar with Google Earth or accessing government websites can discern. The ravages of the mountain pine beetle, and the rapidity of sea level rise were similarly not predicted by the IPCC. Taken together, the underestimated effects of climate warming and the continued procrastination of politicians must cause us to increasingly consider options for adaptation to, rather than prevention of, climate warming.

Few large and important biomes are likely to be as severely affected by climate warming as the Canadian prairies. No 20th century prairie farmer will claim that he ever had too much water to grow his crops. Farming survived on the prairies only by making extreme modifications to the land-use practices brought by European settlers. Even so, farming was abandoned in the driest parts of the prairies. We now know from recent paleo-ecological studies, some by the authors in this book, that the 20th century on the prairies was unusually wet by comparison with most centuries of the last few millennia. Droughts, some lasting a decade or more, haunted the prairies in the distant past, making that of the "Dirty Thirties" look puny by comparison.

Finally, all but two years of the past ten have had lower than normal precipitation for much of the western prairies. Soil moisture levels in some areas are at record lows. Could we be entering one of the mega-droughts of centuries past? It is very possible.

Much of the agriculture in drier parts of the prairies, and most of the larger cities, such as Calgary, Edmonton, and Saskatoon, rely heavily on water from the "water towers" of the Rocky Mountains, where snow packs and glaciers have provided an adequate, if not abundant, supply to support prairie civilization. Yet all signs are that these sources are rapidly disappearing. There is less snow. Shorter and warmer winters allow periodic melting that steals water in midwinter that once accumulated in snow packs, which then that recharged rivers at the very time agriculture needed it. More winter precipitation falls as rain, which seeps away, rather than accumulates. Earlier snowmelts threaten to disconnect the prairies from traditional water sources. Longer, drier summers will see evapotranspiration steal away more of the water, and dwindling glaciers, often the "ace in the hole" that allowed prairie farmers and fish to survive hot, dry, late summer periods, will be gone.

The gloomy background that I describe above makes this book, with its focus on adaptation, look prescient. It is one of a very few works that treats adaptation in a thorough way, and the only comprehensive work on the topic for the western prairie provinces. The authors are some of the most experienced scholars in the small academic community of the prairies.

The collection of papers in this book paint a picture of the prairie world of the next century that would be unrecognizable to people who are alive today. Positive effects of the warming will be few, and disadvantages many. Agriculture will be much diminished, with the advantages of a warmer and longer growing season more than offset by a lack of water. Prairie-forest boundaries will be pushed far to the north, largely the result of increased fires, some following extensive insect outbreaks. Precipitation events will be fewer, but more intense. Transportation by water, and tourism and recreation focused on water resources, will be almost non-existent. New institutional arrangements will be needed to deal with the changes in occupations, health, and other social issues that will face prairie generations in 100 years. The necessary adaptations will have to be aggressive, and unfamiliar enough to be quite risky.

The chapters of the book are lucidly written and understandable to a lay audience. I recommend it highly as a reality check on what we could have largely prevented by simple mitigation measures taken as late as the turn of the millennium, and as a call for actions that must be taken if *Homo sapiens* is to continue to thrive in Canada's western prairies.

Dr. David Schindler
University of Alberta
June 2010

PREFACE

This book is the product of two significant interdisciplinary efforts. The first is the contribution of a group of Prairie scholars to the Government of Canada's report, *From Impacts to Adaptation: Canada in a Changing Climate 2007,* released in March 2008. This major report was the product of Canada's second national assessment of climate change impacts and adaptation. The first assessment, the *Canada Country Study,* was released in 1997. Much of the observation, detection and understanding of climate change and its impacts have occurred since 1997. The second assessment report involved the synthesis and interpretation of a considerable amount of information—more than 3,000 studies, most published since 1997. Because the impacts of climate change, and the appropriate adaptation strategies, differ among regions, the recent national assessment report has a regional format. Chapter 7, "The Prairies," represents the synthesis and expert interpretation of almost 400 studies by 14 authors, and the distillation of a 54-page chapter from a much longer draft. Therefore, following the release of the national assessment, we restored the original draft prairies assessment, updated the content, and invited contributions from more authors. The second interdisciplinary effort was the Institutional Adaptations to Climate Change (IACC) project, an international project that brought together 18 scholars from Canada and Chile. A significant component of the IACC project involved the study of the institutional framework of rural vulnerabilities in the South Saskatchewan River basin, a watershed extending across the three Prairie Provinces. Many of the results of the IACC project are found in several of the chapters of this book, The result is *The New Normal: The Canadian Prairies in a Changing Climate,* a comprehensive and up-to-date overview of climate change in the Prairie Provinces; the impacts on natural resources, communities, human health and sectors of the economy; and the adaptation options that are available for alleviating adverse impacts and taking advantage of new opportunities provided by a warmer climate.

Given the origins of this book, much of it reflects the frameworks and approaches of the national assessment process and the IACC project. Therefore we wish to express our appreciation to the staff of the assessment Secretariat in the Climate Change and Impacts Division of Natural Resources Canada, and in particular the lead author of *From Impacts to Adaptation,* Dr. Don Lemmen. Don and his staff, notably Fiona Warren who coordinated the peer review and technical editing of the prairies chapter, set high standards and are largely responsible for quality of the core content of this book. We also would like to thank our colleagues and research fellows of the IACC project, as well as its staff and technicians for their efforts and contributions, For the quality of the presentation, we have the staff of CPRC Press to thank, and especially Brian Mlazgar, the Publications Manager and editor of this book, and Duncan Campbell, for his cover and text design.

David Sauchyn, Harry P. Diaz and Suren Kulshreshtha
October 2010

REFERENCE

Lemmen, D.S., F.J. Warren, J. Lacroix and E. Bush (eds.). 2008. *From Impacts to Adaptation: Canada in a Changing Climate 2007.* Ottawa: Government of Canada.

THE NEW NORMAL

CHAPTER 1

INTRODUCTION

*Harry P. Diaz, David Sauchyn
and Suren Kulshreshtha*

The present climate crisis shows no signs of abatement. On the contrary, it is becoming more intense and increasing uncertainties and fears about the outcome of the process of climate transformation. The complexity of this transformation is, to a large extent, a product of the intricate relationships between natural and social systems, both of them complex systems on their own. There is a strong consensus that climate change is the product of human activities that raise the concentrations of greenhouse gases (GHG) and, consequently, increase global average temperatures. Global warming has led to regional climate changes, which are increasingly impacting the dynamics of local natural and social systems and their interrelationships. These impacts, which are expected to increase, could bring serious risks and damages to ecosystems and livelihoods, disrupting the precarious balance between people and their environment (Parry et al., 2007).

The natural and social systems of the Canadian Prairies are vulnerable to climate change. Mid-latitude continental interiors have among the highest rates of recent and projected warming and climatically-sensitive dryland ecosystems. The Prairies cover one fifth of Canada's surface area and contain 17% of the national population, extending east from the Rocky Mountains to Hudson Bay. The Prairie Ecozone, the largest natural region in the Prairie Provinces, is the most extensively modified region of the country with more than 80% of Canada's agricultural land and active mining and energy sectors. Throughout Canada's largest dryland, seasonal and prolonged water deficits strongly influence human activities. Temperature records for the region show increases in mean annual temperature since the 1970s, and all the projections indicate that temperatures will continue to increase even more in the future. We already have the capacity to anticipate where this trend is taking us. On the one hand, it could offer significant opportunities, such as a warmer and longer growing season and enhanced productivity of forest, crops and grassland. On the other hand, the trend could also generate significant risks, such as water scarcities due to water loss by evapotranspiration, or more intense extreme events in the form of droughts and rainstorms (Sauchyn and Kulshrestha, 2008). There is, of course, a vaster territory of uncertainties and challenges related to the complexities of climate systems and the intricate dimensions of human responses, where questions of vulnerability and social justice acquire a central relevance (Smit, 2006; Timmons Roberts, 2009).

How are we dealing with this complex climate crisis in the Canadian Prairies? Responses from social actors have been unequal and uneven, as they have been in other parts of the world. To a large extent this attitude has been informed by what Giddens (2009) called a paradox: given the intangibility of the process and the spurious argument that climate change is something of the future, present societies are reticent to do something.

Western Canadians have been slow in responding to the challenge of climate change, but the initial generalized skepticism has diminished considerably. A recent survey of western attitudes revealed that almost 8 in 10 respondents believe that climate change is a serious problem, and define as a priority the reduction of greenhouse gas emissions. This awareness of climate change and of its causes is somehow limited by a desire to maintain the economic path of development: almost 60% of the respondents indicated that the reduction of GHG emissions should not be made at the expense of the economy (Berdahl, 2008).

Clearly, for the majority of Western Canadians, climate change and its impacts are not immediate problems and, as such, there is no urgency in resolving the climate crisis. However, there are sectors with a more direct experience with global change. Anecdotal evidence shows that those more exposed to the uncertainties of climate variability, such as farmers and ranchers, seem to have an increasing awareness and interest in climate change.

Governments' responses have also been limited, failing to provide an effective and comprehensive response to the climate crisis. Despite the fact that Canada signed the Kyoto agreement, it has failed to reduce emissions as agreed. Furthermore, federal and provincial governments are still far from developing a comprehensive climate policy approach that integrates mitigation and adaptation efforts, and even farther from developing a proper climate policy integration that facilitates the mainstreaming of climate policy into a variety of fields, as stated in Chapter 12.

There are, however, indications that a political change is taking place. There are an increasing number of policy-makers and government agency representatives that recognize the risks of the climate crisis and the urgency to do something about it. Some government agencies, such as the Agri-Environmental Services Branch of the federal Ministry of Agriculture (formerly the Prairie Farm Rehabilitation Administration), have assumed a position of leadership in responding to the crisis. A recent initiative is Natural Resources Canada's Regional Adaptation Collaborative (RAC) program, oriented to increase the capacity to integrate climate change into existing planning processes. The process of creating the Prairies RAC has advanced cooperation within and between the three provincial governments towards planning and implementing adaptation to climate change.

The most organized response to the climate crisis has been from a large number of researchers based at universities, government agencies and NGOs. Recognizing the seriousness of the climate crisis and the urgency to ameliorate the associated risks and hazards, prairie scientists have made a significant effort to increase their understanding of the causes of climate change, its trends, the associated impacts, and the capacities that are required to manage the risks and opportunities offered by a new climate regime. Climate research has been the domain of natural scientists, but the increasing recognition that climate change cannot be separated from the dynamics of society has resulted in the engagement of a wider range of scientists. Thus, in recent years, there has been a flurry of activities in numerous disciplines and practice areas to add their efforts to a comprehensive understanding of the climate transformation, expanding the disciplinary approach into the social sciences and the humanities. These efforts have taken place not only at the level of individual scientists, but also in a more institutionalized form. The emergence of informal networks that bring together scientists from different disciplines and the creation of research organizations focused on climate change, such as the Prairie Adaptation Research Collaborative, are indicators of the institutionalization of climate research.

This book represents another step in the process of scientists responding to the climate crisis. It brings together a group of 22 scholars who through 23 chapters present their diverse knowledge and expertise about climate change and its impacts in the Prairie Provinces. These scholars represent various disciplines, ranging from climatology to sociology, reflecting the complexity of the scientific effort in the Canadian Prairies. We believe that this effort is an important contribution to understanding the multiple dimensions of the crisis and seeking solutions to the problems, a belief that has been a central motivation for producing this volume.

THE ORIGINS OF THE BOOK

As indicated in the Preface, this book has its foundations in the work produced by 14 Prairie scholars for Canada's second national assessment of climate change and adaptation (Lemmen et al., 2008). This work involved a significant effort of interpretation, systematization, and synthesis of more than 300 prairie-climate related studies, most published since 1997. The researchers, representing a wide spectrum of disciplines, met in several workshops to discuss the prairies' contribution to the second national assessment and the progress of the work. The process built on the experiences and the leadership of the two project leaders and co-authors of this book: David Sauchyn and Suren Kulshreshtha. The result of this interdisciplinary and collaborative process was the accumulation of a significant amount of material, from which only a small portion was published in the assessment as a 50-page chapter (Sauchyn and Kulshreshtha, 2008). This book presents an updated version of most of the national assessment material produced by this network of scholars, providing systematic information about climate change impacts and climate adaption in Alberta, Saskatchewan, and Manitoba. To complement the core of the book several chapters focused on specific climate relevant issues—many of them completed since the publication of the national assessment—have been added. Many of these additional chapters are the product of several studies carried out in the context of the Institutional Adaptation to Climate Change (IACC) project, an interdisciplinary project focused on the adaptive capacities of formal institutions and their role in the distribution of rural vulnerabilities. The project, supported by the Social Sciences and Humanities Research Council of Canada, followed a vulnerability approach assessing the present and past adaptive capacities of communities and governance networks in the context of the future climate conditions expected for the South Saskatchewan River Basin, a watershed that covers most of the southern parts of Alberta and Saskatchewan (IACC, 2009). The three editors of this book and many of the contributors actively participated in the project. The book is, in this way, the most comprehensive effort to present the 'state of art" of climate research in the Canadian Prairie region.

THE CONTENTS

The 23 important contributions to this book are divided into three sections. Section 1, "The Setting," has three chapters that set the scene for the contributions that compose the other two sections. It establishes the framework that defines the parameters in which the different articles in the next two sections should be understood. We start with the chapter of Suren Kulshreshtha and Harry Diaz (Chapter 2), who provide a socio-economic description of the region, focusing on the present and future aspects of two dimensions that are strongly linked to the climate vulnerabilities of the region: major population shifts brought about by global changes and the path of economic development pursued in the region. Thus, this contribution provides the prairie socio-economic scene where climate change is taking place. The next two chapters provide the climate scene. In Chapter 3, David Sauchyn discusses the natural variability of prairie climate, focusing on the driest region, southern Alberta and southwest Saskatchewan. Because weather station and water level records in the region are from the period affected by anthropogenic global warming, and because they are short relative to some climate cycles, he examines the proxy record of the prairie climate of the past millennium to reveal the natural climate variability that underlies the trends imposed by global warming. The final chapter of the first section, from Elaine Barrow (Chapter 4), provides a brief discussion of climate change scenarios and presents future scenarios for the Prairies. She constructed these scenarios for the two main ecozones in the prairie region, namely grassland and forest, using data from climate change experiments undertaken with seven Global Circulation Models (GCMs) and several Special Report on Emission Scenarios (SRES).

Most of the chapters found in the second section of the book, "Major Themes in Impacts and Adaptation," are the direct product of the work done for the national assessment. They discuss the vulnerabilities of both natural and social systems to the new climate conditions, as well as of the adaptive mechanisms that could be used to reduce these vulnerabilities. The section begins with the chapter about water, the most critical and vital resource in the Prairies. James Byrne, Stefan Kienzle and David Sauchyn offer their perspective on the impacts of climate change on water resources and aquatic ecosystems in the Prairie Provinces (Chapter 5). Their discussion of these impacts on the water cycle includes the effects of changing precipitation and temperature, the consequences for rivers and watersheds, extreme weather events, changes to water quality, and adaptation of water management to cope with the climate change impacts on water resources. Norman Henderson and Jeff Thorpe discuss the impacts of current and future climate change on the ecosystems of the Prairie Provinces (Chapter 6). As they state, climate change has the potential to significantly modify the Prairies' ecosystems, a change that will directly impact on the sustainability of our livelihoods. Following their discussion of the impacts, Henderson and Thorpe present the range of limited adaptation options. In the next chapter, Elaine Wheaton and Suren Kulshreshtha extend the natural resource theme to discuss the impact of climate change on agriculture (Chapter 7), one of the most important economic sectors in the three Prairie Provinces. Given its dependency on climate and the scarcity and variability of water resources, agriculture is highly vulnerable. In this context, the authors present and discuss estimates of the main impacts of future climate change on agriculture in the Canadian Prairies, as well as adaptations and vulnerabilities.

Forestry is also an important economic activity in the three provinces and, as with the case of agriculture, it is strongly affected by climate and climatic variability. Mark Johnston and Tim Williamson (Chapter 8) discuss both the short- and long-term vulnerabilities of forests and forest management and the potential forms of adaptation that could be adopted to ensure the resiliency of forest resources, as well as some of the key uncertainties of the sector. Danny Blair focuses his contribution (Chapter 9) on the transportation network in the Prairie Provinces, one of the most extensive and diverse in Canada. Some components of the network, which include not only roads and rails, but also a substantial number of winter roads, dozens of airports, ferries and boat landings, and a single ocean port, will benefit from warmer temperatures, while others will suffer negative consequences. Blair discusses not only the vulnerabilities of the network, but also the need to improve the existing level of adaptive capacity to face the new challenges.

The focus of the section then shifts to the social dimensions. Debra Davidson, following the vulnerability approach, discusses the exposure, sensitivity and adaptive capacity of prairie human settlements to climate (Chapter 10). Starting with a discussion of the geography of the social landscape, she proceeds to examine vulnerability at different levels: the individual, the household, cities, and small towns, including agricultural and forest communities, resort-based towns and First Nations. Justine Klaver-Kibria (Chapter 11) takes us into an analysis of the expected impacts of climate change on human health. Her contribution addresses four significant issues: first, the impacts from the external environment, as it is modified by climate change, on people's health; second, examining the different degrees of vulnerability of segments of the prairie populations; third, the adaptive capacity of the Prairie Provinces as it pertains to health; and fourth, gaps in knowledge in the area. The last contribution in the second section, from Margot Hurlbert and Darrell Corkal (Chapter 12), deals with the complex relationships between climate as a policy issue and governments. The chapter discusses the role of different federal and provincial government agencies in relation to climate change, the climate policies pursued by these levels of government, and the areas that need to be strengthened to facilitate the development of an adaptive capacity that could build resilience and help society cope with climate change and climate variability.

The last major section of the book, "Studies in Impacts and Adaptation," consists of a variety of studies on the specific impacts of climate on nature and society, and the challenges of adapting to the present and future climatic conditions, providing the reader with a good overview of the present range of climate research that has been carried out by prairie scholars in recent years. The first contribution (Chapter 13), from Johanna Wandel, Jeremy Pittman, and Susana Prado, discusses the issue of rural community vulnerabilities with empirical illustrations selected from a series of case studies conducted by the Institutional Adaptation to Climate Change (IACC) Project, an interdisciplinary project focused on the adaptive capacities of formal institutions and their role in the distribution of rural vulnerabilities. This project included several in-depth community vulnerability assessments in the South Saskatchewan River Basin of southern Alberta and Saskatchewan. This paper brings together insights from the assessments of three rural communities—Taber and Hanna, Alberta, and Outlook, Saskatchewan—focusing on the exposures/sensibilities and adaptive strategies that prevail in these settings. The contribution from Ryan MacDonald, James Byrne and Stefan Kienzle (Chapter 14) deals with the vulnerability of a sensitive watershed. Water supply in the Prairies is highly dependent on snowmelt from the eastern slopes of the Rocky Mountains. The authors assess, in the context of future climate change scenarios, potential changes in mountain snowpack in the St. Mary watershed in southern Alberta. As expected, changes could be significant, and proper adaptive strategies will require changes to water management in the basin. Chapter 15, authored by Suren Kulshreshtha, Virginia Wittrock, Lorenzo Magzul and Elaine Wheaton, brings us back to the dimension of social vulnerabilities. Their contribution focuses on the vulnerability of a First Nation community, the Kainai Blood Indian Reserve, in southern Alberta. This study, also done in the context of the IACC project, analyzes the impact of climate hazards, such as drought and floods, on this Aboriginal community, along with the exposures, sensitivities, and the technical and social adaptive capacities of the members of the community.

The next three chapters are focused on water resources, one of the Prairies' resources most seriously affected by a new climate regime. The work of Elaine Wheaton, Suren Kulshreshtha and Virginia Wittrock (Chapter 16) focuses on a specific extreme climate event, the 2001–2002 drought, and its biological, physical, and socio-economic impacts on the Prairie Provinces. All the projections for the region indicate that climate change increases the future probability of more prolonged and intense droughts. In these terms, the authors' assessment of the 2001–2002 drought shows the potential future impacts of droughts and the gaps in adaptive capacity that must be addressed. One of the forms of adaption that could reduce the impacts of droughts on agricultural activities is irrigation. The contribution from David Sauchyn and Suren Kulshreshtha (Chapter 17) is a review of irrigation activity in the Prairie Provinces. It encompasses the present extent of irrigation in the three provinces, its biophysical and socio-economic impacts, and water use issues, including improving water use efficiency, as well as the relevance of irrigation in the context of the future climate change scenarios. The next contribution from Margot Hurlbert, Darrell Corkal and Harry Diaz (Chapter 18) extends the discussion of water resources in the Prairies to the institutional setting established by government institutions and their water policies in Alberta and Saskatchewan, as they apply to a changing climate. The authors pay particular attention to water policy strategies, drought planning, and the integration of civil society organizations into water governance, complementing the arguments developed in the previous two chapters.

The contribution from Norman Henderson, Elaine Barrow, Brett Dolter and Edward Hogg (Chapter 19) moves the discussion into impacts of climate change on a set of unique and valuable island forests in the northern Great Plains. Based on this impact assessment, the authors outline the options, including recommendations, for the management of these impacts. Danny Blair and David Sauchyn (Chapter 20) discuss the issue of winter roads, which are vital lifelines for the northern communities; these roads are dependent on ice thickness and texture, so the expected increase in temperature is a serious issue for their viability. The chapter is based mostly on a study of the relevance of ice roads for five First Nations in northern Manitoba, showing how the indirect impacts of climate change could affect the quality of life of isolated communities. The next contribution from Jeremy Pittman (Chapter 21) also deals with Aboriginal communities. The chapter is based on a community vulnerability assessment of two First Nations, Shoal Lake and James Smith, both of them in northern Saskatchewan. This assessment, based on a vulnerability approach, discusses the exposure/sensitivities of both groups and their existing coping capacity. The chapter shows the limitations on adaptive capacity imposed by limited economic resources, and the urgent need to improve the institutional support to isolated First Nation communities.

Mark Johnston (Chapter 22) focuses on the impacts of climate change on the areas of tourism and recreation. The author shows how the impacts of climate change on ecosystems could have both positive and negative impacts, redefining many of the activities that are included in tourism and recreation, such as hunting, park visitation, swimming, fishing, boating, canoe-tripping and whitewater activities, in different areas of the Prairie Provinces. The next chapter (Chapter 23), from David Sauchyn, deals with the sensitivity of a significant form of natural capital, prairie soil landscapes, to changes in the climate. The new climate regime may bring increasing erosion and expose the landscape to a process of desertification, with negative consequences for agriculture and ecosystems. The last contribution, from Mark Johnston and Tim Williamson, discusses the potential impacts of climate change on the island forests of central Saskatchewan. The authors argue that these small forests are likely to be affected in the future. The warmer climate will expose them to pests, and increasing soil moisture deficiency could reduce tree growth. Forest management, they argue, could reduce these vulnerabilities, but only to a certain extent.

The contributions in this volume are not exhaustive or comprehensive. There are many other scholars working in the field, whose efforts have not been represented in this book. However, this collection of chapters provides the reader with a fair representation of the variety of interests, concerns, and preoccupations of a wide range of scientists involved in climate research. We certainly expect that the reading of these multiple contributions will stimulate an enlightened response to the climate crisis for the Canadian prairie region.

REFERENCES

Berdahl, L. 2008. *Hot Topics: Western Canadian Attitudes Toward Climate Change.* Canada West Foundation Report (accessed October 12, 2009. Available from the World Wide Web: *www.cwf.ca/V2/cnt/publication_200804101002.php.* 14 pp.

Giddens. A. 2009. *The Politics of Climate Change.* Polity Press: Cambridge.

IACC. 2009, *Integration Report. The Case of the South Saskatchewan River Basin, Canada,* Canadian Plains Research Center Press, Regina. Available at www.parc.ca/mcri/pdfs/papers/into1.pdf

Lemmen, D.S., F.J. Warren, J. Lacroix and E. Bush (eds.). 2008. *From Impacts to Adaptation: Canada in a Changing Climate 2007.* Government of Canada, Ottawa. 448 pp.

Parry, M.L., O.F. Canziani, J.P. Palutikof, P.J. van der Linden and C.E. Hanson (eds.). 2007. "Summary for Policymakers." Pp. 7–22 in *Climate Change 2007: Impacts, Adaptation and Vulnerability. Contribution of Working Group II to the Fourth Assessment Report of the Intergovernmental Panel on Climate Change.* Cambridge University Press: Cambridge.

Sauchyn, D. and S. Kulshreshtha. 2008. "Prairies." Pp. 275–328. in D.S. Lemmen, F.J. Warren, J. Lacroix and E. Bush (eds.). *From Impacts to Adaptation: Canada in a Changing Climate 2007.* Government of Canada: Ottawa.

Smit, B. 2006. "Where from and Where to: Climate Change Impacts and Adaptation Research and Practice." In E. Wall, M. Arsmtrong and S.D. Manityakul (eds.). *Climate Change and Canadian Society. Social Science Research Issues and Opportunities.* C-CIARN Symposium Report, University of Guelph: Guelph.

Timmons Roberts, J. 2009. "Climate Change: Why the Old Approaches aren't Working," in K. Gould and T. Lewis (eds.). *Twenty Lessons in Environmental Sociology.* Oxford University Press: New York.

1. THE SETTING

CHAPTER 2

SOCIO-ECONOMIC DESCRIPTION OF THE PRAIRIE REGION

Suren Kulshreshtha
and Harry P. Diaz

INTRODUCTION

The prairie region of Canada is one of the most diverse regions in Canada. Each of the three Prairie Provinces is diverse, both in terms of population trends as well as socio-economic development status. On account of its heavy dependence on primary production, it is more vulnerable to changes imparted by changing climatic patterns. However, the vulnerability of the region is affected not only by climate conditions, but also by many other internal and external factors. Climate and non-climate conditions affect each other in complex ways, increasing or reducing the exposure and the adaptive capacities of the multiplicity of systems that constitute the Prairies. This chapter focuses on two non-climate factors that are strongly linked to the climate vulnerabilities of the region: major population shifts brought about by global changes and the path of economic development pursued in the region. The impacts of climate change upon the region and its ability to adapt to the new challenges—successfully facing adverse impacts and taking advantage of new opportunities created by climate change—will be affected by the intricate relationships between climate, population patterns, and economic strategies.

In this chapter, demographic changes are reviewed, including projections of the population in the three Prairie Provinces, along with employment patterns and major economic activities. This is followed by an assessment of the nature of vulnerability of the region to climate change based on the present and future changes in population and its composition.

CURRENT DEMOGRAPHIC AND ECONOMIC DESCRIPTORS FOR THE PRAIRIE REGION

In 2006, the prairie region had over 5 million people (approximately 17% of the Canadian population). Prairie population over the last 50 years has doubled (Table 1), but this is not shared equally among the three provinces. This uneven growth has resulted in concentrated population growth in Alberta, while the populations of Manitoba and Saskatchewan have been almost stagnant. Bernard (2007) suggests that although study of population change starts with basic counts (e.g. number of births, deaths, migration), the notion of transition is closely tied to these counts. Understanding these transitions involves understanding that people's lives are longitudinal, multifaceted, linked, and embedded in various social contexts. Thus, population changes, together with economic changes, are major determinants of how transitions and life courses are shaped in the future. These various processes at play in the life course of individuals are, in the aggregate, the causes of population changes.

Particulars	Unit	1951	1975	2001	2006	% Change 2006 over 1951
Total Population	'000	2,549	3,781	5,225	5,539	217.3
Percent of Canada	%	18.2	16.2	16.8	16.9	—
Dependent Population	'000					
Under 15 years		765	1,016	1,158	1,055	137.9
Over 65 years		200	347	614	679	329.5
Total		965	1,383	1,772	1,734	179.7
Dependent Population Percent of Total	%	37.9	36.0	33.9	31.3	—
Net International Migrants	'000	25.1	21.4	11.7	27.2	—
Employment	'000	955	1,757	2,662	2,593	271.5
Real Gross Domestic Product	Bill. $	—	—	$189	$259	—
Real GDP per capita	$	—	—	$36,172	$46,759	—

Table 1. Salient Economic Indicators for the Prairie Economy, Selected Years
Source: Statistics Canada–CANSIMa for GDP, and Statistics Canada; CANSIMb for population.

POPULATION TRENDS

Population in Alberta has nearly tripled since 1951, reaching a total population size of approximately 3.4 million people by 2006. In fact, according to Statistics Canada (2007), the number of Albertans increased by 10.6% between 2001 and 2006, double the national growth rate (+5.4%). However, since 1961, the Manitoba population has increased by an average annual rate of only 0.5%, reaching a total population size of a little over 1 million in 2006. Unlike the other Prairie Provinces, the total population in Saskatchewan has remained at approximately 1 million over the past 50 years, with the strongest population growth of approximately 11% between 1951 and 1961, and decreasing toward the end of the century—a reduction of 4.3% between 1986 and 2006. Alberta now houses almost two-thirds (61%) of the region's population against only slightly over a third in 1951 (Figure 1). These trends are expected to continue into the future.

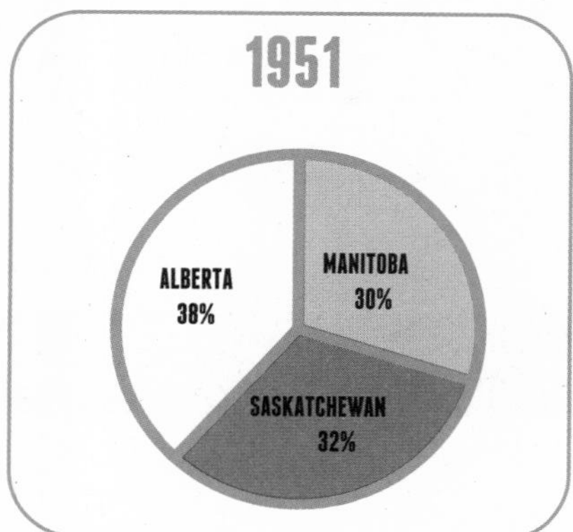

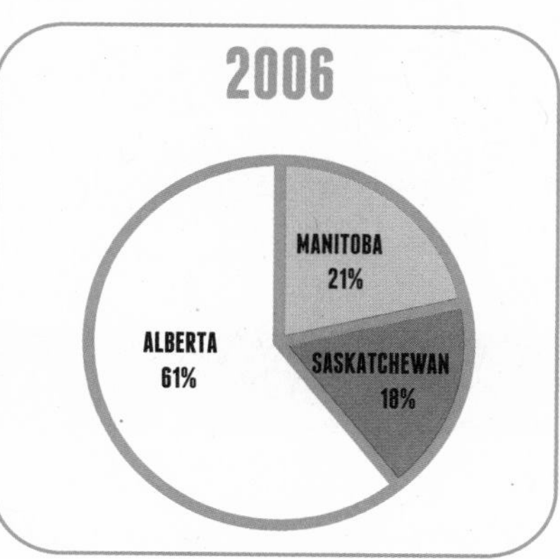

Figure 1. Relative Distribution of Population in the Prairie Region, by Province, 1951 and 2006.

The region has been characterized by a high working force relative to its population, as measured in terms of ratio of dependent to the rest of the population. In 1951, dependent population (those under 14 and over 65 years of age) was almost 38% of the total. By 2006, the region saw a decrease in dependency to 31%. In-migration of young people in certain parts of the region has contributed to this change. Again, these features are not distributed equally among all provinces, as shown in Appendix Table A.1. Alberta had the lowest dependent population in 2006 and has shown a larger decline in such population. This is perhaps indicative of migration of younger working age people in the province.

Components of Population Change (2003–2006)

Changes in the population of the prairie region can be explained through some of the changes in the composition of the population. Table 2 shows that the decline in Saskatchewan's population during this period has resulted from the net inter-provincial out-migration exceeding the small increases obtained through natural population growth and net international immigration. Although Manitoba has also experienced a large portion of net inter-provincial out-migration, this does not exceed the province's natural increase in population growth and net international in-migration. Unlike Saskatchewan and Manitoba, Alberta experienced population growth resulting from increases in all three components of population growth, particularly net inter-provincial immigration which accounted for almost half of its average annual increase in total population size of 2.2% during this period. In fact, Alberta was the only Canadian province where natural increase did not decline between 2001 and 2006 (Statistics Canada, 2007).

Province	Total Change in Thousands (Three Year Total)				Average Annual % Increase
	Natural Growth	Net International Migration	Net Inter-provincial Migration	Total	
Manitoba	11.9	19.9	-18.4	13.0	0.45%
Saskatchewan	8.7	4.5	-23.1	-9.9	-0.13%
Alberta	65.0	38.6	102.1	205.7	2.21%

Table 2. Components of Population Change, 2003–2006. Source: Statistics Canada–CANSIMb.

Age Structure of Population (1996–2006)

Growth in the population of different Prairie Provinces impacts their respective age structure. Although, in 2006, Canada's median age was higher than that for the three provinces (at 38.8 years), Saskatchewan's median age was relatively high (at 37.7 years) within the three Prairie Provinces. Alberta's population in 2006 had the lowest median age at 35.5 years, making its population the youngest in the Prairie Provinces, as well as among the ten Canadian provinces (Table 3).

Region	Median Age (in Years)	Dependant Population (% of Total Population)		Working Age Population (% of Total Population)
		0–19 years	65+ years	20–64 years
Canada	38.8	24.0	13.2	62.8
Alberta	35.5	26.0	10.4	63.6
Manitoba	37.3	26.6	15.3	59.9
Saskatchewan	37.7	27.0	14.9	58.1

Table 3. Age distribution, and Dependent and Working Population, Canada and the Prairie Provinces, 2006. Source: Statistics Canada–CANSIMc.

The aging of a population is nothing new in the Prairies, but the significant "aging at the top" is a phenomenon of the present (Beaujot, McQuillan and Ravanera, 2007). The Prairie Provinces have entered a stage where the baby-boomer cohort moves beyond childbearing and on to retirement, resulting in somewhat significant aging at the top. These changes have implications for health care and pensions, as well as for the potential impacts of warmer weather upon population health.

Further focusing on the age structure of the Prairie Provinces in 2006, Saskatchewan had the smallest working-age population, with 58.1% of its residents between the ages of 20 and 64. Alberta, in contrast, had the largest working-age population, with 63.6% of its population aged 20 to 64. Manitoba had a working-age population of 59.9%. Some of this is a result of working-age migration out of Manitoba and Saskatchewan into Alberta, and lack of immigration relative to other regions of Canada (such as Ontario, Quebec and British Columbia).

Rural-Urban Population Shifts

Population dynamics are also related to the distribution of the population within each one of the provinces. In 2006, approximately 30% of Saskatchewan's population, 28% of Manitoba's population and 17% of Alberta's population lived in predominantly rural regions. Saskatchewan was the only prairie province that suffered a decline (-7.2%) of its population from these rural areas between 1991 and 2006. Alberta, in contrast, reported the highest population gain (+14.7%) in its predominantly rural region, while Manitoba experienced a population increase of 7.5% in this region during the 1991–2006 period.

Looking specifically at the natural component of population change in predominantly rural regions of the Prairie Provinces, both Saskatchewan and Manitoba experienced declines in their natural population growth between 1991 and 2006, while Alberta experienced increases in natural population growth within these regions. Table 4 shows the rural/urban distribution of the population in the three provinces in 2006.

Province	% of Total Population		
	Rural	Total Urban	Census Metropolitan Areas only
Alberta	17.0	83.0	62.0
Manitoba	28.0	72.0	59.0
Saskatchewan	29.7	70.3	43.5

Table 4. Urban and Rural Populations of the Prairie Provinces, 2006. Source: Statistics Canada (2006a).

Aboriginal Population

People with Aboriginal ancestry are one of the fastest growing groups in Canada. Between 1901 and 2001, this group increased tenfold, while the total population of Canada rose by a factor of only six (Statistics Canada, 2003). In 2006, among all Prairie Provinces, Alberta had the largest Aboriginal population (188,365). For that same year, Manitoba had an Aboriginal population of 175,395 and Saskatchewan had the smallest Aboriginal population (141,890) (Table 5).[1] Looking at the proportion of the Aboriginal population in 2006 to the general population in the Prairie Provinces, 15.5% and 14.9% of the population in both Saskatchewan and Manitoba, respectively, was of Aboriginal ancestry. Alberta had the smallest proportion of Aboriginal population—5.8% in proportion to the general population. The rate of growth of the population of people of Aboriginal ancestry was also different for the three provinces, with Alberta experiencing the highest rate of growth, followed by Manitoba, and then Saskatchewan.

Particulars	Manitoba	Saskatchewan	Alberta
Median Age of Aboriginal (Years) in 2006	23.9	21.7	24.8
Median Age of Non-Aboriginal (Years) in 2001	38.5	38.8	35.4
Population of Aboriginal Ancestry in 2006	175,395	141,890	188,365
Aboriginal Population Percent of Total Population in 2006	15.5%	14.9%	5.8%
Aboriginal Population in Major Centres in 2006 [Rank in Canada in 2001]	Winnipeg (63,745) [1]	Saskatoon (19,820) [6] Regina (16,535) [7] Prince Albert (12,140) [9]	Edmonton (38,170) [2] Calgary (24,425) [4]

Table 5. Selected Characteristics of Aboriginal Population, Prairies by Province, 2006 and 2001. Source: Statistics Canada (2003) and Statistics Canada (2006).

The different rates of population increase between the Aboriginal and non-Aboriginal populations have also resulted in a relative difference in the median age of their populations. The people of Aboriginal ancestry are relatively younger compared to the non-Aboriginal people. As shown in Table 5, in 2001, the non-Aboriginal median age in Manitoba was 38.5 years, whereas the Aboriginal median age in 2006 was almost 16 years younger—at 22.8 years.

A portion of the Aboriginal population in the Prairie Provinces resides in cities. Winnipeg has the highest number of people of Aboriginal ancestry (estimated at 63,745 in 2006). In fact, all prairie cities have claimed ranks in the top nine cities in Canada with large numbers of Aboriginal people.

Major Economic Activities

In 2004, the prairie region contributed to the national economy a total of $259 billion worth of value-added activities, as measured by the gross domestic product (GDP).[2] This amounted to $37,637 for every person living in the region in 2004. This compares slightly unfavourably with the 2004 estimate of Canadian GDP of $46,759 per capita.[3] However, this wealth is not shared equally among all provinces. Rate of change in Alberta's GDP has been faster than for Canada as well as for the other two provinces. In fact, per capita GDP in both other provinces is lower than that for Alberta. In 2004, Alberta's per capita GDP was 26% higher than that for Saskatchewan, and 40% higher than that for Manitoba.

For the region as a whole, the largest contributions to the GDP are from the primary resource sectors, as shown in Table 6. Almost a quarter of the total

Industry	% of Total Gross Domestic product			
	Manitoba	Saskatchewan	Alberta	Prairies
Crop and Animal Production	4.6	6.7	1.9	3.1
Mining	2.6	12.4	14.6	12.3
Other Primary Goods Producing Industries	0.4	0.5	0.4	0.4
Manufacturing	11.6	7.4	9.5	9.5
Construction	5.3	5.7	10.6	9.0
Trade	12.7	11.6	11.0	11.4
Utilities	3.1	2.6	1.9	2.2
Public Administration	6.9	5.8	4.0	4.7
Other Services	53.0	47.2	46.1	47.4
Total	**100.0**	**100.0**	**100.0**	**100.0**

Table 6. Distribution of Gross Domestic Product by Industry, 2006, Prairie Region, by Province. Source: Statistics Canada–CANSIMa.

value added is contributed by these sectors. Given that some of these sectors (e.g. agriculture) are largely dependent on climate conditions, they are significantly impacted by extreme events, such as the drought of 2001, when the contribution of the agricultural sector in Alberta and Saskatchewan was slightly lower than it would be under normal years. The next single largest contributing group of economic activity is the service sector which constitutes roughly a fifth of the total economy. In the region, the public administration sector also makes a significant contribution to the regional GDP in all three provinces.

The make-up of the three provinces is similar in some ways, but there are some significant differences. One of the similarities in all three provinces is their large dependence on primary resource production—agriculture, forestry and mining—and, in general, a small manufacturing sector (relative to the rest of Canada). The differences lie in the nature of the primary production sector. In Saskatchewan, agricultural production predominates, while in Alberta, mining activities dominate the scene. Of all the provinces, Manitoba has the largest manufacturing sector (in relative terms). All three provinces are heavily dependent on service industries. This is a major growth sector in the region.

Employment Patterns

Employment patterns in the prairie region basically follow the distribution of economic activities as depicted by the GDP pattern. The only exception is that the service industries are the largest source of employment in the region. The next in terms of importance for creating employment are the trade sectors—both wholesale trade and retail trade industries. These industries contribute around 17% to the total employment in the region as a whole (Figure 2). The primary resources sector is the fourth largest employer group (following manufacturing), with 7% of the total employment in the region. It should be noted that these industries, particularly agriculture, have lost their place from being the major employer in the past. Technological change in these industries, particularly the labour saving devices, has led to this state in more recent times.

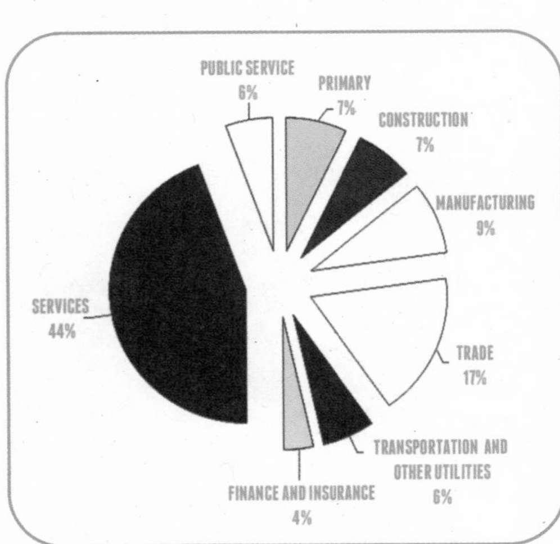

Figure 2. Distribution of Employment in the Prairie Region, by Major Industry Groups, 2006

Regional Economic Profile

From the above discussion it is obvious that the prairie region is not a homogenous entity. The population growth, population composition and economic structure of each province are significantly different from each other. Manitoba, facing a modest growth in population, has been able to maintain a steady employment growth in a diversified economy. In this province, among the goods-producing industries, manufacturing tops the list, with almost half of the GDP from all goods-producing industries. In 2001, agriculture was the fourth largest goods-producing industry, after construction and utilities. Like the other two provinces, farm population in the region has been declining consistently, although the trend has shown a little moderation during the last decade. Average farm size in 2006 was 405 ha, which was the lowest among the region. Cattle farms, as well as grain and oilseed farms, dominate the agricultural landscape. Irrigation in the province is not considered to be a necessity, and is limited to the production of potatoes for processing.

Despite the continued dominance of farming and resources, Saskatchewan's economy has shown promising signs of diversification (Hirsch, 2005a). Manufacturing shipments have increased by 55% in 2004 over the last decade or so. The economy, however, is very much export-oriented. The top export commodities are crude petroleum oil, potassium chloride, spring and durum wheat, and canola. The farm population is declining, although the province still has the highest number of people on farms. The number of farms is decreasing and the average farm size is increasing. In 2006, the average farm size was 586 ha—highest among the three provinces. Grain and oilseed farms and cattle farms dominate the agricultural scene. Major water use in the region is for irrigation, although it is confined to the region around Lake Diefenbaker, and the southwest portion of the province.

The Alberta economy has been booming over the last decade or so, and this has created a great degree of optimism for the future. The gains are sparked by the high energy prices for oil and natural gas (Hirsch, 2005b). In addition, the optimism is also sparked by the oil sands projects, where investment is expected to rise sharply. Other sectors of the economy, such as manufacturing, construction, services and public finance, are also doing well. Farm incomes in recent years have improved following the drought years (2001 and 2002). Major farm types in the province include cattle farms, followed by field crops and wheat farms. Farm size is small (427 ha) relative to neighbouring Saskatchewan. The decline in the farm population is very similar to that observed for other provinces. Major water use in the province is for irrigation. In fact, Alberta has almost two-thirds of all irrigated area in Canada.

FUTURE PROJECTIONS

Predicting a complex set of economic structures for a region as heterogeneous as the Prairies is very risky. Many factors—some internal and many external—

will shape the future economic performance of the region. Dependence of the region on export markets makes any prediction even more risky. In spite of it, certain predictions about the region have been made. Most notably among these are the population projections, which are described in this section.

POPULATION PROJECTIONS

Population projections for the three provinces and for the prairie region along with Canada are shown in Table 7. By 2031, the region's population could be as high as 7 million—an increase of 28.7%. The highest growth is expected to be in Alberta, where population would increase from 3.4 million in 2005 to 4.5 million—an increase of 39.5%. The slowest growth in population is expected for the province of Saskatchewan at 2.9%. Manitoba's population is expected to increase somewhat more slowly than Alberta's population with a 22.8% increase between 2005 and 2031.

		2031 Population (Thousands)*					
Province / Region	2005 Population (Thousands)	Low Growth	Medium Growth (Recent migration)	Medium Growth (Medium migration)	Medium Growth (West coast migration)	Medium Growth (Central-west migration)	High Growth
Manitoba	1,178	1,259	1,375	1,356	1,335	1,378	1,447
Sask.	985	937	967	976	981	1,064	1,023
Alberta	3,376	3,925	4,391	4,145	3,892	4,543	4,403
Prairies	5,539	6,121	6,733	6,477	6,208	6,985	6,873
Canada	32,623	36,261	39,045	39,029	39,015	39,052	41,811
Prairies as a % of Canada	17%	17%	17%	17%	16%	18%	16%

Table 7. Future Population Growth, Prairies and Canada, 2005–2031. Source: Statistics Canada (2005a)

Future Projections of Visible Minority Groups (2001–2017)

The population of visible minority groups (as shown in Table 8) is a relatively small portion of the total population in the Prairie Provinces. In 2001, its proportion varied between 2.7% for Saskatchewan to 11% for Alberta. This population group is projected to increase in the future.

Major increases are projected for Alberta where this population group will almost double, from 330,000 to 656,000 people. By 2017, visible minority

Province	Population (in Thousands)		Proportion of Total Population (Percent)	
	2001	2017	2001	2017
Alberta	330.1	656.4	11.0	17.1
Manitoba	88.2	138.5	7.7	11.8
Saskatchewan	26.8	44.2	2.7	4.7

Table 8. Projected Visible Minority Groups' Population in the Prairie Provinces, 2001 and 2017. Source: Statistics Canada (2005c)

groups will constitute 17% of the Alberta population. Increases for the other two provinces are much smaller: 57% for Manitoba and 64.9% for Saskatchewan. However, relative to Alberta, these groups will constitute a smaller proportion of the total population in these provinces, even in 2017.

Future Projections for Aboriginal Population (2001–2017)

The proportion of the Aboriginal population is projected to increase to 21% in Saskatchewan and 18% in Manitoba by 2017 (Table 9). However, it will continue to be a smaller proportion of the total Alberta population in 2017.

Province	Population (in Thousands)		Proportion of Total Population (Percent)	
	2001	2017	2001	2017
Alberta	167.9	232.6	5.5	6.3
Manitoba	159.4	221.1	13.8	18.4
Saskatchewan	138.3	202.8	13.8	20.8

Table 9. Projected Aboriginal Population in the Prairie Provinces, 2001 and 2017. Source: Statistics Canada (2005b)

Focusing on the proportion of the Aboriginal population aged 0–14 in 2001, 26% of Saskatchewan's population aged 0–14 was of Aboriginal ancestry, which is projected to increase to 37% in 2017. In Manitoba, similarly in 2001, 24% of the population aged 0–14 was Aboriginal. This figure is projected to increase to 31% in 2017. In regards to the population aged 20–29, 17% of Saskatchewan's young adults in 2001 are Aboriginal. This number is projected to increase to 30% in 2017. In Manitoba, 17% of young adults aged 20–29 were Aboriginal in 2001; this is projected to increase to 24% in 2017. These trends suggest that the Prairie Provinces, having a higher Aboriginal representation, will likely be concerned with issues of education and transitions to adulthood (Antal, 2007). These include high school retention and access to post-secondary education, as well as labour market integration, health and housing aspects of these people.

Quality of Life Issues

Much of the focus of public policies is on improving the economic well-being of Canadians. Economic well-being, in general, leads to improved quality of life. However, in spite of this increased awareness of this aspect, the systematic examination of "Quality of Life (QOL)" indicators is a relatively new area of investigation.[4] Only in the last few years has there been a serious inquiry dedicated to the identification of those factors that characterize "QOL." Because this is a relatively new area of investigation, these projects involve mostly qualitative research efforts. For example, the Centre for Health Promotion defines QOL as the degree to which a person enjoys the important possibilities of his (or her) life.[5] As such, no hard numbers for these factors yet exist, and solid projections of quality of life factors in the future have not been undertaken. Williams et al. (2008) suggest that a variety of factors, such as housing tenure, access to basic services, and parks and green spaces, affect people's perception of their QOL.

A recent approach to measure QOL using the economic environment index, social environmental index, and physical environment index has been developed by Natural Resources Canada (see Natural Resources Canada, 2004). The

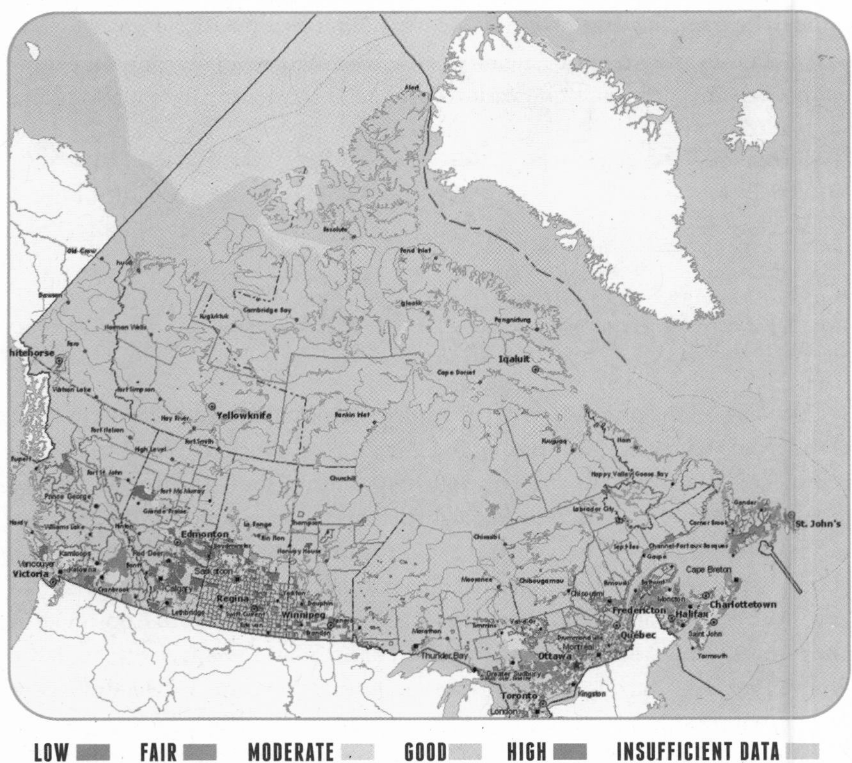

Figure 3. Map of Canada showing Quality of Life Index. Source: Natural Resources Canada (2004)

overall index combines these three indexes. Results are shown in Figure 3, and suggest that people in the three Prairie Provinces enjoy better quality of life than some other regions in Canada.

In the absence of hard numbers, the future quality of life can still be inferred from what is known of the changing demographic character of the region. As such, the implications of demographic trends that are apparent today and reliably projected into the next two decades will likely impact future QOL.

QOL research efforts to date, working in consultation with various groups of citizens, have consistently identified similar sets of priorities and factors that can be said to constitute quality of life. These sets of factors always include and prioritize issues of health, health care and education as being the most important constituents of QOL. Other issues like the environment, income and poverty, crime, and social programs addressing housing and social support are also commonly mentioned. Discussions of each of these factors have led researchers to identify particular groups within the population who can be said to be "at risk" of missing out on each particular QOL factor. The "at risk" segments most often mentioned are those very groups that are at the heart of the demographic changes that are expected in the Prairies in the coming decades: seniors, Aboriginal/First Nations people and recent immigrants. So, for Saskatchewan and Manitoba, quality of life issues will revolve around the well-being of their ever more significant Aboriginal/First Nations populations and of their growing number of seniors. Alberta, though it will continue to be a younger and less Aboriginal population, will still see dramatic increases in both of those groups and will need to address their quality of life issues. In addition, Alberta will experience more international immigration than the other Prairie Provinces and, accordingly, it will also need to address the unique housing and income challenges that face recent immigrants.

The changing proportions of these "at risk" groups in the population allow us to infer the future quality of life on the prairies. Among others, these may include:

- Aboriginal/First Nations are an "at risk" group, particularly "on-reserve" populations. This group is less healthy than the general population and, thus, particularly susceptible to chronic diseases like diabetes.

- Seniors, although previously thought to be healthier than in the past, are an "at risk" group with regard to health as a factor in quality of life. Individual health costs are incurred in the last five years of life regardless of age. The aging population and the increasing number of seniors are not the sole explanation for the coming increase in health care costs; other factors will also drive up costs. For example, the number of "very elderly" seniors, those over 85 years of age, is expected to increase dramatically by 2050. There will be an increase in the availability and demand for expensive health care technologies. Use of pharmaceuticals is expected to increase. New diseases may emerge, or existing ones may become more prevalent. Furthermore, aging baby-boomers will have a higher level of awareness of health issues, thanks to the Internet, and will have higher expectations than did previous generations.

- Seniors, as a group, are responsible for nearly half of the operating health spending nationally, more than any other age group. This trend is expected to become more pronounced in the next two decades. By 2027, it is expected that around 60% of health care spending will be devoted to this age group. (See Table 10).

- Seniors fear crime more than any other group. Thus, the growing number of seniors could result in more perceived crime.

- Aboriginal youth are specifically at risk due to patterns of disadvantage: poverty, housing, etc.

Age Group (Years)	1999/00	2004/05	2009/10	2014/15	2019/20	2024/25	2026/27
0 –14	7.4	6.5	5.8	5.3	5.0	4.6	4.5
15 –44	26.4	24.4	22.3	20.6	18.9	17.3	16.6
45 –64	20.9	22.8	24.2	23.5	21.9	19.8	19.0
65+	45.3	46.3	47.7	50.6	54.2	58.3	59.9
Total	100.0	100.0	100.0	100.0	100.0	100.0	100.0

Table 10. Projected Share of Operating Health Spending by Age Group, Canada 1999–2000 to 2026–2027 (Percent). Source: Provincial and Territorial Ministers of Health (2000).

SUMMARY OF SALIENT FEATURES OF THE REGION

Alberta, with its booming oil-based economy, is where the entire population increase for the region has occurred; and this will continue be the case for the next 30 years. The populations of Manitoba and Saskatchewan are expected to stabilize near their present levels. However, these two provinces are not immune to the challenges of population growth, since the major cities are expected to grow at the expense of the rural population. For some regions, such as Manitoba, population growth will come from immigration (Azmier, 2002), whereas, for Saskatchewan, much of the growth will come from the Aboriginal population.

In addition to the shifting rural/urban balance in Manitoba and Saskatchewan, the composition of those provinces' populations will undergo other shifts in the next three decades. Seniors and Aboriginal/First Nations people, particularly Aboriginal youth, will become ever more significant segments of the populations of those two provinces. And even with its younger population, Alberta will also see higher numbers of seniors and Aboriginal/First Nations people, as well as a large number of recent immigrants. These compositional changes in the populations of the Prairie Provinces will change the face of the region and may alter its priorities.

One of the trends developing in Canada—and the prairie region should be no exception to it—will result in worker shortages in future years (Sauvé, 2003). Immigration would be able to soften this crisis, but much of it depends on factors outside the domain of the Prairie Provinces.

One of the main features of the prairie economy is the amount of change that has occurred over the last few decades. The advances in oil sands recovery techniques, an increase in the film industry, increased agricultural productivity, widespread adoption of computer technology, and consolidation of warehousing and industrial production are indicative of a dynamic economic system.[6] Expansion of irrigation in southern Alberta has softened the adverse effects of weather and climate. These and similar changes will continue to shape the region further in the future.

In the context of the impacts of climate change and adaptation to the new conditions, two main issues are prominent:

a. The Prairies are one of the most vulnerable regions to the impacts of projected changes, especially the southern areas of Alberta and Saskatchewan. The confluence of an increasing aridity of the region, and a concentration of population growth in southern Alberta is a serious challenge from the point of view of water resources.

b. Those population sectors that are defined as potentially "at risk"—as a result of physical or socio-economic limitations—are at the heart of the demographic changes we expect to see on the Prairies in the coming decades: seniors, First Nations people and recent immigrants. In the context of future climate conditions, Saskatchewan and Manitoba quality of life issues will revolve around the well-being of their ever more significant Aboriginal/First Nations populations and of their growing number of seniors. Alberta will continue to have a younger and less-Aboriginal population, but will still face increases in both of those groups, and will need to address their quality of life issues. As well, Alberta will attract more immigrants than the other Prairie Provinces, with the subsequent need to address the unique housing and income challenges that face recent immigrants.

CONCLUSIONS

The future may bring forth major shifts within the prairie region in terms of population composition as well as economic activities.

There is the potential for differential levels of vulnerability among the diversity of regions that constitute the Prairies. These differentials are not only related to an environmental diversity, but also to an uneven distribution of population and resources. A higher proportion of the population lives in the province of Alberta, which also concentrates a significant part of the wealth. Population projections show that this trend will continue, leaving other regions without the human and institutional resources to assist people through proper programming. As climate change brings forth uneven change in the economic activities, major migration patterns may develop, putting more strains on the socio-economic fabric of the region.

There is gradual population shift from rural areas to large urban areas, which reduces the viability of many rural communities; this may put special pressure on cities and create major social problems. Under climate change, this trend would accelerate.

Population trends indicate an increase of sectors of the population that could be highly vulnerable to climate change. These groups include seniors, immigrants and Aboriginal people.

The major economic drivers in the region are agriculture and natural resources. All other sectors are driven by these and other primary goods-producing industries. Climate change would create significant indirect impacts on the region. For example, changes in the amount of precipitation—resulting in decreased soil moisture—would affect agriculture and add to the need for more irrigation development. At the same time, a decrease in surface water could limit such irrigation activities in regions that need it most.

A shift in the ecozone northwards would affect productivity of forests and agricultural production, and bring forth significant shifts in the economic structure of the region. This may also require a major investment in infrastructure, as these regions would have to be built up to face the challenge of new economic development. Energy production, which is a major source of boom to some parts of the region, may also become vulnerable under climate change through reduced water availability.

The northern part of the Prairies is currently inhabited by Aboriginal peoples. These people make up the majority at the present time. With more economic opportunities in the northern part due to climate change (particularly with the ecozone shift), a situation of conflict between the new migrants and the local Aboriginal peoples might emerge.

ENDNOTES

1. One of the limitations of these data is they are based on under-coverage. For some communities, enumeration was not permitted, or was interrupted before it could be completed. These data are not included in these estimates.
2. The gross domestic product is a measure of the region's money value of goods and services becoming available to the people from that region's economic activities.
3. Much of this is explained by lower per capita GDP values in 2004 for Manitoba (at $30,054) and Saskatchewan (at $33,282). In contrast, Alberta had the highest per capita GDP in the region at $41,952 for the same period. Details are shown in appendix Table A.1.
4. Part of the difficulty in this is the fact that perception of well-being is relate to values (Ben-Arieh and Fromes, 2007).
5. These possibilities result from the opportunities and limitations each person faces and reflects the interaction of person and environmental factors. The measurement of this concept can include information related to "Being, Belonging, and Becoming" (Centre for Health Promotion, Undated).
6. For an excellent discussion of economic transformations, see Roach (2005).

REFERENCES

Antal, K. 2007. "Our Population in 2017—Diverse in Our Diversity." *Horizons* 9, no. 4: 32–41.

Azmier, J.J. 2002. *Manitoba in Profile*. Canada West Foundation: Calgary.

Ben-Arieh, A. and I. Fromes. 2007. "Indicators of Children's Well-being: What Should be Measures and Why?" *Social Indicators Research* 84: 249–50.

Bernard, P. 2007. "The Interconnected Dynamics of Population Change and Life-Courses Process." *Horizons* 9, no. 4: 13–16.

Beaujot, R., K. McQuillan and Z. Ravanera. 2007. "Population Change in Canada to 2017 and Beyond." *Horizons* 9, no. 4: 3–12.

Centre for Health Promotion. Undated. *The Quality of Life Model* [online]. Toronto: University of Toronto. [accessed December 24, 2007]. Available from World Wide Web: *http://www.utoronto.ca/qol/concepts.htm*

Hirsch, T. 2005a. *A Soft Landing: Saskatchewan's Economic Profile and Forecast*. Canada West Foundation: Calgary.

——. 2005b. *Firing on (Almost) all Cylinders—Alberta's Economic Profile and Forecast*. Canada West Foundation: Calgary.

Natural Resources Canada, 2004. "Quality of Life." In *Atlas of Canada*. Ottawa.

Provincial and Territorial Ministers of Health. 2000. *Understanding Canada's Health Care Costs: Final Report* [online]. [accessed December 24, 2007]. Available from World Wide Web: *http://www.health.gov.on.ca/english/public/pub/ministry_reports/ptcd/ptcd_doc_e.pdf*

Roach, R. 2005. "Economic Transformations in Western Canada." *Dialogues* [online]. Summer. [accessedDecember 22, 2005] Available from World Wide Web: www.cwf.ca

Sauvé, R. 2003. *Canadian Age Trends and Transitions to 2016*. Sooke, B.C.: People Pattern Consulting.

Statistics Canada. 2003. *Aboriginal Peoples of Canada: A Demographic Profile*. Catalogue No. 96F0030XIE2001007. Ottawa.

——. 2005a. *Population Projections for Canada, Provinces and Territories 2005–2031*. Catalogue No. 91–520-XIE. Ottawa.

——. 2005b. *Projections of the Aboriginal Populations, Canada, Provinces and Territories, 2001–2017* [online]. Catalogue no. 91–547-XIE. Ottawa. [accessed January 12, 2006]. Available from World Wide Web: *http://www.statcan.ca/english/freepub/91–547-XIE/2005001/bfront1.htm*

——. 2005c. *Population Projections of Visible Minority Groups, Canada, Provinces and Regions.* Catalogue No. 91–541-X1E. Ottawa.

——. 2006a. *"Population and Dwelling Counts, for Canada, Provinces and Territories, and Urban Areas, 2006 Ccensus—100% Data."* In Population and Dwelling Count: Highlight Tables, 2006 Census [online]. Catalogue no. 97–550-XWE2006002. Ottawa. [accessed November 28, 2006]. Available from World Wide Web: *http://www12.statcan.ca/english/census06/data/popdwell/Filter.cfm?T=802*

——. 2006b. *Community Profile, 2006—Provinces and Territories in Canada, 2006 Census of Population* [online]. Ottawa. [accessed September 29, 2008]. Available from World Wide Web: (*http://estat.statcan.ca/cgi-win/cnsmcgi.exe?Lang=E&STATFile=EStat\English\SC_RR-eng.htm*).

——. 2007. *Portrait of the Canadian Population in 2006.* Catalogure No. 97–550. Ottawa.

Statistics Canada–CANSIMa. *Table 379–0025—Gross Domestic Product (GDP) at Basic Prices, by North American Industry Classification System (NAICS) and Province, Annual (Dollars)* [online]. CANSIM (database), Using E-STAT (distributor). Ottawa. [accessed November 29, 2007]. Available from World Wide Web: *http://estat.statcan.ca/cgi-win/cnsmcgi.exe?Lang=E&ESTATFile=EStat\English\CII_1_E.htm&RootDir=ESTAT/*

Statistics Canada–CANSIMb. *Table 051–0001—Estimates of Population, by Age Group and Sex, Canada, Provinces and Territories, Annual (Persons Unless Otherwise Noted)* [online]. CANSIM (database), Using E-STAT (distributor). Ottawa. [accessed November 23, 2007]. Available from World Wide Web: (*http://estat.statcan.ca/cgi-win/cnsmcgi.exe?Lang=E&ESTATFile=EStat\English\CII_1_E.htm&RootDir=ESTAT/*).

Statistics Canada–CANSIMc. *Table 051–0011/2—International migrants, by Age Group and Sex, Canada, Provinces, and Territories, Annual (Persons)* [online]. CANSIM (database), Using E-STAT (distributor). Ottawa. [cited November 22, 2007]. Available from World Wide Web: (*http://estat.statcan.ca/cgi-win/cnsmcgi.exe?Lang=E&ESTATFile=EStat\English\CII_1_E.htm&RootDir=ESTAT/*).

Williams, A., P. Kitchen, J. Randall and N. Muhajarine. 2008. "Changes in Quality of Life Perceptions in Saskatoon, Saskatchewan: Comparing Survey Results from 2001 and 2004." *Social Indicators Research* 85, no. 1: 5–21.

APPENDIX

Particulars	Unit	Manitoba				Saskatchewan				Alberta			
		1951	1975	2001	2006	1951	1975	2001	2006	1951	1975	2001	2006
Total Population	'000	777	1,022	1,149	1,178	832	921	1,017	985	940	1,838	3,059	3,376
% of Canada	%	5.5	4.4	3.7	3.67	6.0	4.0	3.3	3.0	6.7	8.0	9.8	10.2
Population under 15 years	'000	223	265	255	228	255	248	231	190	287	503	672	637
Population over 65 years	'000	66	107	156	180	67	102	147	147	67	138	311	352
Dependent Population	% of total	37.2	36.4	35.8	32.9	38.7	38.0	37.2	34.2	37.7	34.9	32.1	29.3
Net International Migrants	'000	—	5.6	3.0	8.7	—	1.8	0.7	2.1	—	14.0	8.0	16.4
Net Interprovincial Migrants	'000	—	-6.0	-4.0	-8.6	—	0.7	-8.4	-9.1	—	23.0	20.0	57.1
Employment	'000	299	461	558	521	302	422	472	411	354	874	1,632	1,661
Unemployment rate	%	4.5	4.4	5.0	4.3	6.8	3.2	5.8	4.7	4.0	4.0	4.6	3.4
Real GDP	Bill. $	—	—	$33	$40	$7	$16	$31	$37	—	$59	$125	$182
Real GDP Per Capita	$	—	—	$28,728	$34,237	$8,575	$17,970	$30,311	$39,001	—	$32,937	$40,694	$54,320
Real International Exports	Bill. $	—	—	10.0	11.3*	1.9	4.0	13.0	14.0*	—	19.7	45.2	—
Exports as a % of GDP	%	—	—	61	64*	44	49	68	70*	—	56	57.9	—

Table A.1. Selected Features of the Prairie Provinces, by Province, Selected Years
* Data for 2004

CHAPTER 3

PRAIRIE CLIMATE TRENDS AND VARIABILITY

David Sauchyn

INTRODUCTION

The climate change scenarios described in this chapter indicate that we can expect the Prairies to become much warmer. Average annual temperatures have been rising since weather stations were first established in the 1880s, as illustrated in Figure 1 for 12 communities spanning the Prairie Provinces. Even though temperatures, and possibly precipitation, are rising, we should not expect warmer and wetter weather throughout the year and every year. Figure 1 shows that, since 1895, mean annual temperatures have deviated each year from the consistently increasing trend. The Prairies have Canada's most variable climate and are among the most variable climates in the world. Because global warming is superimposed on the natural variability, large differences between seasons and years will continue. In fact, simulations of future climate suggest that the short-term variability will be amplified by global warming, increasing the probability of large departures from normal conditions. Thus, we can occasionally expect some colder than average weather, even as average annual temperatures continue to rise.

Prairie precipitation is even more variable between years and decades. In Figure 2, total annual precipitation recorded at Medicine Hat is plotted as positive (blue) and negative (red) departures from the annual average. While the mean value is 384.5 mm, precipitation has ranged from 185.5 mm in 2001 to 689.3 mm in 1927. Precipitation swings from large deficits to large surpluses between years (e.g. 2001 to 2002) and decades (e.g. 1917–24 versus the 1950s). An often used measure of variability is the ratio of the standard deviation for a set of sta-

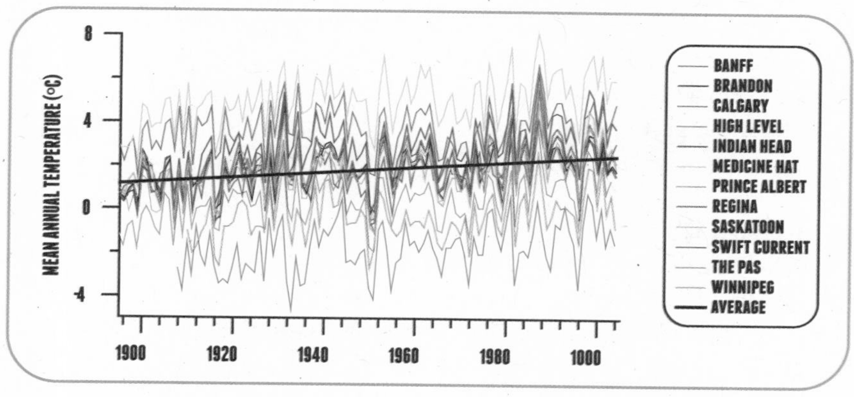

Figure 1. Mean annual temperature records from across the Prairie Provinces for the period 1895–2005. For these 12 long records, there was an average increase of 1.6 degrees from a low of 0.9 at Calgary to 2.67 at Swift Current.

tistics to their mean value. When Longley (1952) determined this coefficient of variation (CV) for precipitation records from across Canada, the highest values (largest inter-annual variation) occurred in the Prairies over an area roughly corresponding to the Saskatchewan River Basin (SRB), Canada's largest dryland watershed. The CV in this region exceeds 25%; at Medicine Hat, it is 28.5%.

This chapter describes the natural variability of prairie climate with a geographic focus on the SRB, the region with the most variable climate and water

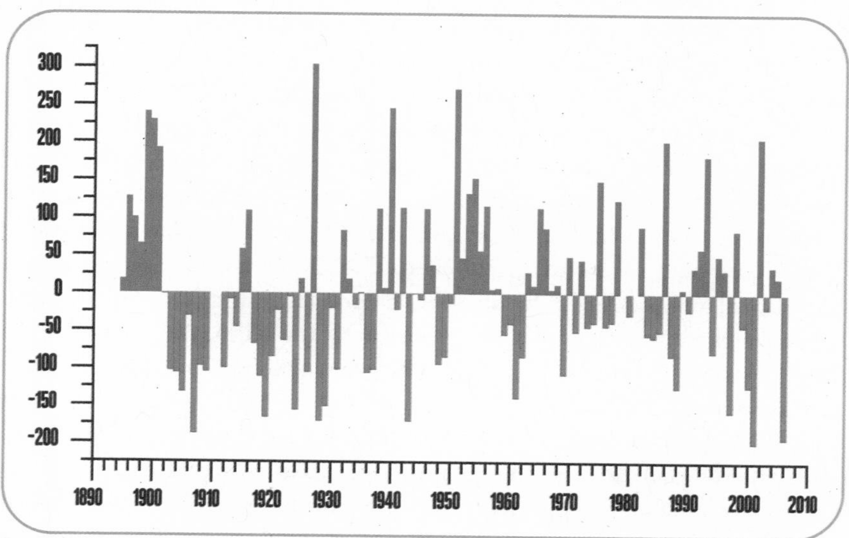

Figure 2. Positive (red) and negative (blue) departures from mean annual precipitation, Medicine Hat, Alberta, 1895–2006.

supplies. Because weather station and water level records are from the period affected by anthropogenic global warming, and because they are short relative to some climate cycles, this chapter examines the proxy record of the prairie climate of the past millennium to reveal the natural climate variability that underlies the trends imposed by global warming.

RECENT CLIMATE TRENDS

Recent trends in the climate of Canada have been the subject of various studies (Akinremi, McGinn and Cutforth, 1999; Beaubien and Freeland, 2000; Gan, 1998; Zhang et al., 2001; Shabbar and Bonsal, 2003; Bonsal et al., 2001; Zhang et al., 2000; Bonsal and Regier, 2007). The trends identified over Western Canada include significant decreases in cold spell frequencies and extreme low temperature during 1900–1998. There are also trends toward more days with extreme high temperature in winter and spring, but these are not as pronounced as the decreases to extreme low values. The number of frost days over most of Alberta and Saskatchewan has decreased, with a few minimal increases scattered throughout the provinces. The frost-free season is longer mostly due to an earlier start and similar ending date. Warmer winter/spring temperatures are correlated with earlier spring flowering in central Alberta; eight days earlier at Edmonton over the last six decades (Beaubien and Freeland, 2000). The blooming of aspen poplar is almost a month earlier than a century ago.

These trends in the frequency of extreme daily temperatures, and analyses of annual and seasonal temperature trends, strongly suggest that Western Canada is not getting hotter, but rather less cold. There has been a greater increase in daily minimum (as opposed to maximum) temperatures, and the largest warming has occurred during winter and early spring. Increases during summer have only been observed for minimum temperature. This observed warming has beneficial effects that include a longer frost-free period and more growing degree days. However, changes to the timing of temperature-related events (e.g. spring runoff) could ultimately have adverse effects on mid- to late-summer water supplies.

Total precipitation has increased over the last 99 years. It increased steadily from the 1920s to 1970, and then levelled off for the annual time series, but not for seasonal trends. Areas in central Alberta during the spring season and southern Saskatchewan during fall have decreasing trends. The ratio of solid to total precipitation has also increased annually over the Prairies with the greatest increase occurring in fall, but there is a decreasing trend in central Saskatchewan. The southern regions of Alberta and Saskatchewan, and some central areas, have experienced summer precipitation decreases. Precipitation trends are much less consistent than for temperature mainly because of the large natural variability and characteristic periodicity at inter-annual to multi-decadal scales, such that the detection of trends depends very much on the period of record examined and the regional response of precipitation patterns to natural and human (global warming) causes of variability.

Historical trends in the summer climate moisture index (CMI = precipitation – potential evapotranspiration) suggest drier conditions of 1–2 mm/yr in southwest and west-central Alberta and an even larger deficit of 1–4 mm/yr in southern Saskatchewan. This drying trend throughout west-central and southern regions of the Prairies is in contrast to increased CMI, i.e. less dry conditions, along the Rocky Mountains and in northwestern Alberta. A recent drying trend reflects the severity of spatial extent and severity of the 2001–2002 drought followed an absence of dry years; 2001 and 2002 were the worst drought years since 1961, and the worst two-year drought since 1929–1930. The most severe and prolonged prairie-wide droughts during the instrumental record occurred in the early part of the 20th century (1915 through the 1930s).

THE CLIMATE OF THE PAST MILLENNIUM

The many studies that have analyzed recent trends in climate and water variables are all made possible by the weather and water gauges observation networks that have been maintained for approximately the past 100 years. While these records enable the detection of trends and variability, the rate of global warming is of concern only because we know that it is unusual in the context of the history of the earth's climate. Long-term trends, gradual regional responses to global climate changes, rare abrupt climate change, and long climate cycles are evident only in paleoclimatic records derived from geological and biological archives. Most instrumental records are too short to capture multi-decadal cycles and the full range of extreme conditions. They also are confined to the period of anthropogenic climate changes. The global climate of the past few millennia provides a historical context for understanding present climate change. The warmer conditions of the Medieval Climate Anomaly (MCA) of the 9th to 14th centuries, to the Little Ice Age (LIA) of the 15th to 19th centuries, have encompassed a large range of variation in temperature (Bradley, 2003). Temperatures inferred from boreholes on the Canadian Plains (Majorowicz, Safanda and Skinner, 2002) and from tree rings at high elevations in the Rocky Mountains (Luckman and Wilson, 2005) show that the warmest climate of the past two millennia is during the 20th century.

In the Prairie Provinces, changes in climate are reflected in the shifting of vegetation, fluctuations in the level and salinity of lakes, patterns in tree rings, and the age and history of sand dunes (Lemmen and Vance, 1999; Vance and Wolfe, 1996). Most of the paleoclimatic records in this region are from the physical (e.g. level), chemical (e.g. salinity) and biological (e.g. biodiversity) characteristics of lake sediments. The frequency and duration of dry periods also has been inferred from the age and history of sand dune deposits (Wolfe, 1997), which are extensive in southwestern Saskatchewan and southeastern Alberta. The regional reactivation of a dune field requires a dry period lasting several years to decades (Vance and Wolfe, 1996). These lake and sand dune records indicate that, early in the Holocene (the postglacial period of the past 10,000 years), the climate was generally warm and dry than today, culminating in the mid-Holocene warm-dry "climatic optimum," when dune activity was so extensive that evidence was not preserved (Wolfe, Ollerhead and Lian, 2002). Pollen records from lakes suggest low water level aspen parkland where today there is coniferous forest (Sauchyn and Sauchyn, 1991; Vetter, Last and Sauchyn, 2000). This period would be warmer than today if not for anthropogenic global warming, which reversed the generally cooler and wetter conditions that have characterized the latter part of the Holocene.

Sediments usually provide coarse climate histories, given limits to the accuracy of determining their age. The reconstruction of annual to decadal climatic variability over the past millennium requires more accurate dating methods and climate archives of higher resolution. From the precise optical dating of quartz grains, Wolfe et al. (2001) identified widespread reactivations of sand dunes about 200 years ago and correlated this geomorphic activity with tree-ring records of prolonged droughts of the mid- to late 18th century. A lag is apparent between peak dryness around 1800 and the onset of dune activity at about 1810. Dune stabilization has occurred since 1890. The droughts of the 1930s and 1980s were insufficient to renew dune activity. High-resolution lake sediment records have been obtained recently with the continuous sampling of sediment cores at fine intervals. The diatom assemblages from Humbolt Lake in Saskatchewan revealed multi-centennial shifts in moisture regime (Michels et al., 2007; Laird et al., 2003). A marked shift to moister conditions occurred about 800 years ago at Chauvin Lake and about 670 years ago at Humbolt Lake, i.e. near the end of the Medieval Climate Anomaly (MCA) and the onset of the Little Ice Age. Using paleoenvironmental information from the Peace-Athabasca Delta, Wolfe et al. (2008) determined that the levels of Lake Athabasca have fluctuated systematically over the past millennium. The lowest levels were during the 11th century, while the highest lake levels coincided with maximum glacier extent during the Little Ice Age.

Whereas these sediment studies provide a historical context for the observed climate changes of recent decades, another climate proxy, tree rings, is the source of both climate information and an absolute annual chronology. The precise measurement of tree-ring width, and calibration of this proxy using instrumental data, enables the reconstruction of the climate variable(s) that limit annual tree growth. Tree-ring and archival records from Manitoba (Ferguson and St. George, 2003; Blair and Rannie, 1994; Rannie, 2006; St. George and Nielsen, 2002, 2003) have highlighted the recurrence of wet years and flooding, and point to a contrast in climate between the western and eastern Prairies. In the dry climate of the western Prairies, tree growth is limited each year by available soil moisture and, therefore, tree rings are a proxy of precipitation and drought. Figure 3 is a bar code representation of annual moisture conditions at Calgary, Alberta, as reconstructed from tree rings collected nearby in the Wildcat Hills near Cochrane. Besides a distinction between wet and dry years with above and below average moisture, negative departures from average conditions are further classified as mild, strong and severe drought. A scan of the bar code reveals that periods of decades or longer that lacked above average moisture (blue years) include the early part of the 20th century, but also dry periods of similar or longer duration before the Prairies were settled. The most severe droughts also are pre-instrumental, including the intense drought of the 1790s associated with sand dune activity in the Great Sand Hills and low water levels in the North Saskatchewan River, such that furs could not be transported from Fort Edmonton (Wolfe et al., 2001; Sauchyn, Stroich and Beriault, 2003). This, the tree rings and other climate proxies suggest that the climate of the 20th century was relatively favourable for the settlement of the Prairies, as it lacked the sustained droughts of preceding centuries that affected sand dune activity, the fur trade, and the health of Aboriginal people (Sauchyn et al., 2002; Sauchyn, Stroich and Beriault, 2003). The short duration of drought since the 1930s may be linked to multi-decadal climate variability, more so than to climate change, which is expected to cause increased aridity and more frequent drought (Kharin and Zwiers, 2000; Wetherald and Manabe, 1999).

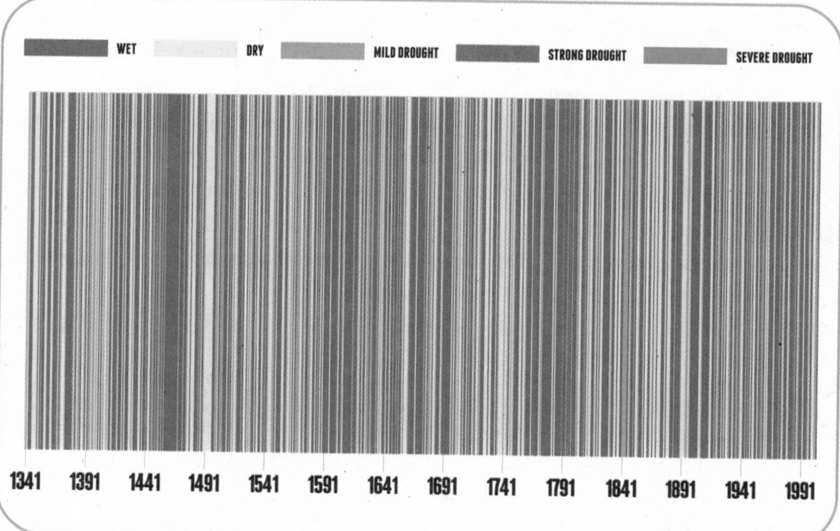

Figure 3. A reconstruction of the annual moisture conditions at Calgary, Alberta, from tree rings collected nearby in the Wildcat Hills near Cochrane.

CONCLUSIONS

Scenarios of future climate in this chapter are derived from global climate models with increased greenhouse gases to simulate the human modification of the atmosphere. Besides the trends in climate variables attributable to anthropogenic global warming, future climate will include the natural variability in the regional climate causing departures from the trends. Trends in mean values and departures from the mean are just two measures or scales of climate variation. For Canada's western interior, there are two relevant and important links between natural climate variability and anthropogenic global warming. First, the large inherent variability from season to season and year to year obscures the trends imposed by global warming and causes perceptions, whether incidental or biased, of no climate change. Second, a warmer, drier average climate will result from a high frequency of dry warm days, seasons and years. Global warming is projected to result in increased climate variability and a larger range of extreme events.

In the context of impacts and adaptation, however, there is another important distinction between trends or shifts in average climate and short-term departures from normal conditions and extreme events. Weather extremes have immediate impacts, and tend to result in adaptations, whereas trends and long-term shifts in mean conditions tend to have potential impacts that are avoided or alleviated though planned adaptation or gradual autonomous adaptations. Typically, prairie communities and economies are most vulnerable to extreme weather events and short-term climate variability, especially drought, and do not recognize climate trends as hazards (Marchildon, 2009). Thus, the future vulnerability of prairie communities will depend on the extent to which these regional climate risks are affected by global warming. The most challenging and risky global warming scenario is likely to be the prospect of severe and or prolonged drought.

REFERENCES

Akinremi, O.O., S.M. McGinn and H.W. Cutforth. 1999. "Precipitation Trends on the Canadian Prairies." *Journal of Climate* 12, no. 10: 2996–3003.

Beaubien, E. and H. Freeland. 2000. "Spring Phenology Trends in Alberta, Canada: Links to Ocean Temerature." *International Journal of Biometeorology* 44: 53–59.

Blair, D. and W.F. Rannie. 1994. "Wading to Pembina: 1849 Spring and Summer Weather in the Valley of the Red River of the North and Some Climatic Implications." *Great Plains Research* 4, no. 1: 3–26.

Bonsal, B. and M. Regier. 2007. "Historical Comparison of the 2001/2002 Drought in the Canadian Prairies." *Climate Research* 33: 229–42.

Bonsal, B.R., X. Zhang, L.A. Vincent and W.D. Hogg. 2001. "Characteristics of Daily and Extreme Temperatures over Canada." *Journal of Climate* 14: 1959–76.

Bradley, R.S. 2003. "Climate Forcing During the Holocene." Pp. 10–19 in A.W. Mackay, R.W. Battarbee, H.J.B. Birks and F. Oldfield (eds.), *Global Change in the Holocene: Approaches to Reconstructing Fine-Resolution Climate Change*. Arnold: London.

Ferguson, G. and S. St. George. 2003[M2]. "Historical and Estimated Ground Water Levels Near Winnipeg, Canada, and Their Sensitivity to Climatic Variability." *Journal of the American Water Resources Association* 39: 1249–59.

Gan, T.Y. 1998. "Hydroclimatic Trends and Possible Climatic Warming in the Canadian Prairies." *Water Resources Research* 34: 3009–15.

Kharin, V.V. and F.W. Zwiers. 2000. "Changes in the Extremes in an Ensemble of Transient Climate Simulations with a Coupled Atmosphere-Ocean GCM." *Journal of Climate* 13: 3760–88.

Laird, K.R., B.F. Cumming, S. Wunsam, O. Olson, J.A. Rusak, R.J. Oglesby, S.C. Fritz and P.R. Leavitt. 2003. "Lake Sediments Record Large-Scale Shifts in Moisture Regimes Across the Northern Prairies of North America During the Past Two Millennia." *Proceedings of the National Academy of Science* [M3]100: 2483–88.

Lemmen, D.S. and R.E. 1999. "An Overview of the Palliser Triangle Global Change Project." Pp. 7–22 in D.S. Lemmen and R.E. Vance (eds.), *Holocene Climate and Environmental Change in the Palliser Triangle: A Geoscientific Context for Evaluating the Impacts of Climate Change on the Southern Canadian Prairies.* Geological Survey of Canada, Bulletin 534.

Longley, R. W. 1952. "Measures of the Variability of Precipitation." *Monthly Weather Review* 80: 111–17.

Luckman B.H., and R.J.S.Wilson. 2005. "Summer Temperature in the Canadian Rockies During the Last Millennium—A Revised Record." *Climate Dynamics* 24: 131–44.

Majorowicz, J., J. Safanda, and W. Skinner. 2002. "East to West Retardation in the Onset of the Recent Warming Across Canada Inferred from Inversions of Temperature Logs." *Journal of Geophysical Research* 107 (B10): pp. 6–1—6–12.

Marchildon, G.P. 2009. *Prairie Forum: A Dry Oasis: Institutional Adaptation on the Canadian Prairies* (Special Issue), 34, no. 1 (Spring).

Michels, A., K.R. Laird, S.E. Wilson, D. Thomson, P.R. Leavitt, R.J. Oglesby and B.F. Cumming. 2007. "Multi-Decadal to Millennial-Scale Shifts in Drought Conditions on the Canadian Prairies Over the Past Six Millennia: Implications for Future Drought Assessment." *Global Change Biology* 13: 1295–1307.

Rannie, W.F. 2006. "A Comparison of 1858–59 and 2000–01 Drought Patterns on the Canadian Prairies." *Canadian Water Resources Journal* 31, no. 4: 263–74.

Sauchyn D.J., E.M. Barrow, R.F. Hopkinson and P. Leavitt. 2002. "Aridity on the Canadian Plains." *Géographie physique et Quaternaire* 66: 247–69.

Sauchyn, M. and D. Sauchyn. 1991. "A Continuous Record of Holocene Pollen from Harris Lake, Southwestern Saskatchewan, Canada." *Palaeogeography, Palaeoclimatology, Palaeocoogy* 88: 12–23.

Sauchyn, D.J., J. Stroich and A. Beriault. 2003. "A Paleoclimatic Context for the Drought of 1999–2001 in the Northern Great Plains." *The Geographical Journal* 169: 158–67.

Shabbar, A. and B. Bonsal. 2003. "An Assessment of Changes in Winter Cold and Warm Spells over Canada." *Natural Hazards* 29, no. 2: 173–88.

St. George, S. and E. Nielsen. 2002. "Hydroclimatic Change in Southern Manitoba Since A.D. 1409 Inferred From Tree Rings." *Qauternary Research* 58, no. 2: 103–11.

St. George, S. and E. Nielsen. 2003. "Palaeoflood Records for the Red River, Manitoba, Canada Derived from Anatomical Tree-Ring Signatures." *The Holocene* 13, no. 4: 547–55.

Vance, R. and S. Wolfe. 1996. "Geological Indicators of Water Resources in Semi-Arid Environments: Southwestern Interior of Canada." Pp. 251–63 in A.R. Berger and W.J. Iams (eds.), *Geoindicators: Assessing Rapid Environmental Changes in Earth Systems*. A.A. Balkema: Rotterdam, Netherlands.

Vetter, M.A., W.M. Last and D.J. Sauchyn. 2000. "Palynology and Lithostratigraphy of Deep Lake, Saskatchewan, Canada (The Other Deep Lake)." 8th International Paleolimnology Symposium. August 20 to August 24, 2000. Queen's University: Kingston, Ontario.

Wetherald, R. T. and S. Manabe. 1999. "Detectability of Summer Dryness caused by Greenhouse Warming." *Climatic Change* 43, no. 3: 495–512.

Wolfe, B.B., R.I. Hall, T.W.D. Edwards, S.R. Jarvis, R.N. Sinnatamby, Y. Yi, and J.W. Johnston. 2008. "Climate-Driven Shifts in Quantity and Seasonality of River Discharge Over the Past 1000 Years From the Hydrographic Apex of North America." *Geophysical Research Letters* 35: 1–5.

Wolfe, S. 1997. "Impact of Increase Aridity on Sand Dune Activity in the Canadian Prairies." *Journal Arid Environments* 36: 421–32.

Wolfe, S., J. Ollerhead and O. Lian. 2002. "Holocene Eolian Activity in South-Central Saskatchewan and the Southern Canadian Prairies." *Géographie Physique et Quaternaire* 56: 215–27.

Wolfe, S.A., D.J. Huntley, P.P. David, J. Ollerhead, D.J. Sauchyn and G.M. Macdonald. 2001. "Late 18th Century Drought-Induced Sand Dune Activity, Great Sand Hills, Southwestern Saskatchewan." *Canadian Journal of Earth Sciences* 38: 105–17.

Zhang, X., K.D. Harvey, W.D. Hogg and T.R. Yuzyk. 2001. "Trends in Canadian Streamflow." *Water Resources Research* 37, no. 4: 987–98.

Zhang, X., L.A. Vincent, W.D. Hogg and A. Niitsoo. 2000. "Temperature and Precipitation Trends in Canada During the 20th Century." *Atmosphere-Ocean* 38, no. 3: 395–430.

CHAPTER 4

CLIMATE CHANGE SCENARIOS FOR THE PRAIRIE PROVINCES

Elaine Barrow

Scenarios in generic terms are "coherent, internally consistent and plausible description(s) of a possible future state of the world" (IPCC, 1994). A climate change scenario in particular refers to the average change in climate between some future time period and a baseline time period (currently 1961–1990). There are many possible future climate states of the world depending on assumptions made about population and economic growth, energy use and conservation, and technology development and transfer, to name but a few of the important factors influencing emissions of greenhouse gases into the atmosphere. Changing the assumptions made about any of these factors may result in a different projection of future greenhouse gas emissions (an emissions scenario), which ultimately leads to a different climate change scenario.

The role that climate change scenarios may play in the assessment of climate change impacts was highlighted in the Third Assessment Report of the Intergovernmental Panel on Climate Change (IPCC, 2001):

- *Illustration of climate change*: by providing information about the range of plausible future climates in a given region.
- *Communication of the potential consequences of climate change*: by examining the impact of a particular future climate (e.g. the effects on species at risk of local extinction). In this way, scenarios may be used as awareness-raising devices.

- *Strategic planning*: by quantifying possible future sea-level and climate changes to design, for example, effective coastal or river flood defences, sewer and storm water systems.
- *Guiding emissions control policy*: by specifying alternative socio-economic and technological options in order to achieve a specific atmospheric composition (e.g. concentration of greenhouse gases). In this case, scenarios may be used to challenge people to think about a range of alternative futures which may sometimes be associated with unconventional socio-economic structures. Scenarios are also vital aids for evaluating options for mitigating future greenhouse gas and aerosol emissions.
- *For methodological purposes*: by determining our knowledge (or ignorance) of a system through, for example, the description of altered conditions, the use of a new scenario development technique, or by evaluating the performance of impact models and determining the reasons for any differences in results.
- *To explore the implications of decisions*: by examining the impacts resulting from specific scenarios of future climate and the actions taken to ameliorate particular harmful impacts associated with the scenarios.

As well as playing a number of different roles in vulnerability, impacts and adaptation studies, climate change scenarios can also be constructed in a num-

Emissions Scenario	Description
A1FI	A future world of very rapid economic growth and intensive use of fossil fuels
A1T	A future world of very rapid economic growth, and rapid introduction of new and more efficient technology
A1B	A future world of very rapid economic growth, and a mix of technological developments and fossil fuel use
A2	A future world of moderate economic growth, more heterogeneously distributed and with a higher population growth rate than in A1
B1	A convergent world with rapid change in economic structures, "dematerialization," introduction of clean technologies, and the lowest rate of population growth
B2	A world in which the emphasis is on local solutions to economic, social and environmental sustainability, intermediate levels of economic development and a lower population growth rate than A2

Table 1: Summary descriptions of the six illustrative SRES scenarios.

ber of different ways, with advantages and disadvantages associated with each individual scenario construction technique. Smith and Hulme (1998) put forward the following four criteria to aid scenario selection to ensure that climate change scenarios are of most use for impact researchers and policy makers:

1. *Consistency with global projections*: Scenarios should be consistent with a broad range of global warming projections based on increased concentrations of greenhouse gases. This range was given as 1.1°C to 6.4°C by 2090–2099 in the IPCC Fourth Assessment Report (IPCC, 2007), relative to 1980–1999 as a period average.
2. *Physical plausibility*: Scenarios should not violate the basic laws of physics, which means that not only should the changes in one region be physically consistent with those in another region and globally, but that changes in the different climate variables should also be physically consistent.
3. *Applicability in impact assessments*: Scenarios should describe changes in a sufficient number of climate variables on a spatial and temporal scale that allows for impact assessment. So, for example, it may be necessary for scenarios to provide information about changes in temperature, precipitation, solar radiation, humidity and wind speed at spatial scales ranging from a single site to global and at temporal scales ranging from daily to annual means.
4. *Representativeness*: Scenarios should be representative of the potential range of future regional climate change in order for a realistic range of possible impacts to be estimated.

Of the climate change scenario construction methods available, scenarios constructed using global climate model (GCM) output generally conform better with the criteria listed above. A complete description of the available scenario construction techniques can be found in IPCC-TGICA (2007).

GCMs are numerical models that represent mathematically the physical processes of, and the known feedbacks between, the atmosphere, ocean, cryosphere, and land surface. These models are used for the simulation of past, present, and future climates, and have undergone considerable evolution since their first appearance about forty years ago, largely due to the substantial advances in computing technology during this time. Most GCMs have a horizontal resolution of between 250 and 600 km, with 10 to 20 vertical layers in the atmosphere and as many as 30 layers in the ocean. This resolution is quite coarse, particularly when considered in comparison to the scales at which most impacts studies are conducted, so that it is impossible to model directly some of the smaller-scale processes (e.g. cloud and precipitation processes) occurring in the atmosphere and ocean. Such processes have to be parameterized, i.e., averaged over larger scales, or related to other variables that are explicitly modelled.

The most advanced GCMs are coupled atmosphere-ocean models, in which three-dimensional models of the atmosphere are linked dynamically with three-dimensional models of the ocean. These transient response models include the ocean circulation and transfers of heat and moisture between the oceanic surface and the atmosphere. Thereby, they are able to simulate the time-dependent response of climate to changes in atmospheric composition. Therefore, they can be used to simulate the climate response to changing atmospheric greenhouse gas and aerosol concentrations and provide useful information about the rate as well as the magnitude of climate change.

The more recent of these transient response GCMs begin by modelling the effects of past changes in radiative forcing, i.e., the effect of historical changes in atmospheric composition (typically from the 18th or 19th century) on the radiation balance of the atmosphere. These are known as "warm start" experiments to distinguish them from the earlier transient experiments that did not have this capability. Simulations are then continued into the future using a scenario of future radiative forcing, which is derived from an emissions scenario.

Obviously, future emissions of greenhouse gases and aerosols into the atmosphere depend very much on factors such as population and economic growth and energy use. For its Fourth Assessment Report, the IPCC (2007) continued to use the six SRES (Special Report on Emissions Scenarios) marker scenarios described in Table 1 (Nakicenovic et al., 2000) for the period 2000–2100, and added a number of experiments in which greenhouse gases and aerosols were held constant at year 2000 levels, CO_2 was doubled and quadrupled, and experiments with greenhouse gases and aerosols held constant after 2100, thus providing new information on the physical aspects of long-term climate change and stabilisation. Many climate modelling centres have undertaken experiments with some or all of these emissions scenarios (see, e.g. data available on the IPCC Data Distribution Centre: *www.ipcc-data.org*).

Given the large number of GCMs and associated data currently available, Smith and Hulme (1998) proposed four criteria to help determine which GCMs are suitable for climate change scenario construction:

1. *Vintage*: Recent model simulations are likely—though by no means certain—to be more reliable than those of an earlier vintage since they are based on recent knowledge and incorporate more processes and feedbacks.
2. *Resolution*: In general, increased spatial resolution of models has led to better representation of climate.
3. *Validation*: Selection of GCMs that simulate the present-day climate most faithfully is preferred, on the premise that these GCMs are more likely—though not guaranteed—to yield a reliable representation of future climate.

4. *Representativeness of results*: Alternative GCMs can display large differences in the estimates of regional climate change, especially for variables such as precipitation. One option is to choose models that show a range of changes in a key variable in the study region.

The IPCC Task Group on Data and Scenario Support for Impact and Climate Assessment (TGICA) established the IPCC Data Distribution Centre (DDC; www.ipcc-data.org) web site to facilitate the provision of GCM output and climate change scenarios to the impacts and adaptation research community. All GCMs and experiments on the DDC must have met the following criteria (Parry, 2002):

- be full 3D coupled ocean-atmosphere GCMs;
- be documented in the peer-reviewed literature;
- have performed a multi-century control run[1] (for stability reasons); and
- have participated in the Second Coupled Model Intercomparison Project (CMIP2; *http://www.pcmdi.llnl.gov/cmip/cmiphome.html*).

In addition, GCMs which have a resolution of at least 3° × 3°, which have participated in the Atmospheric Model Intercomparison Project (AMIP; www.pcmdi.llnl.gov/amip/), and which consider explicit greenhouse gases (e.g. carbon dioxide, methane, nitrous oxide, etc.) are preferred.

Despite the advances in computing technology that have enabled large increases in the resolution of GCMs over the last few years, climate model results are still not sufficiently accurate (in terms of absolute values) at regional scales to be used directly in impacts studies (Mearns, Rosenzweig and Goldberg, 1997). Instead, mean differences between the model's representation of current climate (this baseline period is currently 1961–1990) and some time period in the future

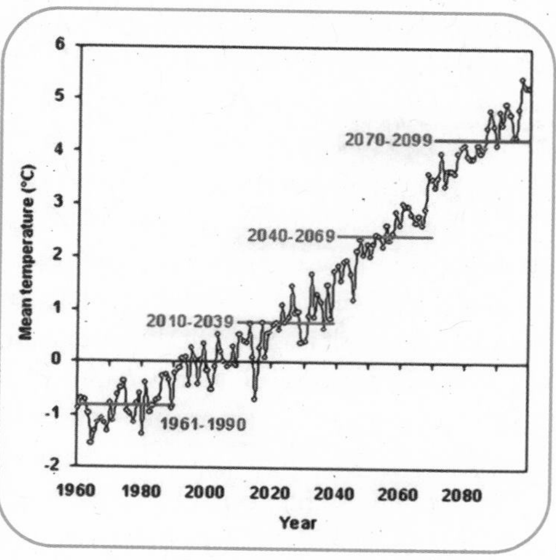

Figure 1: Construction of climate change scenarios from GCM output. The blue line indicates the 30-year mean for the 1961–1990 baseline period, while the red lines indicate the 30-year mean values for the 2020s (2010–2039), the 2050s (2040–2069) and the 2080s (2070–2099). Scenarios are constructed by calculating the difference, or ratio, between the means of the future and baseline periods, depending on the climate variable under consideration (in this case, mean surface air temperature is illustrated).

are calculated (the *climate change scenario*; see Figure 1) and then combined with some baseline observed climate data set to obtain a *climate scenario* (IPCC, 1994). Conventionally, differences (future climate minus baseline climate) are used for temperature variables, and ratios (future climate/baseline climate) are used for other variables such as precipitation and wind speed. Typically, a number of fixed time horizons in the future (e.g. the 2020s (2010–2039), the 2050s (2040–2069), and the 2080s (2070–2099)) are considered in impacts studies.

Most climate change scenarios derived from GCM output are based on changes in monthly or seasonal mean climate, although the greater quantities of model output now being archived by climate modelling centres include daily output and information on certain types of extreme events (e.g. mid-latitude cyclone intensities). Even though model output is being made readily available at finer time resolutions, it does not mean that it is any more meaningful than the output at monthly or seasonal time scales.

CLIMATE CHANGE IN THE PRAIRIES

Climate change scenarios were constructed for the Prairies chapter of *From Impacts to Adaptation: Canada in a Changing Climate 2007* (Sauchyn and Kulshreshtha, 2007), and these are reported in more detail here. Scenarios were constructed for the two main ecozones in the prairie region, namely grassland and forest, using data from climate change experiments undertaken with seven GCMs and several SRES emissions scenarios. These were the most recent scenarios available and were constructed in accordance with the recommendations of the IPCC-TGICA, the international body of experts convened to ensure consistency in climate impacts and adaptation research.

An at-a-glance summary of the changes in mean temperature and precipitation averaged over the grassland and forest regions is provided by the scatter plots in Figures 2 and 3 (see end of chapter for Figures 2 and 3). On an annual basis (Figure 2), it is apparent that the changes in mean temperature are outside of the range of natural climate variability (derived from the long control run of CGCM2) as early as the 2020s and in particular in the 2050s and 2080s. For these three future time periods, the median temperature changes (the vertical blue lines) are consistently outside of the range of natural climate variability (the grey squares) for both the grassland and forest regions. For the grassland region, some of the projected changes in annual precipitation are still within the range of natural climate variability, even by the 2080s, although the median precipitation change (indicated by the horizontal blue line) for this suite of scenarios is outside of the range of natural climate variability by this time period. In contrast for the forest region, median changes in annual precipitation are outside of the range of natural climate variability as early as the 2020s. Median annual temperature changes are similar in both the grassland and forest regions for the

2020s, 2050s and 2080s. However, the median changes in precipitation indicate slightly larger increases in the forest region in all time periods.

Figure 3 illustrates the seasonal changes in mean temperature and precipitation for the 2050s over the grassland and forest regions. For both regions the projected median temperature changes are outside of the range of natural climate variability in all seasons. This is not the case for precipitation, with median changes in this variable being within the range of natural climate variability in summer and fall in the grassland region and in summer in the forest region. Again, the median changes in mean temperature are similar in all seasons in both regions. Median changes in precipitation are similar in winter and spring, but decreases or a very slight increase are indicated in summer and fall, respectively, in the grassland region, while increases are indicated in the forest region. There is more variability in the projections of mean temperature and precipitation change in winter and spring compared to summer and fall in both regions.

One of the uses of this type of scatter plot is to identify those GCMs and experiments which indicate conditions that are more extreme than the median values for the suite of scenarios being considered. For example, there are more than 20 scenarios of climate change illustrated on the scatter plots and most vulnerability, impacts and adaptation studies would find the use of so many scenarios unwieldy to say the least. Examination of the scatter plots, however, can reveal which scenarios indicate cooler and wetter, cooler and drier, warmer and wetter, and warmer and drier conditions than the median values. Thus, the range of projected changes can be captured by considering five scenarios (the four "extremes" and one scenario representing median change values), rather than by attempting to use all of the scenarios illustrated.

To translate the scenarios of climate change from the scatter plots to a map of the Prairie Provinces, all of the scenarios were interpolated onto a common GCM grid (i.e. that of CGCM2) and the minimum, median and maximum changes were then calculated and plotted on a grid-box by grid-box basis (see Figures 4–7 at end of chapter). This means that the values in each individual grid box are not necessarily from the same scenario of climate change. Whereas these maps summarize the potential minimum, median and maximum changes in each grid box, it is not recommended that this information be used directly in an impacts model (e.g. a hydrological or crop-growth model) since the physical consistency between the different climate variables and between grid boxes is not necessarily maintained. To ensure physical consistency, individual climate change scenarios should be used.

Figures 4 and 5 illustrate the geographic distribution of the projected minimum, median and maximum changes in annual mean temperature and precipitation, respectively, for the 2020s, 2050s and 2080s. Minimum and median changes in annual mean temperature are generally homogeneous across the Prairies, but there is a gradient in the maximum changes in the 2050s and 2080s, from the largest changes in the northeast to the smallest in the southwest. Minimum changes in annual mean precipitation indicate that decreases of between 0% and 10% occur across most of the Prairies in all time periods. Median and maximum changes in annual mean precipitation indicate increases across the Prairies in all time periods. The pattern of changes in seasonal mean temperature change for the 2050s (Figure 6) indicate a very similar picture to that of annual mean temperature (Figure 4). The gradient apparent in the maximum change for seasonal mean temperature is most apparent in winter and fall. For seasonal precipitation change for the 2050s (Figure 7), minimum changes indicate predominantly decreases in precipitation, particularly in the summer season where the largest decreases occur in the southwest of the prairie region. These decreases are also apparent in the southwest of the region in summer for both the median and maximum change patterns. Elsewhere, and in the other seasons, increases in precipitation occur which are generally largest in the northeast of the region.

Another way of summarizing the climate change scenario information is to use box-and-whisker plots. These are statistical plots which provide summary information about a data sample—in this case, the climate changes over the grassland and forest regions. The box and enclosed line define the upper and lower quartiles, and median value, respectively. The whiskers are the lines extending from each end of the box to show the extent of the rest of the data. Thus, the box represents the central 50% of the data sample and the whiskers indicate the maximum and minimum data values if there is a dot located on the lower whisker. If there are outliers in the data, indicated by "+" symbols, then the whisker length is 1.5 x inter-quartile range.

Figures 8–11 (see end of chapter for Figures 8–11) are box-and-whisker plots for annual mean changes in mean temperature, precipitation and maximum and minimum temperature, respectively, for the 2020s, 2050s and 2080s. For mean temperature (Figure 8), the box-and-whisker plots are very similar for the grassland and forest regions, although there are more outliers (i.e. more extreme scenario values) in the forest region and in the 2080s, in particular. For precipitation (Figure 9), there is a much larger spread in the scenario values in the grassland region compared to the forest region, although the median changes (indicated by the horizontal line in the box) in precipitation are greater in the forest region. Changes in maximum (Figure 10) and minimum (Figure 11) temperature are similar for the grassland and forest regions, with generally more spread in the scenario values in the 2080s in both regions, and more "extreme" scenario changes in maximum temperature in the forest region.

CONCLUSIONS

This chapter has described the different methodologies and guidelines available for the construction of scenarios of climate change. Scenarios were constructed for the forest and grassland ecozones of the prairie region in accordance with these guidelines.

Averaging scenario changes across these two ecozones and presenting them as scatter plots provides a quick summary of the climate change projections over the prairie region:

- On an annual basis, median changes in mean temperature are similar in both the grassland and forest regions for the 2020s, 2050s and 2080s, while median changes in annual precipitation indicate slightly larger increases in the forest region in all time periods.
- On an annual basis, changes in mean temperature are outside of the range of natural climate variability in both regions as early as the 2020s, but in particular in the 2050s and 2080s.
- Some of the individual scenarios indicate that annual precipitation changes are still within the range of natural climate variability by the 2080s in the grassland region, although the median annual precipitation change is outside of the range of natural climate variability for this time period. For the forest region, the median change in annual precipitation is outside of the range of natural variability in all three future time periods.

Providing a spatial summary of the climate change scenarios is much more difficult and, in this case, all available scenarios were examined and then maps of the minimum, median and maximum changes in mean temperature and precipitation were plotted. Although these maps do not represent physically-consistent scenario changes on a grid-box by grid-box basis, they do give an indication of the probable range of future climate change.

- For annual mean temperature, minimum and median changes are generally homogeneous across the Prairies, indicating increases of between 0°C and 2°C in the 2020s, and between 2 and 4°C (minimum) and 4 and 6°C (median) in the 2080s. Maximum changes in annual mean temperature exhibit a southwest to northeast gradient, with increases between 2 and 4°C in the 2020s, and between 6 and 8°C in the southwest Prairies and between 14 and 16°C in the northeast part of this region in the 2080s.
- Minimum annual precipitation changes indicate slight decreases (of between 0% and 10%) across the prairie region in all time periods. Median and maximum changes indicate increases across this region in all time peri-

ods, with the eastern half of the region generally experiencing larger increases in annual precipitation. Increases are generally 10–20%, but may be as much as 40–50% in the northeastern part of this region.

ENDNOTES

1. For all GCM control runs, the atmospheric composition is set at or near pre-industrial conditions, and there are no changes in forcing for the duration of the run. Output from such a simulation provides valuable information about the stability of the model (e.g. if there are errors in the model formulation, it may drift towards an unrealistic climate over time) and the model's representation of natural climate variability.

REFERENCES

IPCC (Intergovernmental Panel on Climate Change). 1994. *Technical Guidelines for Assessing Climate Change Impacts and Adaptations*. Prepared by Working Group II [T.R. Carter, M.L. Parry, H. Harasawa and S. Nishioka (eds.)] and WMO/UNEP. CGER-IO15–94. University College: London, U.K. and Centre for Global Environmental Research, National Institute for Environmental Studies: Tsukuba, Japan, 59 pgs.

——. 2001. *Climate Change 2001: The Scientific Basis*. Contribution of Working Group I to the Third Assessment Report of the Intergovernmental Panel on Climate Change. (J.T. Houghton, Y. Ding, D.J. Griggs, M. Noguer, P.J. van der Linden, X. Dai, K. Maskell and C.A. Johnson, [eds.]). Cambridge University Press: Cambridge, New York, 881 pgs.

——. 2007. "Summary for Policymakers." In S. Solomon, D. Qin, M. Manning, Z. Chen, M. Marquis, K.B. Averyt, M. Tignor and H.L. Miller (eds.). Climate *Change 2007: The Physical Science Basis*. Contribution of Working Group I to the Fourth Assessment Report of the Intergovernmental Panel on Climate Change. Cambridge University Press: Cambridge, New York.

IPCC-TGICA. 2007. *General Guidelines on the Use of Scenario Data for Climate Impact and Adaptation Assessment*. Version 2. Prepared by T.R. Carter on behalf of the Intergovernmental Panel on Climate Change, Task Group on Data and Scenario Support for Impact and Climate Assessment, 66 pgs.

Mearns, L.O., C. Rosenzweig and R. Goldberg. 1997. "Mean and Variance Change in Climate Scenarios: Methods, Agricultural Applications, and Measures of Uncertainty." *Climatic Change* 35: 367–96.

Nakicenovic, N., J. Alcamo, G. Davis, B. de Vries, J. Fenhann, S. Gaffin, K. Gregory, A. Grübler, T.Y. Jung, T. Kram, E.L. La Rovere, L. Michaelis, S. Mori, T. Morita, W. Pepper, H. Pitcher, L. Price, K. Raihi, A. Roehrl, H-H. Rogner, A. Sankovski, M. Schlesinger, P. Shukla, S. Smith, R. Swart, S. van Rooijen, N. Victor and Z. Dadi. 2000. *Emissions Scenarios. A Special Report of Working Group III of the Intergovernmental Panel on Climate Change*. Cambridge University Press: Cambridge and New York. 599 pgs.

Parry, M. 2002. "Scenarios for Climate Impact and Adaptation Assessment." *Global Environmental Change* 12: 149–53.

Sauchyn, D. and S. Kulshreshtha. 2007. "Prairies." Pp. 275–328 in D. Lemmen, F. Warren, J. Lacroix, and E. Bush (eds.), *From Impacts to Adaptation: Canada in a Changing Climate 2007*. Government of Canada: Ottawa.

Smith, J.B. and M. Hulme. 1998. "Climate Change Scenarios." Pp. 3–1 to 3–40 in J.F. Feenstra, I. Burton, J.B. Smith and R.S.J. Tol (eds.), *UNEP Handbook on Methods for Climate Change Impact Assessment and Adaptation Studies*. United Nations Environment Programme: Nairobi, Kenya and Institute for Environmental Studies: Amsterdam.

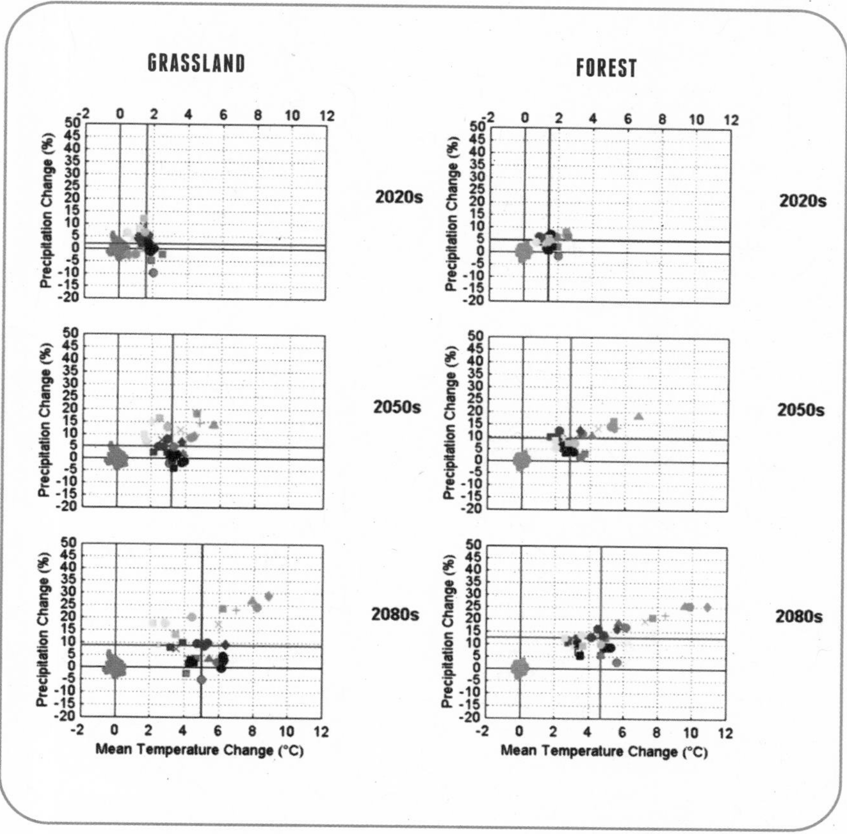

Figure 2: Scatter plots illustrating the annual changes in mean temperature (°C) and precipitation (%) averaged over the grassland and forest regions of the Prairies. Each colour and symbol represents a different GCM and climate change experiment, respectively: Grey squares—natural climate variability as modelled by CGCM2. Emissions scenarios: ♦ - A1FI; + - A1T; ▲ - A1; ★ - A1B; ● - A2; ✕ - B1; ■ - B2. Colours: CGCM2—black; HadCM3—dark blue; CCSRNIES—green; CSIROMk2—magenta; ECHAM4—red; NCARPCM—yellow; GFDL-R30—cyan.

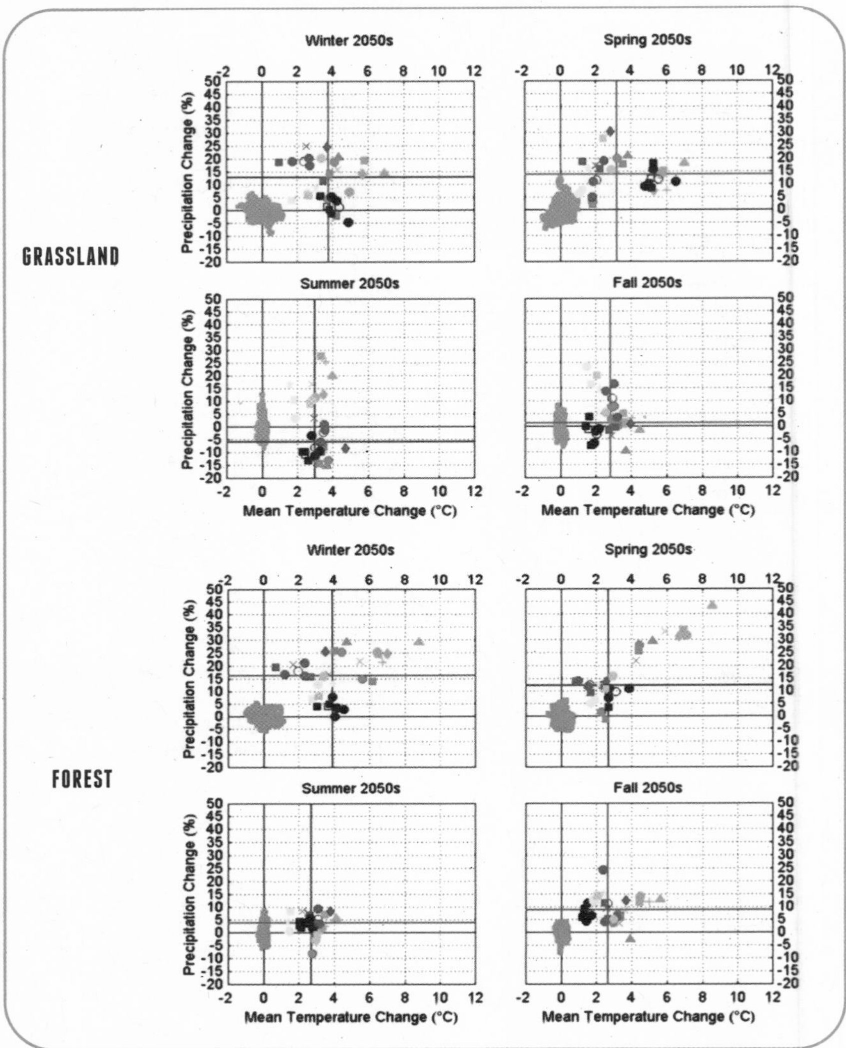

Figure 3: Scatter plots illustrating the seasonal changes in mean temperature (°C) and precipitation (%) averaged over the grassland and forest regions of the Prairies for the 2050s. Each colour and symbol represents a different GCM and climate change experiment, respectively: Grey squares—natural climate variability as modelled by CGCM2. Emissions scenarios: ♦ - A1FI; + - A1T; ▲ - A1; ★ - A1B; ● - A2; ✕ - B1; ■ - B2. Colours: CGCM2—black; HadCM3—dark blue; CCSRNIES—green; CSIROMk2—magenta; ECHAM4—red; NCARPCM—yellow; GFDL-R30—cyan.

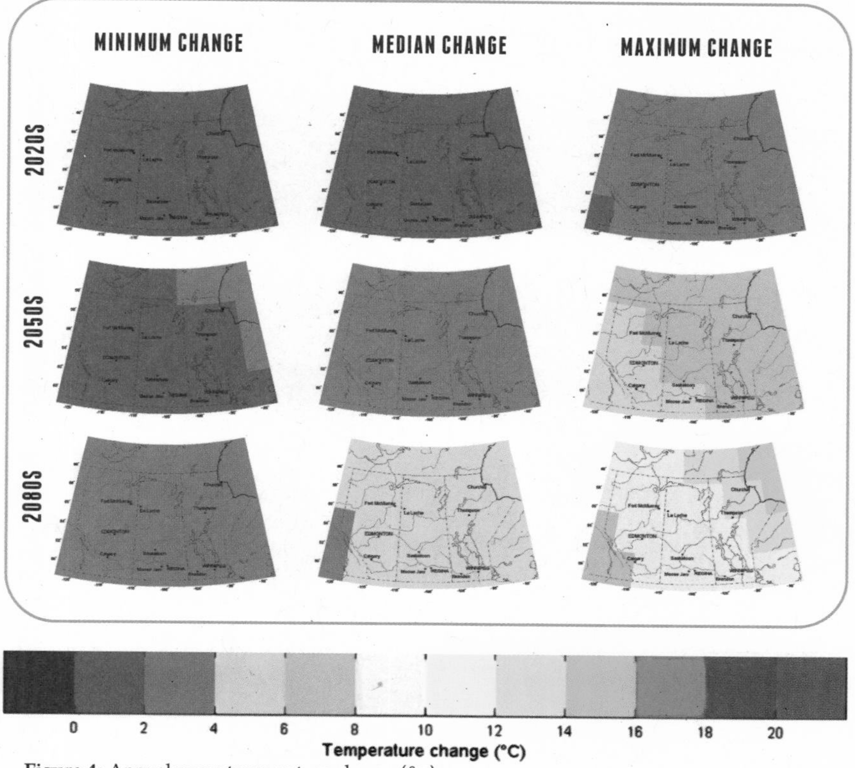

Figure 4: Annual mean temperature change (°c).

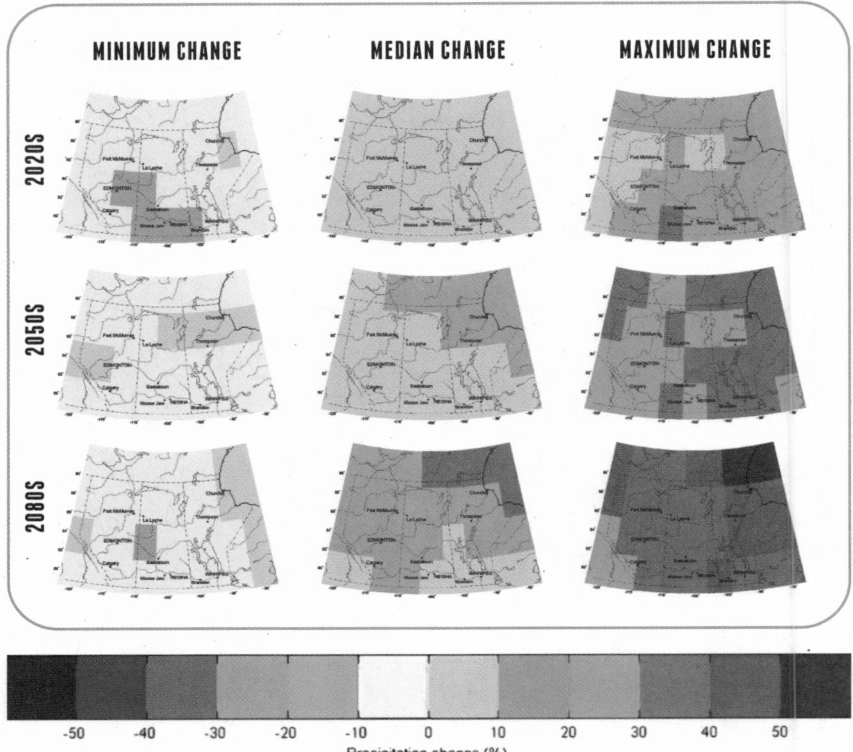

Figure 5: Annual mean precipitation change (%).

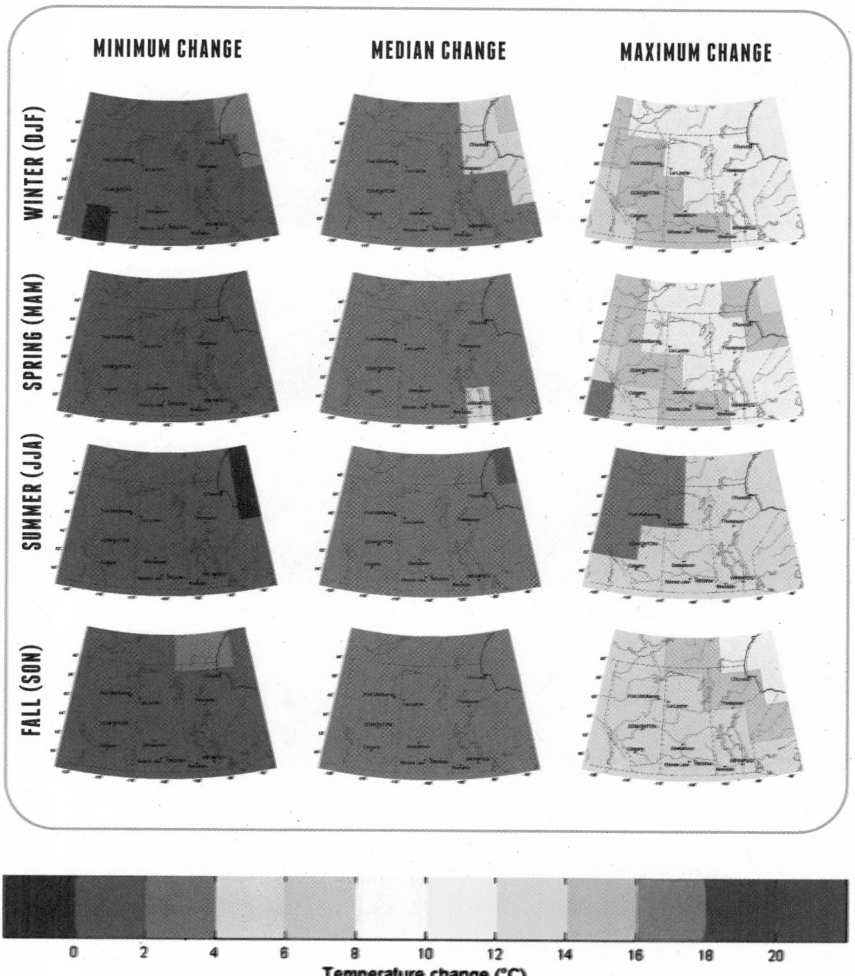

Figure 6: Seasonal mean temperature change (°c) for the 2050s.

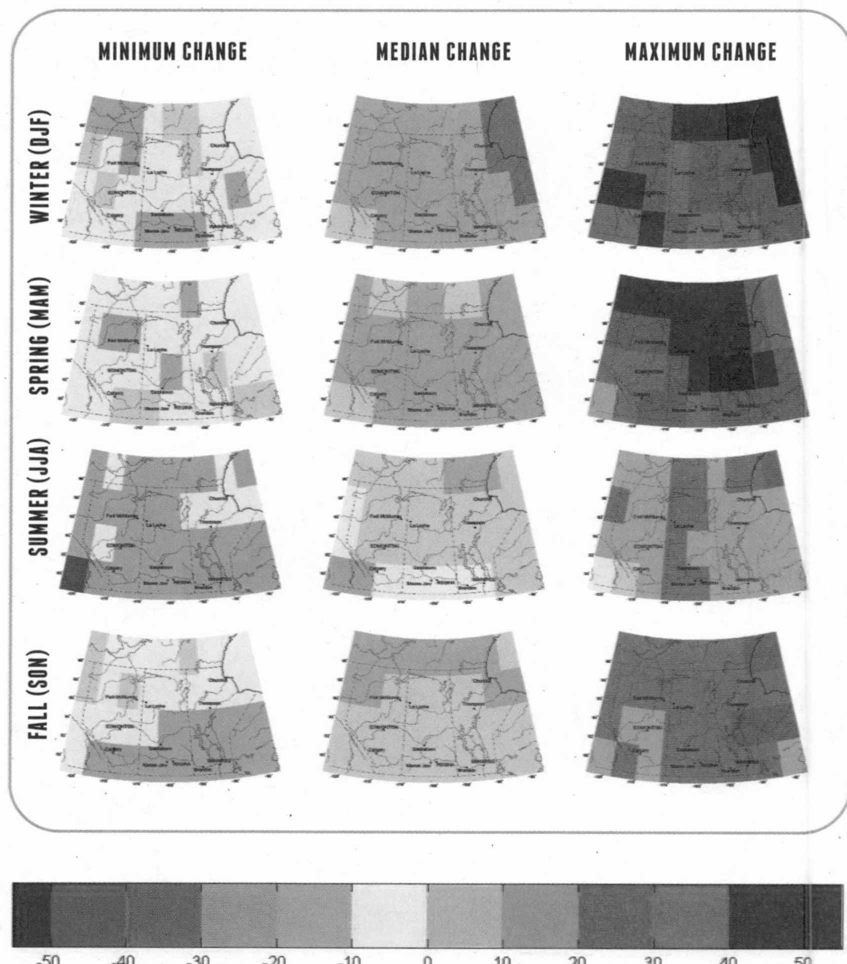

Figure 7: Seasonal mean precipitation change (%) for the 2050s.

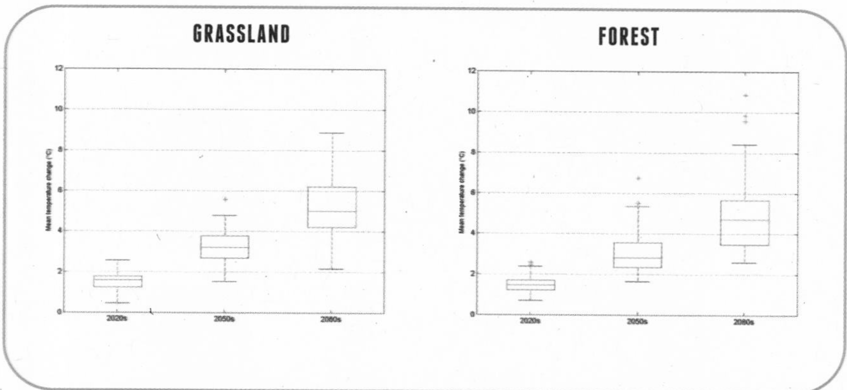

Figure 8: Box-and-whisker plots for annual mean temperature change (°c) for the 2020s, 2050s and 2080s, for the grassland and forest region of the Prairies.

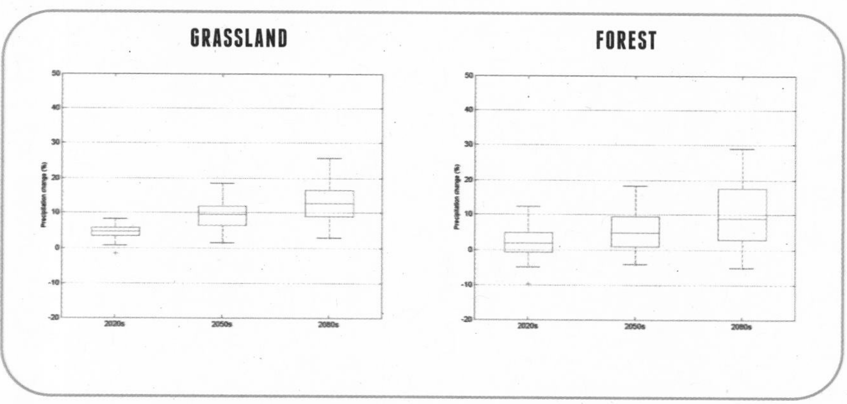

Figure 9: Box-and-whisker plots for annual mean precipitation change (%) for the 2020s, 2050s and 2080s, for the grassland and forest region of the Prairies.

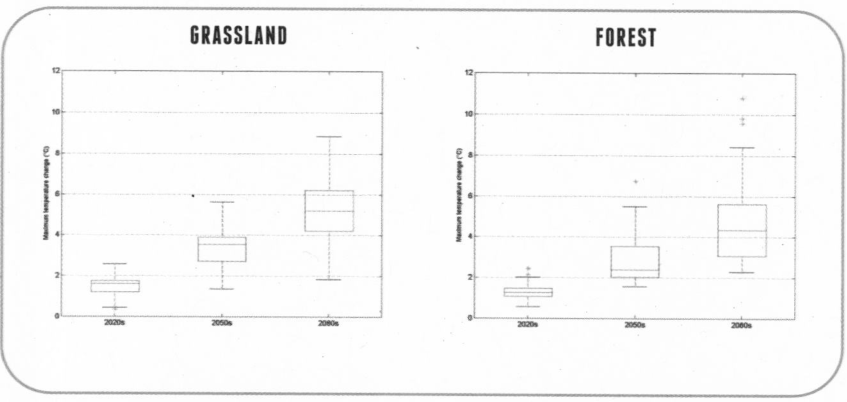

Figure 10: Box-and-whisker plots for annual maximum temperature change (°c) for the 2020s, 2050s and 2080s, for the grassland and forest region of the Prairies.

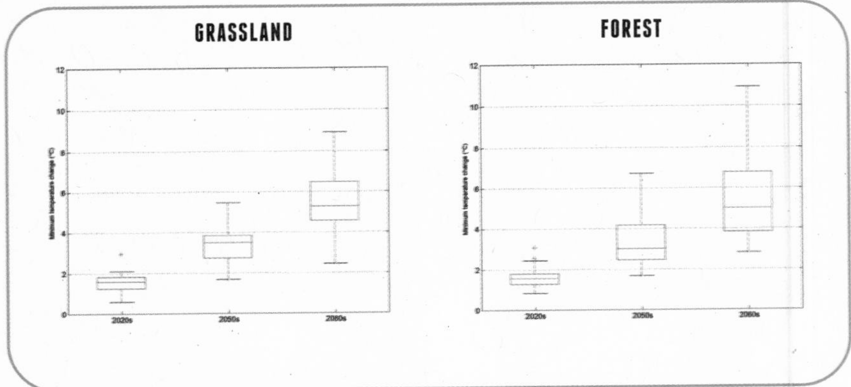

Figure 11: Box-and-whisker plots for annual minimum temperature change (°C) for the 2020s, 2050s and 2080s, for the grassland and forest region of the Prairies.

2. MAJOR THEMES

in Impacts and Adaptations

CHAPTER 5

PRAIRIES WATER AND CLIMATE CHANGE

James Byrne, Stefan Kienzle and David Sauchyn

Water is a critical component and key variable in the functioning of ecosystems and a vital resource for society. Water resources define the nature of human interaction with ecosystems. They are sensitive to fluctuation in climate, especially in dry environments like the Canadian Prairies. Some of the greatest societal stresses endured in Western Canada have been directly related to extremes in the hydroclimate system that impact every sector of society. Drought is Canada's most costly weather event. The historical record of cyclical droughts, associated with water shortages and a decline in agricultural productivity, clearly suggests that the Canadian Prairies are negatively impacted by even modest climate change. As global atmospheric change continues, our water resources will come under increasing stress. The Prairie Provinces' water resource systems have already undergone substantial changes due to human activities, including hydroelectric dams and reservoirs, water impoundment and diversion for extensive irrigation development, industrial utilization, coolant water for power plants and other industry needs, extensive diversion for use in oil sands, and for supplying water to cities and towns.

The water balance defines the nature and distribution of the ecosystems of the Prairie Provinces. A balance of precipitation to water loss by evapotranspiration separates the forested area from the grassland ecoregions, where snowmelt and rainwater are consumed in most years by evapotranspiration, and only upland areas with island forests generate runoff. Over much of the northern half of the Prairie Provinces, short cool summers and long cold winters, combined with more precipitation than in the grasslands, support boreal forest. Here, runoff occurs in all seasons, as there are extensive wetlands that store snowmelt and rainfall and release it slowly. Between the grasslands and boreal forest, the parkland—a mix of woodlots and grasslands—is a transition zone where, historically, prairie fires in dry periods limited the expansion of forest, which expanded over grasses during wet periods. These regions typically feature a large snowmelt streamflow pulse in the spring, and highly variable rainfall runoff. This ecosystem transition, which takes place over a thousand kilometres from southeastern Alberta to northern Manitoba can be experienced in a long day hike in the Rocky Mountains. The temperature declines rapidly and precipitation increases at a similar rate with increasing elevation in mountain environments. In approximate terms, a 300-metre change in elevation is equivalent to about 300 kilometres of latitude change in climate (Ahrens, 2005). Climbing from the edge of the grasslands to the mountain peaks in Waterton Park, Alberta, is the equivalent of a climate transition from the southern Prairies to the arctic tundra.

The hydrology of Western Canada is dominated by cold weather processes: snow accumulation and melt, frozen soils and glacier runoff (Pomeroy, de Boer and Martz, 2007; Toth et al., 2009). Therefore, the prairie surface water levels and processes have marked seasonality. Water is mostly stored as snow and ice in fall and winter. In early spring, the rapid release of water from melting snowpacks lying over frozen ground produces high surface runoff. Snow is about a third of the annual precipitation, but accounts for about 80% of the runoff and for most of the water that fills lakes, sloughs, dugouts and reservoirs, and recharges soil moisture and groundwater storage. In late spring and early summer, water is stored as soil moisture, and very little rainfall ever runs off. Plant water use rapidly depletes soil water and dry soils absorb most summer rains. The exceptions are intense summer convective storms that track over narrow swaths of land. There can be substantial local flooding from these storms, particularly in cities and towns, but such random events are not dependable supplies of surface water. Generally, the period from mid-summer to fall is characterized by the least rainfall and, therefore[m1], the least soil moisture, actual evaporation and runoff. Evapotranspiration usually is close to rainfall in less cool years, and especially in the eastern and northern agricultural zone.

The Rocky Mountains capture large snow accumulations that shed runoff into the Saskatchewan-Nelson River system in the southern half of the Prairie Provinces and the Peace and Athabasca rivers that flow north to the Arctic Ocean. Most of the population and irrigated agriculture depends on water from the North and South Saskatchewan rivers, and thus mountain snowmelt. Beyond the mountains and forested foothills, in the rainshadow of the Rocky Mountains, low annual precipitation in the range of 300–400 mm generates very little runoff. Within the flat grasslands, there are large areas of internal drainage that do not contribute to the major river systems. Numerous small, closed depressions (sloughs, wetlands and dugouts) tend to be disconnected from the stream network, but can be important local recharge for shallow groundwater aquifers. Local prairie water resources are limited and very sensitive to changes in climate and land cover. Management of these limited and variable water supplies is challenging in periods of extreme drought. Thus, few prairie streams can be managed for substantial water consumption or irrigation, and groundwater and mountain-fed rivers are the most reliable water supplies.

This chapter illustrates some concerns that the scientific community has regarding the impacts of climate change on water resources and aquatic ecosystems in the Prairie Provinces. We review a wide range of science literature that has reported on historical observations and expected changes to the climate and water cycle under global climate warming. Our discussion of the various climate warming impacts on the water cycle includes the effects of changing precipitation and temperature, the consequences for rivers and watersheds, extreme weather events, changes to water quality, and adaptation to water management to accommodate the response of water resources to climate change.

CHANGING TEMPERATURES IN THE PRAIRIE PROVINCES

All other things being equal, warmer climates lead to greater water use by ecosystems, and declining water supplies. The historical climate record demonstrates substantial warming has already occurred. For 10 climate stations in the Prairie Provinces, Schindler and Donahue (2006) reported increases in mean annual temperatures of 1.0 to over 4.0°C since the early 20th century. Zhang et al. (2000) reported a spring warming trend in Manitoba and northern British Columbia. Gan (1998) determined that, in the past 50 years, the Prairies were experiencing an extensive warming and relatively significant drying trend. He detected a significant warming in January, March, April, and June, and concluded that the statistical results were strong enough to suggest that the warming experienced in the Prairies is likely attributed to global warming. Caprio et al. (2009) have shown that minimum temperatures are rising rapidly at locations in British Columbia and Montana, with the greatest warming trend in late winter, early spring and in midsummer. Gameda et al. (2007) reported an opposite trend for midsummer in agricultural regions of the Prairies. They found that mid-June to July maximum temperatures have declined in the cultivated regions. This very specific midsummer cooling is attributed to strong regional influences of large-scale declines in summer fallow land management practices. However, analyses by Millett, Johnson and Guntenspergen (2009) of temperature trends for a series of climate stations located across the entire southern Prairies and into the Dakotas reported average daily air temperatures have risen about 1°C in 95 years compared to about 0.6°C of global warming in the same period. Millett, Johnson and Guntenspergen (2009) also report greater winter warming overall due to substantial increases in minimum temperatures. They further report that historical trends in the Palmer Drought Severity Index (PDSI) values reflected a wetter water balance for the far southeastern Prairies, but a general drying trend in the western Canadian Prairies.

The data reported by Gameda et al. (2007) suggest that the general warming trend observed over the Prairies has been mitigated by cooling in summer associated with changes in agricultural summer fallow practice. Regional warming may be greater than reported by Gan (1998), Caprio et al. (2009), and Millett, Johnson and Guntenspergen (2009), as land-use changes may be mitigating overall global warming effects. But any cooling benefits of the reduction in summer fallow have likely been negated because summer fallow is no longer a widespread practice; so continued warming of the Prairies is likely. Overall, the literature discussed above supports the contention that warming for the Prairie Provinces will exceed the global average, exacerbating our global warming challenges.

PRECIPITATION

Changes in precipitation in the historical record are difficult to define due to the extreme variability in precipitation events. Generally, climate scientists expect higher global precipitation under a warmer climate due to an increase in sea surface evaporation. However, the distribution of precipitation on the Prairies is governed by other geographic factors. A brief discussion of these factors is helpful to understanding likely changes in precipitation distribution as discussed in the recent science literature.

The southwestern prairies region is dry because the Cordilleran Mountains to the west exert a strong rainshadow effect. Pacific air masses rise and cool over the mountains, and the water vapour condenses as snow and rain on the western slopes. These Pacific airflows descend onto the plains, now much warmer and drier, in the form of Chinook winds. The southern prairies' summer is hottest and driest when the desert air masses of the United States southwest expand over the western part of the continent, often bringing extreme heat and very dry conditions. The southeastern Prairie Provinces are much wetter primarily due to the influence of much wetter air masses originating in the Gulf of Mexico. The polar air masses that dominate the prairie provinces' winter weather will weaken under global warming, allowing the hot dry air of the American southwest and the hot humid air of the American southeast to more routinely affect the Prairie Provinces. The scientific literature suggests historical declines in mean annual precipitation have occurred for the western Prairie Provinces, but the opposite may be happening for eastern regions. Millett, Johnson and Guntenspergen (2009) examined 20th-century precipitation trends and found that annual precipitation is declining in some parts of Alberta, but an opposite trend was detected for stations in southeastern Saskatchewan and Manitoba. Grundstein (2009) used a moisture index to assess historical conditions over the continental United States for 1910–99. For the western states, relatively close to southern Alberta and southwestern Saskatchewan, the index shows a general drying of the climate. But for the eastern states, close to southeast Saskatchewan and southern Manitoba, the historical trend shows a more humid climate. Hughes and Diaz' (2008) work on the continental United States found similar regional historic trends in states near the southern Canadian Prairies. Schindler and Donahue (2006) reported declines in annual precipitation of 14 to 27% since early in the 20th century at six western prairie province climate stations with reliable long term records. Older work by Gan (1998), using the period 1949–89, suggests that the southern half of the Prairies Provinces had become warmer and somewhat drier. Gan's work uses more western than eastern climate stations, which may explain a bias towards general warming and drying. Overall, the concerns about declining precipitation are worrisome for all regions. However, the greatest concern is for regions further west. Global climate model projections (Chapter 4) suggest that most future declines in precipitation will occur in summer, when natural and human systems have the highest demand for water.

The concerns over precipitation trends are exacerbated by changes in storm behaviour. One study (Akinremi, McGinn and Cutforth, 1999) determined that the total number of precipitation events on the Canadian Prairies increased from 1920–95. Over half of the days with precipitation in that time frame had recorded depths of less than 5 mm, and that precipitation increased in the period 1956–95 due to a higher frequency of low-intensity events. Small precipitation events are captured in the vegetation, resulting in more water being lost by evapotranspiration, and less water reaching the soil, rivers, lakes, wetlands, and the water table. Other more recent studies (e.g. Stone, Weaver and Zwiers, 2000; Kharin et al., 2007), however, show recent trends and model projections of increasingly intense precipitation. If precipitation is delivered in fewer and more extreme events, the intermittent weather will tend to be drier and, thus, both drought and unusually wet conditions are expected to be more common with climate change.

RIVERS AND STREAMS OF THE PRAIRIE PROVINCES

The major rivers of the Prairie Provinces are shown in Figure 1. The Athabasca River originates primarily from snow and ice melt in the Rocky Mountains of Jasper National Park. The Peace River has similar roots in the Rocky Mountains of northern British Columbia and the northern regions of Alberta's eastern slopes. Together, these large northern rivers form the Slave River that flows northward into the Mackenzie River that flows to the Arctic Ocean. The Saskatchewan River systems drain the Rocky Mountains from Banff National

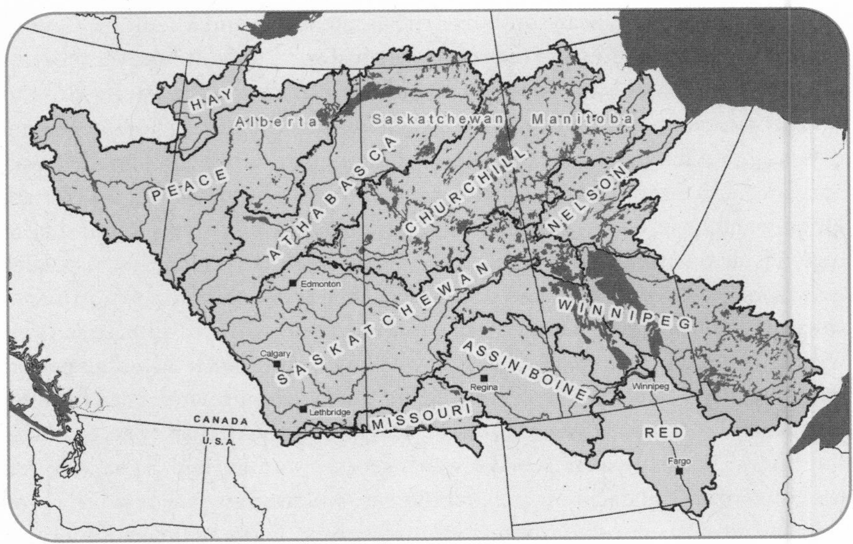

Figure 1. The major river basins of the Prairie Provinces.

Park to Waterton-Glacier International Peace Park on the Alberta-Montana border. The Saskatchewan River system flows into the north basin of Lake Winnipeg. Other major rivers that flow into Lake Winnipeg include the Red River draining northward into Manitoba from the Dakotas and Minnesota, the Winnipeg River that drains off the Precambrian Shield of Ontario and northern Minnesota, and the Assiniboine River that flows out of southern Saskatchewan. The Nelson River flows from the north end of Lake Winnipeg north to Hudson Bay. The other major river in the Prairies is the Churchill River that drains large parts of the wetlands and bogs of the northern boreal forest of all three provinces. These major watersheds include countless small tributaries and, in many cases, closed sub-basins that never contribute to a water body beyond their own borders. The following discussion describes how past and future climate did—and likely will—link to the water balance for the major rivers of the Prairie Provinces.

CLIMATE AND WATER: THE PATH WE ARE ON

Discussions of our changing water balance indicated that both temperatures and precipitation play a major role in the water supply. Spring melting of snow accumulations in the Rocky Mountains is the source of streamflow to most of the population on the Prairies. Stories in the media have made much of the decline of the glaciers in the Rocky Mountains, sometimes implying a state of constant drought will follow the demise of the last glaciers. Such is not true, but it does not mean glacier melt is unimportant. One study (Demuth and Pietroniro, 2003) concluded: "The reliability of water flow from the glaciated headwater basins of the upper North Saskatchewan River Basin has declined since the mid-1900s. Hydrologic and ecological regimes dependent on the timing and magnitude of glacier-derived meltwater may already be experiencing the medium-long-term impacts of climate change discussed by the IPCC." The area studied is in close proximity to the headwaters of the Athabasca and Saskatchewan River systems. Both rivers are fed by valley glaciers originating in the Columbia Ice Fields. Comeau. Pietroniro and Demuth (2009) defined the contribution of glacier melt to the North and South Saskatchewan rivers at Edmonton and Calgary, respectively, as less than 3% of annual regulated flow. However, they estimated that glacier melt might represent up to 10% of annual flow for many headwater streams in the Rocky Mountains. Obviously, the loss of glaciers will not have a large affect on the average annual water balance for major rivers, but in warm, dry years, glacier melt is a very important source of water in late summer. Overall, the demise of the Rocky Mountain Ice Fields will seriously challenge our adaptive capacity by removing a viable streamflow source in the driest years.

The loss of the mountain glaciers is an indicator of the combined impacts of declining winter snow accumulations and longer melting periods. Exploring the combined impacts of warmer winters, longer, warmer summers, and changes in the snowfall regime are critical to understanding the likely response of the larger water supply curve for the Prairie Provinces under climate warming. Specific studies for the Prairie Provinces on streamflow and climate change all suggest a clear trend in declining water supplies. However, there is much additional literature that addresses climate change impacts on water over the western mountain regions of North America, which confirms the trends for Canadian watersheds. Work in the 1980s (Revelle and Waggoner, 1983; Gleick, 1987a and 1987b; WMO, 1989; Byrne, 1989) was forecasting substantial declines in streamflow supplies and/or water demands for locations in western North America, and more recent work supports those early predictions. Low stream flows were observed with increasing frequency on the Canadian Prairies, and low flow magnitude was declining for the period 1912–93 (Yulianti and Burn, 1998). Burn (1994) investigated the spring snowmelt and subsequent runoff events as indicators of climate change in 84 unregulated river basins across Western Canada. He found that rivers located at higher latitudes experienced greater advances in the timing of the peak spring runoff event. Thirty percent of the unregulated rivers showed a statistically significant change to earlier spring runoff. Earlier spring runoff will result in a decline in late season streamflow, as more heat units associated with longer, warmer summers increase evapotranspiration potential, effectively transpiring soil water and increasing potential soil water storage for summer precipitation that does occur. Zhang et al. (2001) found that mean annual streamflow in Canada decreased in the latter 50 years of the last century.

Studies with a broader geographic region report results that confirm the general decline in streamflow in the west. Leung and Ghan (1999) looked at climate change impacts on the Pacific Northwest of the United States. Their studies suggest that the combined effects of surface temperature warming and precipitation changes due to a doubling of global GHGs will result, on average, in snow cover reductions by up to 50%, causing large changes in the seasonal runoff. Leung et al. (2004) used output from a series of GCMs. They found that relatively small changes in temperature and precipitation would result in large changes in snow accumulation, and associated volume and timing of runoff. Mote (2003) analyzed a wide range of spring snow time series of varied lengths for the Pacific Northwest, and showed substantial declines in April 1 snowpack at most locations. These results were supported by the work of Hamlet et al. (2005), which found that historical April spring snow accumulations were declining in the same region for the period of record, and that of Mote (2006), which found consistent declining trends for 1950–97 snowpacks for much of western North America, including the Rocky Mountains of Montana and Alberta.

Trend analysis studies of the historical streamflow records found prominent flow declines of rivers in Alberta and those in adjacent states and provinces (Rood et al., 2005). Overall, rivers draining from the triple divide located in Montana displayed a reduction in discharge of 0.22% per year for the record period available. Further trend analysis work on the historical streamflow record (Rood et al., 2008) concluded that winter flows (especially March) were increasing slightly, spring run-off and peak flows are occurring earlier in the year, and summer and early autumn flows (July–October) were considerably reduced. This trend is indicative of the expected shift of the hydrological balance under climate warming scenarios (Zhang et al., 2001). The increase in the March streamflow may be attributed to an increase in spring precipitation as rain, combined with an earlier snowmelt (Zhang et al., 2001; Burn, 1994; Gan, 1998; Leith and Whitfield, 1998).

All scenarios of future hydroclimate and water supplies consistently project an earlier onset of the spring snowmelt, a tendency towards a more rainfall-dominated hydrograph, and reductions in the annual and spring flow volumes in the 2050s and 2080s. For the South Saskatchewan River Basin, reductions in annual flow, and a shift of the peak annual flow from summer to spring and winter, are projected for the major mountain tributaries (Lapp, Sauchyn and Toth, 2009). The median scenario for the South Saskatchewan River at Medicine Hat is a reduction of 8.5% in mean annual flow by mid-21st century. Lapp et al. (2005) modeled historical and future snowpack for the Oldman Basin in southern Alberta, predicting shorter winters and declining snow accumulations in the low- and mid-mountain elevations. Their results suggested an average decline in spring snow runoff of around 40% by the period 2020–50. Merritt et al. (2006) found similar results for simulated runoff for the Okanagan Basin in southern British Columbia. Finally, MacDonald (2008) forecast declines in spring snowpacks for the St. Mary River in northern Montana under a range of climate scenarios. Larson (2008) predicted that hydroclimate and water supplies in the St. Mary headwaters would lead to substantial declines in spring runoff. However, the work of MacDonald included a scenario where GHG emissions were well controlled in this century. In that case, minimal declines in water supply were predicted. These results are very encouraging as the work suggests humanity can minimize global warming impacts on water if we substantially reduce global GHG emissions.

IMPLICATIONS OF CLIMATE CHANGE FOR MAJOR WATERSHEDS OF THE PRAIRIE PROVINCES

The literature discussed herein strongly supports the notion that regional climate—and related regional water resources challenges—will intensify under global climate warming. Overall, scientists expect larger departures from average conditions and more frequent extreme events. Specific concerns include:

- Much of the western half of the continent is showing historical trends that suggest an increasing influence of the dry tropical climate of the American Southwest. Consequently, we can expect negative impacts on all watersheds originating in the Rocky Mountains and on the western Prairies. The magnitude of the change in these watersheds is uncertain, but modelling results suggest it will be substantial. The fact that so many researchers have already identified statistically significant trends towards lower runoff supplies supports the contention that streamflow reductions will be significant.
- Longer warmer winters will reduce snowpack accumulations in the mountains and on the southern Prairies, leading to lower overall annual runoff in the major rivers, with the reductions more pronounced in the southwestern regions.
- Changes in streamflow timing will result from earlier melting of snow across the Prairie Provinces. More runoff will occur in March, April and May. This will reduce streamflow supplies in summer, fall and into winter. The Peace, Athabasca and Saskatchewan river systems will suffer under this scenario. Stress will build for ecosystems and society, as reduced streamflows and increased water demands lead to routine supply shortfalls.
- Warmer and longer summers will increase water demand for agriculture generally, stressing rain-fed crops across the southern Prairies. Water use, including around 500,000 hectares of irrigation developments, has essen-

tially been fully allocated the Bow and Oldman rivers in southern Alberta. Without further adaptation and management of water demand, reductions in streamflows could jeopardize the irrigation of tens to hundreds of thousands of irrigated hectares in those basins. It is estimated that a 10 to 20% expansion in irrigation in the Bow and Oldman basins is possible with water conservation (Alberta Environment, 2003). Further north and east, rivers are not so heavily allocated, and some irrigation development may occur here in response to warmer and drier conditions. Potential for significant increased human uses (including irrigation) exists in portions of the SSRB in Saskatchewan.

- The Brace Report (PFRA, 2005) suggests that a 500% irrigation expansion is possible in Saskatchewan, with most of the water coming from the South Saskatchewan River. However, these adaptations are not likely to proceed without balancing the environmental, social and economic impacts.

A further problem that may alter both water supply and quality is the occurrence of dramatic forest fires that have occurred in recent years over many regions of western North America. Large-scale forest fires create flash runoff that erodes soils and degrades water quality. Some studies express concern over increasing forest fire risk due to climate change. (Leung et al. 2004) and two prominent studies (Running, 2006; Westerling et al., 2006) explicitly link recent dramatic fire seasons in the western United Sates to climate change. The occurrence and impacts of forest fires are closely connected to climate and weather (Carey et al., 2006).

For the eastern Prairies, warmer temperatures are not likely to create drought stress in most years, as adequate additional precipitation may occur thanks to the humid air that dominates the eastern half of the continent. However, the boundary region between the wet and dry zones that lies over eastern Saskatchewan has shown significant shifts over the years. This region will likely endure much greater interannual variation between intensified wet and dry cycles.

Overall, the greatest risk to the Red and Winnipeg rivers due to climate change may be an increase in flood potential. The Red River has had several high flow years recently, and in April 2009, a cool winter with large snow accumulations and heavy ice cover has once again caused severe flood stress for North Dakota and southern Manitoba.

Climate change will impact the boreal forest and far northern taiga that make up the Churchill River drainage system. Studies forecast increasing climate extremes and probable alterations in local and regional hydrologic conditions due to melting of permafrost under shorter winters and warmer summers (ACIA, 2004; Majorowicz, Skinner and Safanda, 2005; Pearce, 2005). More severe and persistent droughts will cause declining soil water in many years (Sauchyn, Stroich and Beriault, 2003), which will likely increase forest fire extent and net areas burned. Stocks et al. (1998) forecast large increases in the areal extent of extreme fire danger for boreal forest in Canada and Russia under a doubling of atmospheric CO_2. Flannigan et al. (2005) forecast the area burned in Canada will increase 74–118% by the end of this century in a 3× CO_2 scenario. In recent history, during extreme droughts, organic soils have dried and burned with forests, resulting in almost total loss of vegetation and soil cover and, subsequently, the ability to store water locally. Under these conditions, runoff events are less buffered, with subsequent instantaneous responses and the risk of causing flash floods. The drought of 2001–03 extended much further north than expected, bringing drought and a bad fire season to the western boreal forest.

CHANGES FOR SMALL WATERSHED SYSTEMS

Most of the research on climate change and prairie water resources, as described above, has focused on the large rivers that flow from the eastern slopes of the Rocky Mountains and supply most of the population of the Prairie Provinces with water. The smaller rivers and streams that flow from prairie uplands support important riparian ecosystems and are the water supply for much of the rural populations and some larger communities. The few studies of the climate and hydrology of these prairie-sourced streams include long-term research in central Saskatchewan (Pomeroy, de Boer and Martz, 2007). Recent studies in these watersheds (Fang and Pomeroy, 2008) suggest that, in the near term, projected climate change could produce an increase in annual runoff. With further climate warming, however, annual runoff would begin to decline, as winter snow cover would become discontinuous.

In the prairie region, groundwater is the dominant source of late season streamflow in the form of baseflow. If groundwater levels decline due to reduced recharge, prolonged drought or over-pumping, the baseflows will decline to levels where water supply and quality will be adversely affected. This occurred in central Alberta during the 2001–03 drought (Kienzle, 2006), with dire consequences for some water users, such as a collapse of streamflow of the upper Battle River, resulting in severe water restrictions for domestic users, or the shutdown of power plants due to the lack of available cooling water.

Increased rainfall in early spring and late fall will enhance soil moisture and contribute to groundwater recharge when soil water levels are high. Soil moisture is essential for ecosystem and crop productivity. With a warmer climate imposing higher evapotranspiration rates, soil moisture will be utilized, and declining ground moisture retention will lead to a decrease in groundwater recharge and a slow but steady decline in the water table.

EXTREME EVENTS

No single extreme weather event may be explicitly connected to climate change. However, the scientific literature discussed above supports the contention that extreme events will become more common in the future under a changing climate, and what might have been an extreme event in the past may become a commonplace challenge of the future. Recent flood and drought events (Environment Canada, 2007, 2008, 2009) demonstrate the increasing pressures on water resources management and what can be expected in the future according to climate change scenarios that include extreme hydroclimatic events that will increase in severity and frequency. Therefore, these weather extremes are understood to be harbingers of what will in the future be much more common:

- Drought persisted in east-central Alberta and west-central Saskatchewan. From July 1, 2008 to the end of June 2009, total precipitation at Edmonton was about 234mm, less than half the average expected total of 482 mm.
- The summer 2009 drought ended in typical fashion with intense summer rains that caused, for example, flash floods and closure of the Yellowhead Highway near Denholm, Saskatchewan on July15.
- In 2007, Alberta farmers filed over 4,700 crop damage claims—the most ever. Saskatchewan farmers filed 14,000 claims, slightly less than in 2006, when they filed a record number of insurance claims. Manitoba was hit by a massive thunderstorm that includes hailstones as large as grapefruits, resulting in write-offs of many vehicles in Dauphin;
- Canada's first-ever recorded F5 tornado, the most intense such storm possible on the Fujita intensity scale, occurred on June 22, 2007, in Manitoba. A day later, multiple tornadoes passed through southern Manitoba.
- In 2006, a record 221 hailstorms occurred in the three Prairie Provinces, including golf ball-sized hail in Calgary on July 6 and storms almost as severe on July 7 (Environment Canada, 2007).
- The number one weather event of 2004 in Canada, the Edmonton thunderstorm, dropped 150 mm of rain in under an hour on a city already saturated by earlier storms. Losses were estimated at $175 million (Environment Canada, 2004b).
- There was unprecedented flooding of the western reaches of the South Saskatchewan River in 1995 and similar water levels again in 2004. The Red River Floodway was nearly breached, critical flood protection measures were undertaken, and the floodway managed to protect the areas of greater Winnipeg in 1997. However, there was significant flooding damage inflicted on rural communities, businesses and agricultural land.
- In spring 2004, the Canadian Wheat Board reported Alberta and Saskatchewan fields were drier than at any time in recorded history.

- Environment Canada (2004b) characterized the 2002 crop year as "the worst ever for farmers in Western Canada."
- The drought of 2001–03 in central Alberta was the most severe on record (see Chapter 16). The Palmer Drought Severity Index (PDSI) in 2002 at Lacombe, Alberta, was -3.6. This is an all-time low for almost 90 years (Kienzle, 2006). The most severe impacts on soil moisture and groundwater levels are from multi-year droughts.
- In 2001, the St. Mary River Irrigation Project in southern Alberta had insufficient water to meet annual allocations—farms were only provided with 60% of their water needs.

CLIMATE CHANGE IMPACTS ON WATER MANAGEMENT

Climate change will exacerbate threats to clean water and aquatic ecosystems—our most essential water resources. The natural health and wealth of the Prairie Provinces is intimately linked to the quality and quantity of the water resources. Environment Canada (2001, 2004a) documented the threats to water quantity and quality across the country, including many problems facing the Prairie Provinces. Diversions from one watershed to another, dams to store water—mainly for irrigation—as well as chemical, biological and thermal pollution have already seriously altered water quality and aquatic ecosystem health in most prairie regions.

With the potential for higher rates of soil erosion from intense rainfall, especially when vegetation cover is reduced by drought or fire (Chapter 23), climate change may produce sedimentation problems and add nutrients to local water systems, leading to eutrophication of local water bodies; runoff may also carry waste products and could lead to increased pathogen loading in streams in summer. Pathogen loads are already an issue in many prairie watercourses (Hrudey and Hrudey, 2004; Hyland et al., 2003; Johnson et al., 2003; Little, Saffran and Fent, 2003). Urban runoff and sewage effluent contribute additional chemicals, nutrients and microbial populations.

Green masses of cyanobacteria, often called algal blooms, routinely cover much of Lake Winnipeg (Casey, 2006) in summer. These blooms may cover so much of the lake that they are best documented from space (see Figure 2).

Dilution capacity will decline under climate change as streamflows decrease and lake residence times increase accordingly. Therefore, global warming is further reason to change the long- standing view that "dilution is the solution to pollution." This attitude and cumulative pollution loads that are gradually destroying freshwater ecosystems could, when amplified by climate change, push aquatic ecosystems beyond a threshold of ecological collapse (Couillard et al., 2008). Recovery from ecological collapse can take many human lifetimes. A warmer climate regime will likely increase the risk of impact to both ecosys-

Figure 2 On the left, is a satellite image of Lake Winnipeg during the algal bloom of August 7, 2008, just ten days before the algal bloom washed up on Connaught Beach. The two images on the right are algal bloom buildup on Connaught Beach in August 2008. All photos courtesy of Dr. Greg McCullough, University of Manitoba.

tems and human health if watercourses have lower volumes of water, and likely warmer water that will enhance algal growth and alter ecosystems evolved for cooler temperatures.

Declining water supplies and increasing water demands will lead to calls for increased storage, which may support the construction of more dams and reservoirs. However, reservoirs are not greenhouse gas-neutral (St. Louis et al., 2000), and such structures disrupt and segment ecosystems. Impacts of dams and diversions on water include (Environment Canada, 2001):

- thermal stratification within the reservoir and modification of downstream water temperatures;
- eutrophication;
- promotion of anoxic conditions in hypolimnetic water and related changes in metal concentrations in outflow;
- increased methylation of mercury;
- sediment retention;
- associated changes in TDS, turbidity and nutrients in the reservoir and discharged water; and
- increased erosion/deposition of downstream sediments and associated contaminants.

Flow diversions can also produce major changes in water quality. The most dramatic shifts result from mixing of waters from disparate hydro-ecological systems (e.g. across major hydrologic divides or from freshwater to estuarine environments), resulting in changes in chemistry, temperature and sediment. In addition, the transfer of fish, parasites and pathogens can accompany such mixing.

The combined effects of changes to water quantity and quality will require changes to land- and water-management practices. Currently accepted best management practices may no longer fully address the expected changes from a future climate. New water management strategies may be required (e.g. earlier releases to irrigators; balancing water competition between agricultural, energy and urban uses). The timing of spring seeding and fall harvesting may shift earlier in each season, and changes to pest management and soil moisture management may be required (Chapter 7). With potentially increased frequency of extreme events and wider hydrologic variability (drought, flood), new adaptation strategies will likely be necessary to help dryland agriculture and irrigated agriculture cope with a more volatile climate. These examples suggest that research and planning is necessary to determine future adaptive practices to help local decision-makers address new challenges and opportunities posed by a different climate and water regime.

CONCLUSIONS

In this chapter, we reviewed a wide range of scientific studies that have reported recent and expected changes to the climate and water cycle of the Canadian Prairies under global warming. These studies present various tendencies in observed and projected precipitation and water levels. Generally, differences among studies do not represent disagreement among the scientists, but rather the variability and complexity of the water cycle, such that the monitoring and modelling of precipitation and water levels can reveal different trends for different time periods and watersheds. Dry climates, like in the Canadian Prairies, have the most variable precipitation and runoff from year to year. Streamflow also varies at other time scales, from a few years to decades to multiple decades. In Western Canada, this low-frequency variability in the hydroclimate has been linked to oscillations in surface temperature in the Pacific Ocean. As a result of these wet and dry cycles, what seems like conflicting trends in water levels can in fact be the response of watersheds to climate forcing at different rates. Therefore, the analysis and interpretation of streamflow trends depend very much on the length of gauge records and the period of time modeled or observed (Sauchyn and St. Jacques, 2009).

Despite this complex behaviour of hydrological systems, there are some common tendencies in the response of prairie water resources to climate change. Among the more consistent scenarios is the shift in precipitation and runoff from late spring and early summer to winter and early spring. Less surface and soil moisture can be expected in the mid-to-later stages of the longer and warmer summers. Rainfall will be more concentrated in time, with larger amounts in fewer storms. As a result, we can expect some unusually wet conditions, but also long dry spells between the less frequent rainstorms. The net result of these hydrological changes is potentially greater risk of more extreme events and more variability in the distribution of precipitation, affecting ecosystems and human needs. These more extreme conditions, and a wider range of water levels and moisture conditions, likely will determine much of the impact of climate change in the Prairie Provinces. Adaptation to avoid the most adverse impacts on prairie ecosystems, communities and economies must include integrated and adaptive water resource planning and policy to manage a range of variability and extremes that exceeds our past experience.

REFERENCES

ACIA. 2004. *Impacts of a Warming Arctic: Arctic Climate Impact Assessment* [online]. Cambridge University Press: Cambridge, New York. Available from World Wide Web: *www.acia.uaf.edu*

Ahrens, C. Donald. 2005. *Essentials of Meteorology*. Thomson Brooks/Cole.

Akinremi, O.O., S.M. McGinn and H.W. Cutforth. 1999. "Precipitation Trends on the Canadian Prairies." *Journal of Climate* 12, no. 10: 2996–3003.

Burn, D. 1994. "Hydrologic Effects of Climate Change in West-Central Canada." *Journal of Hydrology* 160: 53–70.

Byrne, J.M. 1989. "Three Phase Runoff Model for Small Prairie Rivers: I. Frozen Soil Assessment." *Canadian Water Resources Journal* 14, no. 1: 17–28.

Caprio, Joseph M., Harvey A. Quamme and Kelly T. Redmond. 2009. "A Statistical Procedure to Determine Recent Climate Change of Extreme Daily Meteorological Data as Applied at Two Locations in Northwestern North America." *Climatic Change* 92: 65–81.

Cary, G., R. Keane, R. Gardner, S. Lavorel, M. Flannigan, I. Davies, C. Li, J. Lenihan, T. Rupp and F. Mouillot. 2006. "Comparison of the Sensitivity of Landscape-Fire-Succession Models to Variation in Terrain, Fuel Pattern, Climate and Weather." *Landscape Ecology* 21: 121–37.

Casey, Allan. 2006. "Forgotten Lake." *Canadian Geographic* 126, no. 6: 62–78.

Comeau, Laura E.L., Al Pietroniro and Michael N. Demuth. 2009. "Glacier Contribution to the North and South Saskatchewan Rivers." *Hydrol. Process.* 23: 2640–53.

Couillard, Catherine M., Robie W. Macdonald, Simon C. Courtenay and Vince P. Palace. 2008. "Chemical–Environment Interactions Affecting the Risk of Impacts on Aquatic Organisms: A Review with a Canadian Perspective—Interactions Affecting Exposure." *Environ. Rev.* 16: 1–17.

Demuth, M.N. and A. Pietroniro. 2003. *The Impact of Climate Change on the Glaciers of the Canadian Rocky Mountain Eastern Slopes and Implications for Water Resource-Related Adaptation in the Canadian Prairies.* Prairie Adaptation Research Collaborative: Regina, Saskatchewan. Project P55. 111 pgs.

Environment Canada. 2001. *Threats to Sources of Drinking Water and Aquatic Ecosystem Health in Canada.* NWRI Scientific Assessment Report Series No. 1. National Water Research Institute: Burlington, Ontario. 72 pgs.

——. 2004a. *Threats to Water Availability in Canada.* NWRI Scientific Assessment Report Series No 3 and ACSD Science Assessment Series No 1. National Water Research Institute: Burlington, ON, 129 pgs.

——. 2004b. *Canada's Top Ten Weather Stories for 2004* [online]. Available from World Wide Web: *http://www.msc-smc.ec.gc.ca/media/top10/2004_e.html*

——. 2007. *Canada's Top Ten Weather Stories for 2006* [online]. Available from World Wide Web: *http://www.ec.gc.ca/doc/smc-msc/m_110/toc_eng.html*

——. 2008. *Canada's Top Ten Weather Stories for 2007* [online]. Available from World Wide Web: *http://www.ec.gc.ca/doc/smc-msc/m_110/s10_eng.html*

——. 2009a. "Water Pollution" website [online]. Available from World Wide Web: *http://www.ec.gc.ca/water/en/manage/poll/e_poll.htm*

——. 2009b. *Canada's Top Ten Weather Stories for 2008* [online]. Available from World Wide Web: *http://www.ec.gc.ca/doc/smc-msc/2008/toc_eng.html*

Fang, X. and J.W. Pomeroy. 2008. "Drought Impacts on Canadian Prairie Wetland Snow Hydrology." *Hydrological Processes* 22, no. 15: 2858–73.

Flannigan, M.D., K.A. Logan, B.D. Amiro, W.R. Skinner and B.J. Stocks. 2005. "Future Area Burned in Canada." *Climatic Change* 72: 1–16.

Gameda, S., B. Qian, C.A. Campbell and R.L. Desjardins. 2007. "Climatic Trends Associated with Summerfallow in the Canadian Prairies." *Agricultural and Forest Meteorology* 142: 170–85.

Gan, T.Y. 1998. "Hydroclimatic Trends and Possible Climatic Warming in the Canadian Prairies." *Water Resources Research, American Geophysical Union* 34, no. 11: 3009–15.

Gleick, P.H. 1987a. "The Development and Testing of a Water-Balance Model for Climate Impact Assessment: Modeling the Sacramento Basin." *Water Resources Research* 23, no. 6: 1049–61.

——. 1987b. :Regional Hydrologic Consequences of Increases in Atmospheric Carbon Dioxide and Other Trace Gases." *Climatic Change* 10, no. 2: 137–61.

Grundstein, Andrew. 2009. "Evaluation of Climate Change Over the Continental United States Using a Moisture Index." *Climatic Change* 93: 103–15.

Hamlet, Alan F., P.W. Mote, M.P. Clark and D.P. Lettenmaier. 2005. "Effects of Temperature and Precipitation Variability on Snowpack Trends in the Western United States." *J. Climate* 18: 4545–61.

Hrudey, S. and E. Hrudey. 2004. *Safe Drinking Water: Lessons from Recent Outbreaks in Affluent Nations.* International Water Association Publishing, 486 pgs.

Hughes, M.K. and H.F. Diaz. 2008. "Climate Variability and Change in the Drylands of Western North America." *Global and Planetary Change* 64: 111–18.

Hyland, Romney, James Byrne, Brent Selinger, Thomas A. Graham, James Thomas, Ivan Townshend and Victor P.J. Gannon. 2003. "Spatial and Temporal Distribution of Faecal Indicator Bacteria within the Oldman River Basin of Southern Alberta, Canada." *Water Qual. Res. J. Canada* 38, no. 1: 15–32.

Johnson, J.Y.M., J.E. Thomas, T.A. Graham, I. Townshend, J. Byrne, B. Selinger and V.P.J. Gannon. 2003. "Prevalence of Escherichia coli 0157:H7 and Salmonella spp. in Surface Waters of Southern Alberta and its Relation to Manure Sources." *Canadian Journal of Microbiology* 49: 326–35.

Kharin, Viatcheslav V., Francis W. Zwiers, Xuebin Zhang and Gabriele C. Hegerl. 2007. "Changes in Temperature and Precipitation Extremes in the IPCC Ensemble of Global Coupled Model Simulations." *Journal of Climate* 20: 1419–44.

Kienzle, Stefan W. 2006. "The Use of the Recession Index as an Indicator for Streamflow Recovery After a Multi-Year Drought." *Water Resources Management* 20: 991–1006.

Larson, Robert. 2008. "Modelling Climate Change Impacts on Mountain Snow Hydrology, Montana-Alberta." Master of Science thesis, University of Lethbridge. 136 pgs.

Lapp, S., J. Byrne, I. Townshend and S. Kienzle. 2005. "Climate Warming Impacts on Snowpack Accumulation in an Alpine Watershed: A GIS Based Modeling Approach." *International Journal of Climatology* 25, no. 3: 521–36.

Lapp, Suzan, Dave Sauchyn and Brenda Toth. 2009. "Constructing Scenarios of Future Climate and Water Supply for the SSRB: Use and Limitations for Vulnerability Assessment." *Prairie Forum* 34, no. 1: 153–80.

Leith, R.M. and P. Whitfield. 1998. "Evidence of Climate Change Effects on the Hydrology of Streams in South-Central b.c." *Canadian Water Resources Journal* 23: 219–30.

Leung, L.R. and Ghan, S.J. 1999. "Pacific Northwest Climate Sensitivity Simulated By a Regional Climate Model Driven By a GCM. Part II: 2 x Co2 Simulations." *Journal of Climate* 12, no. 7: 2031–53.

Leung, L. Ruby Yun Qian, Xindi Bian, Warren M. Washington, Jongil Han and John O. Roads. 2004. "Mid-Century Ensemble Regional Climate Change Scenarios for the Western United States." *Climatic Change* 62: 1–3.

Little J.L., K.A. Saffran and L. Fent. 2003. "Land Use and Water Quality Relationships in the Lower Little Bow River Watershed, Alberta, Canada." *Water Quality Research Journal of Canada* 38, no. 4: 563–84.

MacDonald, Ryan J. 2008. "Modelling the Potential Impacts of Climate Change on Snowpack in the St. Mary River Watershed, Montana." Master of Science thesis, University of Lethbridge, 93 pgs.

Majorowicz, J.A., W.R. Skinner and J. Safanda. 2005. "Ground Surface Warming History in Northern Canada Inferred from Inversions of Temperature Logs and Comparison with Other Proxy Climate Reconstructions." *Pure & Applied Geophysics* 162, no. 1: 109–28.

Merritt, W.S., Y. Alila, M. Barton, B. Taylor, S. Cohen and D. Neilson. 2006. "Hydrologic Response to Scenarios of Climate Change in Sub Watersheds of the Okanagan Basin, British Columbia." *Journal of Hydrology* 326: 79–108.

Millett, Bruce, W. Carter Johnson and Glenn Guntenspergen. 2009. "Climate Trends of the North American Prairie Pothole Region 1906–2000." *Climatic Change* 93: 243–67.

Mote, P.W. 2003. "Trends in Snow Water Equivalent in the Pacific Northwest and Their Climatic Causes." *Geophysical Research Letters* 30, no. 12: 1601, doi: 10.1029.

Mote, P.W. 2006. "Climate-Driven Variability and Trends in Mountain Snowpack in Western North America." *J. Climate* 19: 6209–20.

Pearce, F. 2005. "Climate Warning as Siberia Melts." *New Scientist* 187, no. 2512: 12.

Pomeroy, J.W., D. de Boer and L.W. Martz. 2007. "Hydrology and Water Resources." Pp. 63–80 in B. Thraves, M.L. Lewry, J.E. Dale and H. Schlichtmann (eds.). *Saskatchewan: Geographic Perspectives*. Regina: Canadian Plains Research Center.

PFRA (Prairie Farm Rehabilitation Adminsitration). 2005. *Analysis of Issues Constraining Sustainable Irrigation in Canada and the Role of Agriculture and Agri-Food Canada*. Brace Centre for Water Resources, Agriculture and Agri-Food Canada, Prairie Farm Rehabilitation Administration: Regina, Saskatchewan, 2005.

Revelle, R.R. and P.E. Waggoner. 1983. "Effects of a Carbon Dioxide Induced Climatic Change on Water Supplies in the Western United States." Pp. 419–32 in *Changing Climate,* National Academy Press.

Rood, Stewart B., Jason Pan, Karen M. Gill, Carmen G. Franks, Glenda M. Samuelson and Anita Shepherd. 2008. "Declining Summer Flows of Rocky Mountain Rivers: Changing Seasonal Hydrology and Probable Impacts on Floodplain Forests." *Journal of Hydrology* 349: 397–410.

Rood, S.B., G.M. Samuelson, J.K. Weber and K.A. Wyrot. 2005. "Twentieth-Century Decline in Streamflows from the Hydrographic Apex of North America." *Journal of Hydrology* 306: 215–33.

Running, Steven W. 2006. "Is Global Warming Causing More, Larger Wildfires?" *Science* 313 (August 18): 927–28.

Schindler, D.W. and W.F. Donahue. 2006. "An Impending Water Crisis in Canada's Western Prairies" [online]. Proceedings of the National Academy of Sciences of the United States–Early Edition, 7 pgs. [accessed June18, 2007] Available from World Wide Web: *http://www.pnas.org/cgi/doi/10.1073/pnas.0601568103*

Sauchyn, Dave and Jeannine St. Jacques. 2009. "Security of Western Water Supplies under a Changing Climate." Proceedings of the 18th Convocation of the International Council of Academies of Engineering and Technological Sciences, July 13–17: Calgary, Alberta.

Sauchyn, D.J., J. Stroich and A. Beriault. 2003. "A Paleoclimatic Context for the Drought of 1999–2001 in the Northern Great Plains." *The Geographical Journal* 169, no. 2: 158–67.

St. Louis, V.L., C.A. Kelly, É. Duchemin, J.W.M. Rudd and D.M. Rosenberg. 2000. "Reservoir Surfaces as Sources of Greenhouse Gases to the Atmosphere: A Global Estimate." *BioScience* 50: 766–75.

Stocks, B. J., M.A. Fosberg, T.J. Lynham, L. Mearns, B.M. Wottoni, Q. Yang, J-Z. Jin, K. Lawrence, G.R. Hartley, J.A. Mason and D.W. McKenney. 1998. "Climate Change and Forest Fire Potential in Russian and Canadian Boreal Forests." *Climatic Change* 38: 1–13.

Toth, Brenda, Darrell R. Corkal, David Sauchyn, Garth Van der Kamp, Elise Pietroniro. 2009. "The Natural Characteristics of the South Saskatchewan River Basin, Climate, Geography and Hydrology." *Prairie Forum* 34, no. 1: 95–127.

Stone, Dáithí A., Andrew J. Weaver and Francis W. Zwiers. 2000. "Trends in Canadian Precipitation Intensity." *Atmosphere-Ocean* 38, no. 2: 321–47.

Westerling, A.L., H.G. Hidalgo, D.R. Cayan and T.W. Swetnam. 2006. "Warming and Earlier Spring Increase Western U.S.Forest Wildfire Activity." *Science* 313: 940.

WMO. 1989. Proceedings of the Conference on the Changing Atmosphere. Toronto, ON, June 1988.

Yulianti, J.S. and D.H. Burn. 1998. "The Impact of Climate Change on Low Streamflow in the Prairies Region of Canada." *CWRJ* 23, no. 1: 45–60.

Zhang, X., K.D. Harvey, W.D. Hogg and T.R. Yuzyk. 2001. "Trends in Canadian Streamflow." *Water Resources Research* 37, no. 4: 987–98.

Zhang, X., L.A.Vincent, W.D. Hogg and A. Niitsoo. 2000. "Temperature and Precipitation Trends in Canada During the 20th Century." *Atmosphere-Ocean* 38, no. 3: 395–429.

CHAPTER 6

ECOSYSTEMS AND BIODIVERSITY

Norman Henderson and Jeff Thorpe

THE IMPORTANCE OF ECOSYSTEMS AND BIODIVERSITY

Biodiversity is the range of species and genetic resources that maintain the resilience of natural biological systems. Healthy ecosystems are the home of biodiversity. The resilience and productivity of ecosystems depend on the diversity and health of flora and fauna. Changing climate could reduce this biodiversity and result in reduced resilience and adaptive capacity within ecosystems.

In a global analysis, climate change is rated as second only to land use in importance as a factor expected to determine changes in biodiversity during the current century (Sala et al., 2000). Climate change affects biodiversity directly because living organisms and their natural ecosystems can be sustained only within certain environmental tolerances. Changes in environmental means and variability alter the conditions that allow specific species and systems to survive or prosper.

The sensitivity of ecosystems to climate change increases the vulnerability of human populations to climate change. Current economic advantages of biodiversity may be lost as ecosystems are degraded or simplified by changing conditions. For example, biodiversity supports genetic resources and species that could yield new crops, drugs, or pest control products. As plant and animal species change in response to a warmer and drier climate in the Prairies Provinces, tourism, agricultural and recreational activities may be negatively affected, as detailed in other chapters of this book.

More fundamentally, ecosystems and biodiversity ultimately provide us with all of our material needs, from foods, to fuel, to building materials. Ecosystems also provide us with assimilative capacity to process human waste and pollution (our excessive demands on this capacity are themselves major contributors to climate change). Damage to ecosystems and losses of biodiversity reduce the availability of natural resources to us, sometimes irreversibly. Physically, we cannot live outside or apart from the biosphere.

Additionally, the natural world provides us with psychological and spiritual benefits that are often hard to quantify, but which are undoubtedly very real and important. We need other living species, and the home environments of those species, for us to survive mentally, as much as physically. One of the great unknowns of climate-change-induced landscape shifts will be the psychological impact on people of the rapid loss of familiar and valued landscapes such as woodlands, lakes and rivers (Henderson, 2006).

Some also argue that ecosystems have intrinsic value in addition to their human use values (Vilkka, 1997). However analyzed, it is clear that ecosystems and biodiversity are vital—the very foundation of our health, prosperity and survival. Climate change presents a major change driver to ecosystems around the world and the Prairie Provinces are no exception. Indeed, Prairie residents are stewards of a major portion of one of the most climate-change-threatened biomes on the planet, the boreal forest (Soja et al., 2007; Lenten et al., 2008). Both worldwide and locally, we need to act aggressively to reduce greenhouse gas emissions to limit the danger to our ecosystems, as well as to adapt to the inevitable changes underway. A discussion of current and potential future climate change impacts on the ecosystems of the Prairie Provinces (Figure 1) and our adaptation options form the rest of this chapter.

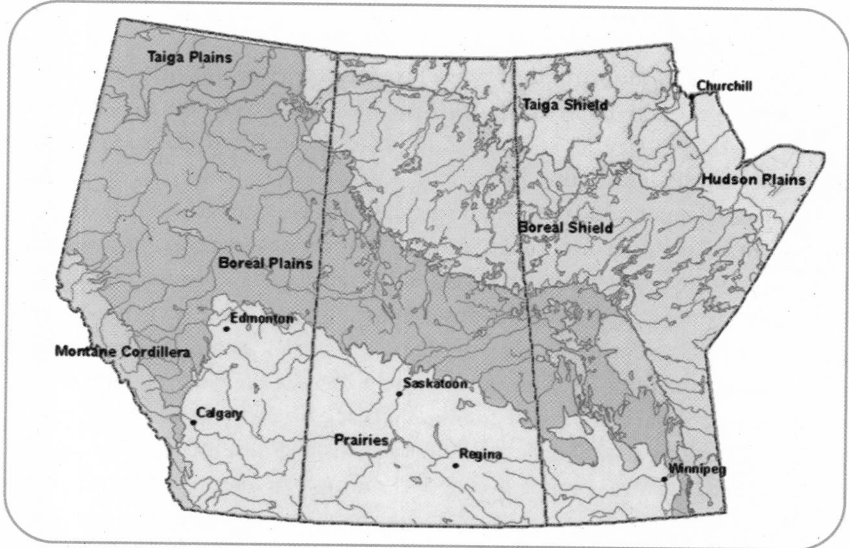

Figure 1. Ecozones of the Prairie Provinces.

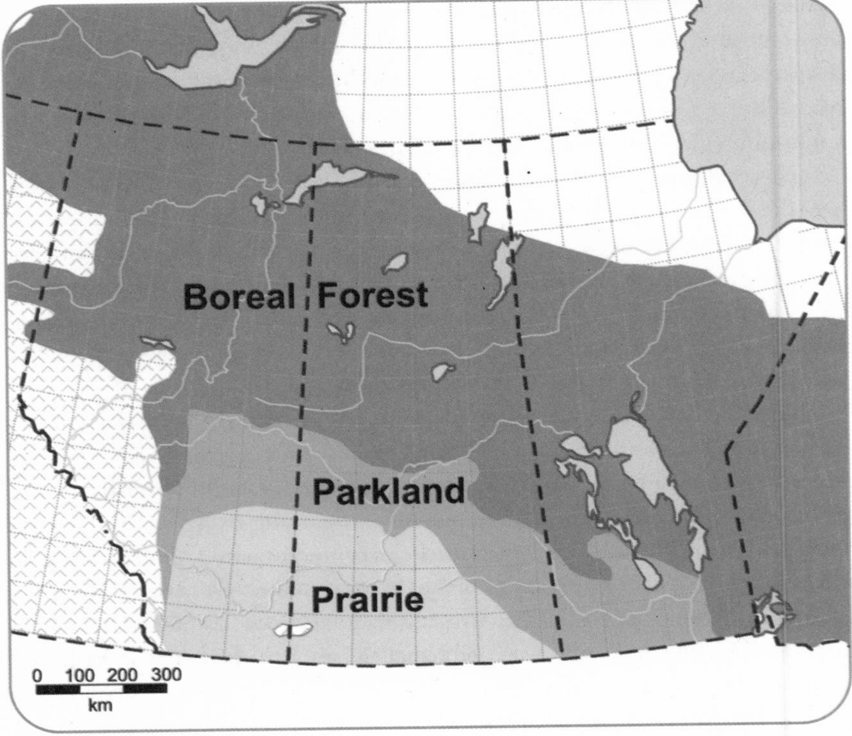

Figure 2. The Prairie, Parkland and Boreal Forest regions of the Prairie Provinces. Credit: E. Hogg

VEGETATION RESPONSE TO CLIMATE CHANGE

The generalized outlook for the Prairie Provinces is a warmer, drier climate. As ecosystems productivity in the Prairie Provinces is often already moisture constrained in the grassland, parkland, and southern boreal ecoregions (Figure 2), increasing aridity is a serious concern.

Climate-change-induced vegetation and faunal changes are already ongoing worldwide (Parmesan and Yohe, 2003). Based on observations of the timing of the flowering of key wild plants, Beaubien and Freeland (2000) found that, in Alberta, a 26-day shift to earlier onset of spring has already occurred over the last century. This represents a major shift in timing. In a comprehensive study of the circumpolar boreal forest biome (including the Prairie Provinces' boreal component), Soja et al. (2007) concluded that evidence of significant change, including more frequent and intense fires and insect infestations, is already evident—indeed, change was proceeding faster even than many model predictions.

Yet in some ways intensive human management of the Prairie Provinces' ecosystems over the past century has confused vegetation evidence of a climate signal. For example, European settlement has resulted in the widespread suppression of fire in both forest and grassland biomes. In some cases this has resulted in tree and shrub expansion and survival where we might otherwise have expected to see retreat in the face of increased aridity. Most of the grasslands have been converted to intensively managed agricultural systems, where annual farm management decisions impact the vegetation more obviously than long-term climate trends. But Aboriginal peoples note the impacts of climate change in their traditional lands, which have often been less intensively managed over the past century. At the 2004 Prince Albert Grand Council Elders' Forum, elders reported more frequent extreme weather events, deterioration in water quality and quantity, changes in species distributions, changes in plant life, and decreasing quality of animal pelts (Ermine et al., 2005).

There is significant change predicted to both the future extent of forested areas and to the nature of future Prairie Provinces forests (Chapter 8). In the absence of moisture limitations or other constraints, plant productivity should rise in the boreal forest with an increase in temperature, growing season length, and atmospheric CO_2 concentrations (Boisvenue and Running, 2006). Increased photosynthetic activity for much of Canada over the period 1981–91 has been attributed to a longer growing season (Myneni et al., 1997). Therefore, the northern extremes of the boreal forest will likely show increased productivity, as heat is the limiting parameter to forest growth at the northern boundary. However, the rate of northward extension of the forest is uncertain and will proceed over many decades, as trees respond to variations in soil temperatures and permafrost and because of uncertain seed dispersal and establishment (Lloyd, 2005).Parisien et al. (2005) predicted increased fire frequency and intensity in the Saskatchewan boreal forest as the result of new climate conditions. Thornley and Cannell (2004), modelling increased fire frequency in coniferous forests for the climate of Prince Albert, predicted long-term decline in net primary productivity. Hogg and Bernier (2005) suggested increasing vulnerability to drought, insects and fire in Canada's western boreal forest. Drought conditions weaken trees' defences to more virulent pathogens (Saporta, Malcolm and Martell, 1998). As conditions become more xeric, the lifespan of conifer needles is reduced, placing conifers under increasing stress (Gracia, Sabaté and Sanchez, 2002). Increased fire disturbance may favour hardwoods, such as aspen, over conifers, such as white spruce (Johnston, 1996). Changes in the timing and intensity of freeze-thaw events, diurnal temperature patterns (Gitay et al., 2001), and storm and wind stress events may also influence vegetation distribution or survival, especially of various tree species (Macdonald et al., 1998), but the detail of how this will occur is not known.

At the southern edge of the boreal forest, extending across the central parts of the Prairie Provinces, forest productivity will probably decrease. In this region, there is a transition from boreal forest in the north, through a belt of aspen parkland (a mosaic of forest and grassland patches), to native grassland in the warmer and drier climates of the south. Hogg (1994) showed that this transition is well defined by a climatic moisture index (annual precipitation minus annual potential evapotranspiration), with positive values in the forest and negative values in the grassland. Results from applying a climate change model based on CO_2 doubling to the climatic moisture index show a substantial northward shift in the boreal forest/aspen parkland boundary (Hogg and Hurdle, 1995). Much of the southern boreal forest could shift to a climate suitable for aspen parkland. Hogg and Schwartz (1997) found that conifers are unable to regenerate naturally in the grassland zone and are only slightly more successful in the aspen parkland. In other parts of the world, cold-climate conifers have shown dieback which has been related to historic warming (Hamburg and Cogbill, 1988). Carr, Weedon and Coutis (2004), Henderson et al. (2002), Herrington, Johnson and Hunter (1997), and Scholze et al. (2006) all expected considerable disruption in the boreal forest's southern boundary region via droughts and associated large-scale fire events. Historical evidence has also shown that extreme periods of drought have led to increased frequency of fire and insect infestations, which have in turn led to increased aspen dieback at the southern limit of the boreal forest (Hogg, 1997).

Average winter temperatures are expected to rise to a greater extent than average summer temperatures, leading to greater overwinter survival of pathogens and increased disease severity (Harvell et al., 2002). The current outbreak of mountain pine beetle, which has killed an unprecedented number of lodgepole pine trees in British Columbia, is thought to be at least partially a result of climate-change-related warmer winters in that province. Warmer winters mean less winterkill of the beetle. The beetle has crossed the Rockies into Alberta, attacking lodgepole pine there, and is now found as far east as jack pine stands in Saskatchewan's boreal forest (Natural Resources Canada, 2006). The eventual impacts on boreal jack pine stands are not known.

Lenten et al. (2008) concluded that the circumpolar boreal forest is one of the large-scale components of the earth's biosphere that may pass a tipping point and be heavily impacted by climate change, with biome conversion and grassland invasion the expected outcome. According to a study of potential range shifts for extant and exotic tree species in the current boreal forest and parkland fringe under one climate change scenario (CGCM1 A1), there could be major changes in species representation in Saskatchewan's boreal forest by 2080 (Carr, Weedon and Cloutis, 2004). Henderson et al. (2002) noted two pathways to forest decline. One possibility is slow and cumulative decline: aspen break-up as trees pass maturity, forest subject to pathogen attack, and conditions too dry for regeneration. Alternatively, the change mechanisms may be spectacular and catastrophic, such as a major fire.

For the national parks specifically, Scott and Suffling (2000: xix) note that climate change represents "an unprecedented challenge for Parks Canada" and "current ecological communities will begin to disassemble and 'resort' into new assemblages." All the prairie-parkland national parks (Elk Island, Prince Albert and Riding Mountain) and Wood Buffalo National Park may undergo a shift to another forest formation type, including grassland invasion, as a result of climate change (Scott and Suffling, 2000). These parks can expect increases in forest fire frequency and intensity and forest disease outbreaks and insect infestations, as well as loss of boreal forest to grassland and temperate forest (de Groot et al., 2002).

The existing Prairie Provinces grassland biome has a vegetation zonation pattern derived in large part from climate. Aspen parkland and fescue prairie (highly productive grassland dominated by plains rough fescue [*Festuca hallii*]) in the moister northern and western fringe give way to mixed prairie (a mixture of mid-sized grasses and shortgrasses) in the warmer and drier south. Vandall, Henderson and Thorpe (2006) modelled the shifts in Saskatchewan vegetation zones resulting from three climate change scenarios for the 2050s (CGCM2 A21, CSIROMK2B B11 and HadCM3 B21). Vegetation zones in the Great Plains of the United States were used as analogues for the warmer future climates projected for Canada. Results for one of these scenarios are shown in Figure 3 (other scenarios gave similar results). Most of the boreal forest up to 54° latitude is replaced by aspen parkland. Most of the aspen parkland is replaced by mixed prairie. Most of the Canadian mixed prairie is replaced by US mixed prairie (i.e. the kind of mixed prairie found in Montana, Wyoming and the Dakotas). The driest area, in southwestern Saskatchewan and southeastern Alberta, shifts to shortgrass prairie, currently found from Colorado southward.

Mixed prairie is dominated by cool-season species (i.e. plants with the C_3 photosynthetic pathway). However, warm-season species (i.e. plants with the C_4 photosynthetic pathway) are relatively more abundant in US mixed prairie than in Canada, and they are dominant in the shortgrass prairie (Vandall, Henderson and Thorpe, 2006). Because higher temperatures are required for warm-season grasses to start growing in spring, cool-season grasses have a competitive advantage in cooler climates with shorter growing seasons. Climatic warming will allow warm-season grasses to develop earlier in spring, which may increase their ability to compete with cool-season grasses in northern areas (Long and Hutchin, 1991). Climate change modelling has shown increasing proportions of warm-season grasses, and in some cases a shift to warm-season dominance, in the northern part of the Great Plains (Coffin and Lauenroth, 1996; Epstein et al., 2002). Epstein et al. (2002) suggest that the higher water-use efficiency of warm-season species could lead to higher productivity and/or reduced transpiration at a given level of precipitation. However, changes in the seasonality of moisture availability may also affect the proportion of warm-season grasses

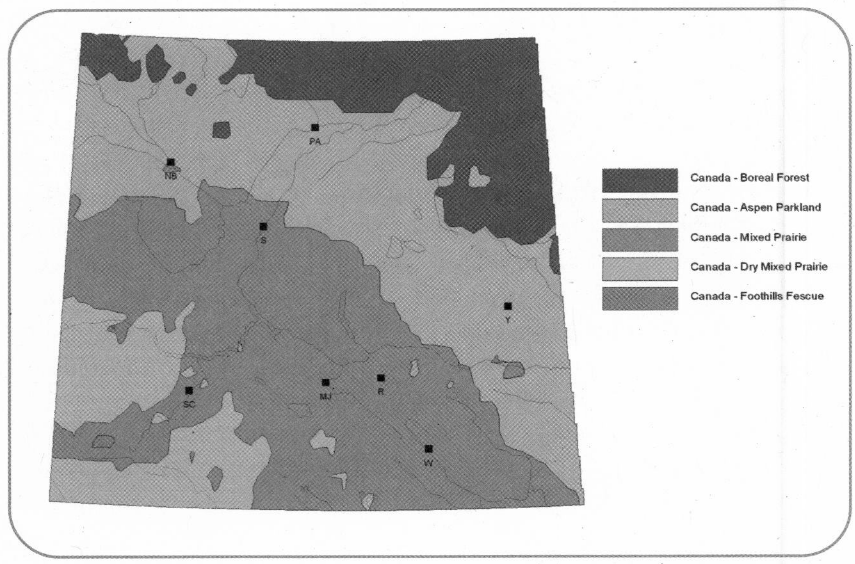

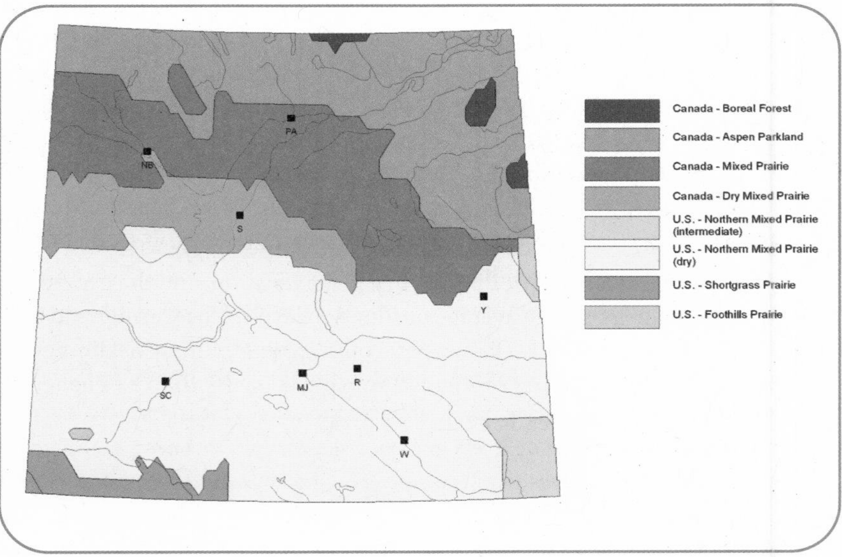

Figure 3. Vegetation zonation in southern Saskatchewan as predicted by ecoclimatic models (Vandall et al., 2006). The upper map shows the zonation resulting from 1961–90 climatic normals. The lower map shows the predicted zonation resulting from the HadCM3 B21 scenario for the 2050s.

(Paruelo and Lauenroth, 1996; Winslow, Hunt and Piper, 2003). Most scenarios project a lower proportion of precipitation in summer, and this could moderate the increase in C_4 grasses favoured by rising temperatures.

Some change could occur by shifts in the abundance of species already present. Warm-season grasses such as blue grama (*Bouteloua gracilis*), sand reed grass (*Calamovilfa longifolia*), sand dropseed (*Sporobolus cryptandrus*), and little bluestem (*Schizachyrium scoparium*) are already widespread in the Prairie Provinces grassland biome, and they could expand relative to the cool-season dominants such as wheat grasses (*Agropyron* spp.) and spear grasses (*Stipa* spp.). However, some of the change could occur by migration into Canada of species that are currently absent or rare. Wolfe and Thorpe (2005) compared sand dunes in the Canadian Prairies with their warmer-climate analogues in Colorado and Nebraska. While many species are common to both areas, there are also many (mostly C_4 grasses) that are found in the south but not in the north (e.g. sand bluestem [*Andropogon hallii*], sandhill muhly [*Muhlenbergia pungens*], and switchgrass [*Panicum virgatum*]). Future vegetation on dunes could be a mixture of species we already have with migrants from the south. The shift in zonal vegetation from mixed prairie to shortgrass prairie shown in Figure 4 could occur initially by an increase in blue grama—the northward migration of buffalo grass (*Buchloe dactyloides*), the other major component of US shortgrass prairie, will likely be a much slower process. The likely source of migration would be adjacent patches of native grassland further south. For this reason, the current fragmentation of grassland by cultivated fields, roads, etc. will be an impediment to migration of new species.

Grassland productivity is clearly limited by moisture supply, as shown by the year-to-year fluctuations in production in a given location, and the geographic variation across the region in average production levels. The warmer and drier climate forecast for the Prairie Provinces would appear to imply declining production. However, modelling of this relationship suggests that the changes in the medium term (i.e. up to the 2050s) are likely to be modest (Thorpe, Houston and Wolfe, 2004; Thorpe, 2007). Figure 4 shows the predicted change in production on loamy-textured soils based on the HadCM3 B21 climate change scenario (Thorpe, 2007). While the forecast increases in temperature and potential evapotranspiration would have a negative effect on productivity, the most important factor is actual evapotranspiration, which is closely tied to annual precipitation, a variable that is not predicted to change dramatically (Thorpe, 2007). Other factors that may moderate the effect of a drier climate on grassland productivity include reduced competition from shrubs and trees and increases in warm-season grasses with higher water-use efficiency.

Modelled shifts in climatic zonation do not specify the exact composition of future vegetation. Connectivity, or the lack of same, will play a major role in

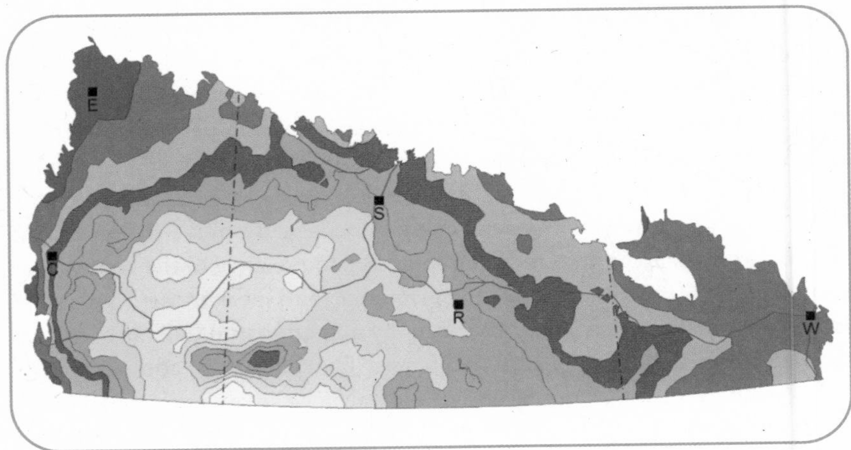

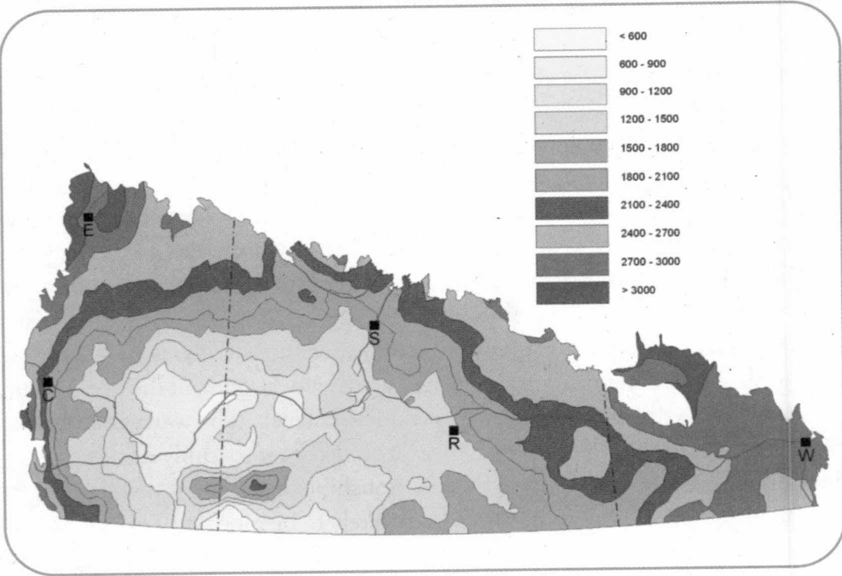

Figure 4. Grassland production (kg/ha) on loamy soils in the Prairie Ecozone of Canada as predicted by ecoclimatic models (Thorpe, 2007). The upper map shows the production resulting from 1961–90 climatic normals. The lower map shows the predicted production resulting from the HadCM3 B21 scenario for the 2050s.

determining change. James et al. (2001) modelled the theoretical dispersal capabilities of prairie grassland species under climate change in two Saskatchewan watersheds. They concluded that there will be significant changes in the distribution and abundance of many prairie species and that habitat corridors will be valuable in allowing some extant species to survive. There will certainly be lags in migration of some species—in fact, if connectivity is not present, a given climatically suitable organism may never arrive at a theoretically newly suitable location. The shifts in the boundaries between historic ecological regions will not be smooth, as they will be driven largely by encroachment of species into previously unsuitable areas that are becoming climatically more tolerable, but which for many years will retain soils generated under the previous climatic regime.

The rate and success of migration will vary among species because of differences in climatic tolerance, genetic variation and dispersal rates (Peters, 1992). Many wind-dispersed species can easily migrate fast enough to adapt to climate change, but there are others for which slow dispersal or restriction to particular habitats will limit their rate of range expansion (Malcolm and Pitelka, 2000). The northward migration of plant species following the retreat of the Wisconsin glaciation is well documented, but Davis and Shaw (2001) pointed out that the predicted climate change for the 21st century will require much faster range shifts (300–500 km/century, compared to pasts shifts of 20–40 km/ century). For example, Rehfeldt et al. (1999) noted that the post-glacial migration of lodgepole pine across British Columbia was much slower than would be required to keep pace with projected future shifts in climate zones. However, evidence of relatively rapid migration in response to recent warming is also beginning to appear. In California's Santa Rosa Mountains, the average elevation of dominant plant species rose by 65 m between 1977 and 2007 (Kelly and Goulden, 2008). In northern Europe, Walther, Berger and Sykes (2005) documented northward expansion in the range of holly (*Ilex aquifolium*), which kept pace with the increase in temperature from 1944 to 2003. In addition to the requirement for dispersal, there may also be a requirement for genetic adaptation. While seeds from more southerly climates may be somewhat pre-adapted to the new environment, there will also be genotype selection for new combinations of photoperiod and temperature response (Davis and Shaw, 2001). This selection pressure could further differentiate species. Differential migration rates among species imply that future communities could be unlike any found under current climates.

Predicting timing and precise ecological thresholds is difficult, as vegetation responds after the fact to climate change, i.e., it is natural for a given ecosystem to be "behind" environmental conditions to some degree, a condition termed ecological inertia. Climate change impacts on mature trees may not be noticeable until biological thresholds are reached and dieback results (Saporta, Malcolm and Martell, 1998). Solomon (1994: 7) notes: "Loss of a cohort may not be evident in change to forest structure for decades or even a century as the established trees continue to survive under increasing environmental stress." Pielou (1991: 99) adds: "Many plants have an astonishing ability to persist in unfavorable habitats." However, such persistence in the face of major environmental change may only be a prelude to sudden and catastrophic change. Anderson et al. (1998) warn that ecosystems can sometimes absorb stresses over long periods of time, but then cross a critical threshold leading to rapid ecosystem and landscape modification.

There are many uncertainties in the above scenarios. One limitation lies in the nature of the models themselves. Most vegetation zonation modelling is static and based mainly on the current positions of ecoregions, i.e., on the existing ranges and distributions of species (e.g. Davis and Zabinski, 1992; Carr, Weedon and Cloutis, 2004; Henderson et al., 2002; Hogg, 1994; Vandall, Henderson and Thorpe, 2006). Models based on plant growth and population dynamics are more realistic in portraying ecosystem transformations, i.e., how we get from A to B. Unfortunately, such process modelling efforts can be very complex undertakings and require many assumptions.

The CO_2 fertilisation effect is a major uncertainty (Wheaton, 1997; Aspen FACE Experiment, 2009). Atmospheric CO_2 concentrations are rising, and will certainly continue to do so for many decades, and possibly for centuries. This increases the water-use efficiency of some plant species (Lemon, 1983; Long et al., 2004) and can possibly make such species more resistant to dry conditions. The fertilisation effect provides some grounds for hope that increasing aridity may not be as damaging to some terrestrial ecosystems as would otherwise be the case. Wang, Chhin and Bauerle (2006) reported a positive CO_2 enrichment effect on the growth of open-grown natural white spruce in southwestern Manitoba. Johnston and Williamson (2005) found increased tree productivity and drought tolerance in northern Saskatchewan under a high CO_2 future.

One major problem in predicting CO_2 enrichment impacts is that the impact occurs on all vegetation simultaneously. It is not enough to know the CO_2 response of one species, rather one needs to know the relative growth advantage, if any, gained by all vegetation species competing for resources at a given site. Work by Long et al. (2004) and Norby et al. (2005) found that trees respond to increased CO_2 concentrations more than other vegetation. But there is still substantial truth to Gitay et al.'s (2001) argument that it is not yet possible to predict overall changes in forest CO_2 uptake and storage, independent of interspecies variations. In their metastudy, Boisvenue and Running (2006: 874) concluded "there is no clear answer as to whether rising CO_2 concentrations will cause forests to grow faster and store more carbon."

In grasslands, Parton et al. (1996) and Ojima et al. (1996) used an ecosystem model, CENTURY, to simulate the effect of doubling CO_2 concentrations on sites around the world, and found increases in production. Similar results were obtained by Baker et al. (1993), who used the SPUR model to simulate CO_2 doubling without climate change in US rangelands. Field experiments in grassland have shown increases in production ranging from 0 to 30% with doubling of ambient CO_2 concentrations (Campbell and Stafford Smith, 2000). However, some research has shown a reduction in forage quality with CO_2 fertilization (Campbell and Stafford Smith, 2000).

Increasing CO_2 concentration could also affect the species composition of grassland. Theoretically, CO_2 fertilization provides a greater relative benefit to C_3 grasses than to C_4 grasses (Long and Hutchin, 1991; Parton, Ojima and Schimel, 1994; Collatz, Berry and Clark, 1998). However, ecosystem experiments have shown that under the dry conditions typical of grasslands this advantage tends to be reduced or eliminated (Nie et al. 1992; Wand et al. 1999; Campbell and Stafford Smith, 2000). Winslow, Hunt and Piper (2003) modelled the impacts of climate change on proportions of C_3s and C_4s on the basis of changes in water availability during the growing season for each group, and concluded that these changes may mask any benefit of rising CO_2 concentrations to C_3s. Morgan et al. (2007) performed a CO_2 enrichment study on shortgrass prairie in Colorado and determined that rising atmospheric CO_2 levels may favour shrub invasion of grasslands. Over the course of the five-year experiment, fringed sage (*Artemisia frigida*), a common subshrub of North American grasslands, increased 40 times in above-ground biomass. Such encroachment would disfavour grazing by livestock. However, differential CO_2 enrichment impacts on Canadian Plains natural prairie grassland systems are largely untested.

Several important anthropogenic-origin atmospheric changes other than increases in CO_2 levels are ongoing as well. UV-b radiation and ground-level ozone levels are increasing. These can be expected to impact negatively on vegetation in general and to possibly counteract any positive CO_2 enrichment effect, but the detail is not known and there are many uncertainties (Henderson et al., 2002). Nitrogen deposition from industrial activity may be affecting species growth and competitive interactions, even in Prairie Provinces locations far from industrial centres (Kochy and Wilson, 2001). Sulphur emissions are known to be rising from tar sands extraction and processing in Alberta, and are impacting soils in northern Alberta and into Saskatchewan. All these impacts occurring simultaneously reinforce our expectation of change, but complicate prediction in detail.

Not only the amount, but also the pattern of future precipitation events may be important. Climate models have less certainty around precipitation than around temperature futures, but a general expectation is that we may experience more variability and more extreme precipitation events (e.g. Kharin and Zwiers, 2000) within a trend towards increased aridity. Laporte, Duchesne and Wetzel (2002) hypothesized that plant productivity could decline where monthly precipitation levels (and other environmental factors) remain constant, but where the precipitation falls in fewer events. Their empirical test results in a northern Ontario grassland ecosystem support this possibility.

Trends are more predictable than exact processes, timing or end points. With respect to the southern half of Saskatchewan (and likely generalizable to adjacent Manitoba and Alberta), Vandall, Henderson and Thorpe (2006) suggest the following vegetation trends over the period from 2006 to the 2050s:

- Reduction in tree growth in moisture-constrained forest regions to the slower growth and lower maximum size typical of the aspen parkland;
- Regeneration failure in dry years in the forest region;
- Gradual reduction in tree cover and expansion of grassland patches in the moisture-constrained forest regions;
- Shrinking of aspen groves in the aspen parkland;
- Reduced invasion of grassland patches by shrubs and poplar sprouts in the aspen parkland;
- Decreasing shrub cover in aspen parkland and mixed prairie;
- Decreases in animal species dependent on woody cover and increases in species dependent on open grassland in the prairie ecozone;
- Decreases of midgrasses and increases of shortgrasses in the grasslands;
- Decreases in cool-season (C3) grasses and increases in warm-season (C4) grasses; and
- Gradual introduction of flora and fauna currently found only in the U.S.

CLIMATE CHANGE IMPACTS ON FAUNA

As with vegetation, wildlife migration patterns and population size have already been affected by recent climate trends (Inkley et al., 2004). In northern regions, earlier dates of disappearance of snow and increasing average temperatures have resulted in earlier nesting and hatching of geese (LaRoe and Rusch, 1995).

Aquatic ecosystems everywhere in the Prairie Provinces can be expected to be stressed by warmer and drier conditions (James et al., 2001; Schindler and Donahue, 2006), with a possible increase in water salinity and turbidity. James et al. (2001) noted that a large number of prairie aquatic species are at risk of extirpation. Many fish species, for example, are sensitive to small changes in temperature, turbidity, salinity or oxygen regimes. Melville (2001) expects climate change to stress lake environments throughout the boreal plains area of the Prairie Provinces, and advocates reducing allowable fish catches as an adaptive measure. For the Prairies, Schindler and Donahue (2006) expect larger algal blooms, accelerated eutrophication, and serious impacts on fish species, owing to a combination of climate change, increasing nutrient runoff, and increasing human use pressures on natural water systems. Baulch et al. (2005) simulated climate change by warming a littoral zone of a boreal lake and examining the impacts on benthic communities, particularly on the epilithon (the biofilm, consisting of algae, bacteria, fungi and invertebrates present on submerged rocky surfaces). They predicted increased metabolic rates of epilithons, but note many uncertainties about benthic impacts.

Warming in Hudson Bay may cause problems for sea-ice-dependent species such as polar bears and ringed seals. Body condition of southern Hudson Bay polar bear populations has been in steep decline since the mid-1980s (Obbard, 2006). Ferguson, Stirling and McLoughlin (2005) anticipate declining ringed seal populations in western Hudson Bay as warming advances.

The prairie pothole region of central North America is the single most productive habitat for waterfowl in the world (Johnson et al., 2005); the Canadian Prairies produce 50% to 80% of North American duck population (Clair et al., 1998). It is likely that increasing aridity in the prairie grasslands will negatively impact migratory waterfowl populations (Poiani and Johnson, 1993; Johnson et al., 2005). Waterfowl numbers decrease in response to drought and habitat loss (Bethke and Nudds, 1995). Fluctuations in weather during the breeding season account for more than 80% of the variation in population growth rate of mallards and other ducks (Hoekman et al., 2002).

Peterson (2003) comparatively modelled climate change impacts on North American mountain-resident bird species versus Great Plains bird species, and found the impacts on the latter more severe. On a global scale, Wilcove (2008) has concluded that migratory species of all kinds are at risk from climate change and other disturbances, as they are vulnerable to disruption at the end points of their journeys as well as along the way. The majority of Prairie Provinces bird species are migratory, and there have been notable declines in the numbers of many migratory bird species in North America.

Reptiles and amphibians, often sessile, are also at risk worldwide, though there is little climate-change-related data for herptiles available for the Prairie Provinces. Nor have impacts on insects and still smaller biota been well studied for the Prairies. It can be assumed there will be major impacts here as well. In particular, the timing of the life cycles of many insects is largely determined by spring temperatures and by the condition of their plant hosts. Spring already occurs substantially earlier than a century ago (Beaubien and Freeland, 2000).

Range distribution changes of almost every animal seem likely. For example, Pitt, Larivière and Messier (2008) concluded that winter severity is the principal constraint on the northern range limit of raccoons; therefore, this animal can be expected to expand northward under milder winter conditions. McCarty (2001) documented rapid northward range expansions of some animal species and local extinctions at southern edges of ranges owing to recent climate change. All fauna can be expected to show impacts in some manner as vegetation change accelerates.

CONTEXT AND PERSPECTIVES ON ADAPTATION

In geologic terms the Prairie Provinces' ecoregions are very young environments. Comparing northern Plains Holocene environmental conditions across millennia, or even across only centuries, indicates a highly variable landscape. We know that the region has a Holocene history of frequent environmental change. Rapid environmental change is not necessarily unnatural. Around 10,000 years ago, rapid climate change in mid-latitude North America led to major forest species conversion at some sites in less than a human lifetime (Watts, 1983). Over time, it is natural, perhaps even positive, that some ecosystems are lost, while completely new ones emerge. Nonetheless, the speed, the scope and the scale of modern ecosystem impacts we are facing in the Prairie Provinces are daunting. Adaptive action appears necessary on many fronts.

How we actually react to climate change impacts on natural systems depends in large part on whether we view climate change as a natural phenomenon or not. If we understand climate change as a natural event, then it logically follows that vegetation community changes that ensue from it are also natural. There is a long-standing and strong tradition in North American nature conservation philosophy that areas of high natural values, and particularly national parks and designated wilderness areas, are best represented where human intervention is least practised and least evident (Henderson, 1992). The preservation of native or natural communities, and of natural ecosystem functions, is the touchstone of most North American conservationists (Nash, 1982). This philosophy is the basis, for example, of policies to allow wildfires to burn (barring endangerment of human life or significant infrastructure) in national parks and some other conservation areas. If climate change is just another natural force like fire, then many would conclude that resultant vegetation changes are not, in principle, to be resisted.

Factually this viewpoint is unsustainable. Research (IPCC, 2007) has demonstrated beyond reasonable doubt that the majority of current climate change is being forced by anthropogenic emissions. Many apparently "natural" phenomena, such as wildfire, increased pathogen incidence, forest succession, and the decrease and increase in various species on the Prairie Provinces' landscapes, will to an ever-greater degree be driven by climate change. Any superficially natural landscapes that result from such phenomena will actually be the by-products of anthropogenic emissions (Henderson et al., 2002).

While climate change impacts cannot be avoided, we are not helpless. We have the choice of a suite of landscape ecosystems viable within given climate parameters. As a matter of conservation policy we can aim to extend ecological inertia, have no impact on it, or reduce it. Vegetation associations that are most "in tune" with the evolving climate will require the least maintenance, i.e., the least degree of human intervention. Conversely, those vegetation ensembles increasingly outside of their natural climate norms can be expected to require increasingly intensive and active human intervention and management to survive. However, with a commitment to a high degree of human intervention, it will be possible in some sites to maintain vegetation (and associated fauna) assemblages that would otherwise certainly disappear.

The EPA (2002: 99) *U.S. Climate Action Report—2002* gives a practical example: "For forests valued for their current biodiversity, society and land managers will have to decide whether more intense management is necessary and appropriate for maintaining plant and animal species that may be affected by climate change and other factors ... One possible adaptation measure could be to salvage dead and dying timber and to replant species adapted to the changed climate conditions."

THE ADAPTATION POLICY CHALLENGE

A failure to incorporate climate change impacts within strategic planning is typical of conservation management throughout the northern Plains region. For example, Montana's forestry policy management strategy for state lands (MDNRC, 1996) postulates a return to historic forest landscapes, which climate change almost certainly makes an impossible objective. Manitoba's "Protected Areas Initiative" (Manitoba Conservation Department, 2000), which aims to protect representative ecosystems of the province, is based on a division of the province into 18 natural regions and sub-regions defined by physiography and common climate, and does not reference climate change. Yet, while physiographic factors may be reasonably enduring, the current climate is not. Some of Manitoba's climate-physiography combinations may shrink in extent, some may disappear entirely, and entirely new ones may arise. Flora and fauna (i.e. the ecosystems and biodiversity the Protected Areas Initiative is intended to protect) will also change accordingly. Saskatchewan's "Representative Areas Network" is based on a division of the province into 11 distinct ecoregions, distinct from those in Manitoba, but also defined in part by climate, as well as by geology, soils, plants, and animals; the latter three are ultimately dependent on climate. Both Manitoba and Saskatchewan ecoregions are now transient features, as a changing climate will change their nature and boundaries. Alberta's "Special Places Program" is a comparable biodiversity protection initiative which also does not incorporate climate change in strategy design or implementation.

Of course, the fundamental problem is not just a regional one. Scott and Lemieux (2005), in their examination of protected area policy, note that the standard approach in Canada and worldwide for the preservation of biodiversity is the protection of representative parcels of a suite of ecoregions. The assumption that underlies the assembling of these protected areas is that they are biogeographically stable. However, Scott and Lemieux (2005: 697) note: "A growing body of scientific research indicates that global climate change would render this assumption untenable in the 21st century." They concluded that biodiversity protection planning must now aim to protect "a moving target of ecological representativeness."

To its credit, Parks Canada has built up a body of research activity in response to the threats climate change may pose to the nation's national parks system (Welch, 2005). However, this research has not led to a resolution of the central challenge that climate, and, therefore, flora, fauna, hydrology, and soils will not be static over this century, and a conservation strategy based on trying to maintain the ecological status quo by protecting selected landscapes from human impacts other than climate change will not succeed. In the context of Canada's national parks, Scott and Suffling (2000) agreed that a landscape maintenance strategy is materially impossible, whatever its philosophical merits or demerits. It simply becomes too difficult to repress or counteract every disturbance, whether gradual or catastrophic, that acts to shift the landscape towards a new equilibrium more closely attuned to the changing climate.

Other researchers have reached similar conclusions that forest managers need to abandon the idea of managing a steady-state system, given that we will be experiencing a changed disturbance regime and climate future (Kurz and Apps, 1993). Stability may not be possible, and aiming to build resilience into the ecosystem, and keeping our ecological options open, may be more appropriate. Pernetta (1994: viii) argued: "A static approach, will ultimately result in protected areas being over-taken by events, they may well exist in areas no longer suitable for the maintenance of the species and ecosystems they were originally designed to conserve." Natural disturbances will have to be managed, exogenous stresses will need to be controlled, and habitat modifications may be necessary to "reconfigure protected areas to new climatic conditions" (Halpin, 1997: 831).

POTENTIAL ADAPTATION MEASURES

In light of the inevitable impacts, what practical measures can be recommended? One way to build adaptation capacity is to increase public awareness of adaptation issues and options. Increasing public involvement is now especially critical, as the public has little knowledge of the probable impacts of climate change on natural systems, and managers have no certainty as to where, within the bounds of possible future landscapes, the public might wish them to steer.

There are many methods to engage the public. Planners could ask the public to evaluate and rate photomontages, or artists' drawings, or computer-generated images of possible landscape futures. This visual survey approach has been successfully employed in British national parks to help determine the public's preferred target landscapes (O'Riordan, Wood and Shadrake, 1993). Planners could employ standard question-and-answer social surveys about landscape preferences. Display areas within protected areas to demonstrate various management activities and outcomes could also be useful for public input. Methods such as these can help determine the public's preferences among alternative possible future landscapes and also provide information about public preferences among alternative management methods to reach preferred landscapes. We can speculate that increased public knowledge will increase pressure for active management to counter the risks of catastrophic change and will also increase support for climate change mitigation efforts.

There is in fact an important link between adaptation and mitigation. Some possible adaptation decisions (such as letting forests burn) have negative mitigation implications, in that such decisions increase CO_2 emissions (at least temporarily). Impacts on CO_2 balances are likely to become increasingly important considerations in future vegetation management everywhere. Grissom et al. (2000) discuss the policy implications (i.e. that net CO_2 emissions be minimized) with respect to the boreal forest, while de Groot et al. (2002) provide estimates of the impact on CO_2 balances of different fire management regimes for Elk Island, Prince Albert, Riding Mountain, and Wood Buffalo National Parks.

Not only the public, but policy-makers too, need to become better informed. At the best of times, it is a challenge to effectively communicate issues, data, research, and management options and consequences between researchers and decision-makers or implementers. When the issues are complex, and political as well as scientific, the challenge is especially great. Jacobs (2002) gives practical advice as to how to bring the scientific and policy-makers' worlds together in a productive exchange.

Aboriginal views are particularly important, as Aboriginal peoples with significant economic, traditional or spiritual connections to natural landscapes are at particular risk from climate change impacts. Management and economic adjustments in the touring, trapping, hunting, and guiding industries may be required as ecosystems evolve. In response to specific declining species populations we may need to implement, where relevant, reduced hunting quotas, increased species stocking, increased habitat protection measures, or other species or ecosystem support measures. A reduction in grazing capacity may require monitoring and reduced stocking for rangeland (Thorpe, 2007).

The views of Aboriginal and other residents of the relevant ecosystems will be important in these decisions. Through historic settlement and appropriation for agriculture of the grassland biome of the Prairie Provinces, European settlers largely destroyed the possibility of a traditional lifestyle for grassland-based Aboriginal peoples. It is discouraging that unintended climate impacts in the boreal biome may have similar negative impacts on Aboriginal peoples resident there.

One important practical adaptation measure will be effective environmental and ecological monitoring. We need to follow trends and developments in the health, numbers and distribution of keystone species such as aspen across the region. We also need to follow the status of many increasingly threatened species. Both an individual species and a holistic ecosystems monitoring approach is important. We need to pay particular attention to relatively understudied and under-monitored areas of climate impacts on ecosystems. Historically, we have had a research focus on harvested species, such as big game, waterfowl, and commercial trees. However, we also need to pay equal attention to threatened elements, such as amphibians, declining migratory songbirds, terrestrial invertebrates, and vulnerable trophic networks in sensitive aquatic ecosystems.

Programs like Plantwatch, which tracks the effects that climate variability and trends have on wild vegetation, will be invaluable. Resulting data can provide an index of the expected and realized impacts of shifts in climate (Beaubien, 1997). Equally important is long-term monitoring of keystone species. An admirable example is the Canadian Forest Service's Climate Change Impacts on the Productivity and Health of Aspen (CIPHA) network of aspen study sites, which has already provided valuable data (Hogg, Brandt, and Kochtubajda, 2001).

At a network level, our selection of protected areas should now emphasize site heterogeneity and habitat diversity at least as much as "representativeness." For example, high relief terrain, such as the Cypress Hills landscape, can always be expected to provide a range of habitats and ecosystems different to the surrounding plains and, therefore, will always contribute to biodiversity, even as the nature of these protected habitats and ecosystems changes over time. However, a low relief landscape, such as Prince Albert National Park, which is mandated to protect fescue grassland, aspen parkland and southern boreal forest within the national system, may fail to preserve these representative landscape elements over time, just as Wapusk National Park may fail to protect polar bear habitat, its mandated *raison d'être*.

At the site-specific scale, promoting landscape heterogeneity—species and ecosystem diversity, but also successional stage diversity—is a good way to increase resilience in the face of climate change. Promoting heterogeneity is a general principle advocated by the US National Assessment Synthesis Team, which also explicitly recommends active management, if necessary, to achieve heterogeneity (Joyce et al., 2001).

Zoning, which is already employed as a management technique in many protected areas, will be a critical intra-site management tool. Zoning can facilitate a differentiated response to climate change and a shift towards multiple target landscapes within a protected area, with some zones proactively managed in response to climate change, while other zones are managed passively or more traditionally. A multiple-target landscape approach would promote landscape diversity and ecosystem robustness and also have scientific and monitoring value (Henderson et al., 2002).

Maintaining or enhancing connectivity (natural corridor linkages or migration routes for particular species populations) between natural ecosystems is an important adaptation response, and commonly recommended as one method of coping with climate change (e.g. Malcolm and Markham, 2000; Scott and Suffling, 2000). The US National Synthesis Team (Joyce et al., 2001: 209) recommends that managers ensure "high levels of connectivity in aquatic and terrestrial systems" in response to climate change. However, connectivity will affect species differentially (Scott and Lemieux, 2005). While some—but not all—species may be able to migrate, other resident species will be threatened by the arrival of new competitors or by the pathogens that increased connectivity supports. Landscape connectivity is therefore a double-edged sword which may hasten unwanted change in some ecosystems.

A ROLE FOR EXOTIC INTRODUCTIONS?

Managers can supply artificial connectivity via, for example, programs to find and introduce drought-resistant seed sources of extant forest and grassland species. Breeding of more drought-resistant varieties might also be useful. More radical connectivity could be supplied by the introduction of new species to increase ecosystem resiliency. This response to climate change is recommended by Ledig and Kitzmiller (1992) in a commercial forestry context. Williams (1997) argued that exotic species can contribute positively to the structure and function of a particular ecosystem, and that exotic species are more likely to have ecological value in human-altered systems than in natural systems.

Aquatic species, particularly those locked into small basins or east-west linkages, may have little ability to move to suitable habitat as climate conditions evolve. The assisted migration of fish species, for example, is a possible adaptation response. However, in their analysis of climate change impacts on global freshwater fisheries, Ficke, Myrick and Hansen (2007: 602) cautioned that the introduction of fishes is "not a decision to be made lightly because of the possible negative consequences for organisms already in that environment." Isolated basin systems, like isolated grasslands or forests, are examples of disjunct systems where effective species migration will typically not be possible without human intervention.

In the case of species with small and shrinking ranges, and threatened by climate change or other factors, there is a contentious debate ongoing as to the merits and demerits of assisted species migration. In some such situations it may be possible to start a new population of the threatened species by translocation, perhaps of seed, to what is hoped to be a favourable environment. The objective is to save a particular species. McLachlan, Hellmann and Schwartz (2007) provided a taxonomy of the various viewpoints on this. To take no management action in some such situations may imply disregarding Leopold's (1949) injunction to "save all the parts."

The principal Prairie Provinces challenge is, however, quite different—not to preserve particular extinction-threatened species so much as to preserve entire threatened systems. In the case of prairie forests, the assisted migration of quite common species to stabilize intended host forests must be considered where extant trees look unlikely to endure and where trees potentially more suited to the new emerging prairies climate regimes exist elsewhere.

Henderson et al. (2002) explored the management options for threatened prairie forest landscapes, focussing on climate change impacts on five isolated "island" forests of the southern Prairie Provinces and immediately adjacent North Dakota and Montana. These forests are refugia of trees and tree-dependent species isolated in a sea of grass and at significant risk from climate change. They are marginal or ecotone systems, borderline between grassland and forest ecosystems, and therefore sensitive to relatively small changes in environmental conditions. As they are relatively small ecosystems, island forests may exhibit reduced genetic diversity and greater vulnerability to catastrophic disturbance, such as wildfire, pathogen attack or severe drought.

Henderson et al. (2002) used a range of climate scenarios derived from three GCMS (HadCM3, CGCM2, and CSIROMk2b) to describe the future climate of the southern Prairie Provinces. Working from the climate scenarios data, future moisture regimes were derived for the five forests (Cypress Hills, Sweet Grass Hills, Moose Mountain, Turtle Mountain, and Spruce Woods), and the implications of these moisture regimes for the dominant trees were considered. Both precipitation and temperature impact on moisture levels. Increased temperatures will have a powerful evaporation effect, such that soil moisture balances will decline substantially. The island forests will therefore suffer serious challenges to tree survival and ecosystem integrity. The details of this research are presented as a case study in Chapter 19.

Henderson et al. (2002) concluded that intrusive management is likely necessary to preserve some type of forest cover at these sites. Management that aims simply to retain existing vegetation, or to restore historical vegetation distributions and ecosystems, will fail as the climate moves farther away from recent and current norms. Possible adaptation actions included managing (reducing) fuel loads; harvesting trees; maintaining a diversity of age stands in the forest; responding aggressively to pathogen disturbances; actively regenerating the existing forest; and, most radically, introducing exotic tree species better adapted to the new climate parameters.

There seem to be no legal prohibitions against the introduction of exotic tree species. Indeed, exotics are frequently planted on freehold land in Canada. However, current provincial policies do not favour exotic tree introductions on Crown forest lands. In Saskatchewan, after forest harvest, the responsible forestry company must typically restock with seed collected in the relevant ecoregion. A Manitoba Forest Renewal Program objective is to ensure that all harvested forests are satisfactorily regenerated to maintain the existing mosaic of forest ecosystem stand types (Manitoba Conservation Department, 2005). In Alberta, forestry policy is "to reproduce the composition and structure of the current forest" (Alberta Reforestation Standards Science Council, 2001). Alberta's *Standards for Tree Improvement in Alberta* is the most detailed document to deal with reforestation "tree improvement" issues yet written amongst the Prairie Provinces with respect to forested provincial Crown lands (Alberta Sustainable Resource Development, 2005). The document's emphasis is clearly on the preservation of existing biodiversity. Genetically modified organisms are not approved for reforestation. The key assumption across the three provinces is that maintenance of the existing forest mosaic is possible.

Thorpe, Henderson and Vandall (2006) evaluated ecological and policy issues related to the possible introduction of exotic tree species into the western boreal forest of the Prairie Provinces as an adaptation to climate change. They concluded that western conifers such as Douglas fir and ponderosa pine, as well as hardwoods of the southern prairies such as Manitoba maple and green ash, may be suited to future climates of the western boreal. However, native boreal species were expected to shift northward in distribution and probably decline in the southern parts of their current range. Red pine, lodgepole pine, ponderosa pine, Scots pine, blue spruce, and Siberian larch are all potentially viable exotic conifer introductions into island forests just south of the southern boundary of the Saskatchewan boreal forest. Bendzsak (2006) recommended trial plantings of all these species within these island forests, in part as a strategy to adapt to climate change.

To propose the introduction of exotic tree species to save prairie forests seems a desperate measure—and indeed it is. There have been numerous cases around the world in which introduction of exotic tree species has led to undesirable consequences, including economic losses, spread of diseases, genetic impacts on native species, site degradation, loss of aesthetic values, and invasion of adjacent ecosystems (Thorpe, Henderson and Vandall, 2006). Globally, invasive behaviour has been the largest threat associated with exotic introductions. Assessments of threats to native biodiversity show exotic invasion to be one of the most important, contributing to the decline of almost half of all endangered species in the United States (Wilcove et al., 1998). According to Haysom and

Murphy (2003), most introduced trees do not become invasive, but "there are now several well-documented studies that show the hazards that can result from an introduced tree or woody shrub becoming invasive . . ." In the Canadian Prairies, examples include Siberian elm (*Ulmus pumila*), European buckthorn (*Rhamnus cathartica*), and caragana (*Caragana arborescens*). Therefore, the ultimate cold-hardy and drought-hardy trees derived from central Asia (Henderson et al., 2002) are also the most likely to cause invasive problems in a Prairie Provinces environment.

Research has shown that one of the best predictors of invasive species is prior invasive behaviour elsewhere (Thorpe, Henderson and Vandall, 2006). In a system developed in the United States for screening proposed introductions, exotic species from other continents are considered to pose a greater threat than those that are native to other parts of North America (Reichard and Hamilton, 1997). Thorpe, Henderson and Vandall (2006) recommended that exotic tree species proposed for introduction in the Canadian Prairies be subject to a standardized assessment process to evaluate benefits versus risks. The value of the species in adaptation to climate change would be one of the benefits considered. Limited planting trials, with appropriate monitoring and evaluation, should precede widespread planting. Assessments should vary with the type of land management (e.g. protected areas versus commercial forest versus private farmland) and the potential for controlling unforeseen problems. This approach legitimizes the concept of introducing tree species for adaptation to climate change, but adds process and oversight to the current unregulated situation, which historically has allowed the introduction of some problem species.

The challenge is to find the balance between regenerative success and undesirable invasion. If we are introducing exotics with the aim of stabilizing an at-risk ecosystem, the implication is that we want the introduction to be able to successfully regenerate in the new landscape—our objective is actually assisted colonization (Hunter, 2007). Otherwise we would be doomed to continuous replanting, which would not be practical in very large biomes, such as the boreal forest. However, species that regenerate too successfully become the problem invasives that exclude other species and threaten biodiversity. The introduction of species from other parts of North America, which might be expected to eventually migrate to our region, likely presents fewer risks than movement of species between continents (Thorpe et al., 2001).

If we are seriously considering the introduction of exotic tree species, logically we must also consider the practicality and desirability of introducing associated midstory (shrubs and bushes) and understory (ground flora) exotics (Carr, Weedon and Cloutis,, 2004; Henderson et al., 2002). Frelich and Puettmann (1999: 507), in the context of restoration ecology, noted: "it is often assumed that understory vegetation will establish over time ('plant trees and the rest will come'), but natural invasion may not automatically bring back all species desired." Unfortunately, while we know little enough about exotic tree introductions in prairie forests, we know even less about mid- and understory introductions. What research we do have has been on trees, and primarily on trees as a source of commercial timber.

It must be understood that all forest management options have the potential to be irreversible. If we continue with present policy, and restocking or natural regeneration fails, it may become increasingly difficult to regenerate some forest environments as the climate tracks warmer and drier. Forest loss could be irreversible if we adapt only slowly or reactively.

MANAGEMENT MODELS FOR RESPONDING TO CLIMATE CHANGE

Henderson et al. (2002) identified three possible natural systems management models as a response to climate change impacts. Carr et al. (2004) added a fourth, which differs from the first three in that it is not designed to be sustainable:

PASSIVE MANAGEMENT MODEL

In this model we assume that protected landscapes are best left untouched by human management so far as possible. The approach is *laissez-faire*, except that where we have interrupted natural processes in the past (e.g. by suppressing fire), we may seek to reinstate these processes or replicate their effects in so far as it is practicable.

The advantages of this model are that it is traditional, initially inexpensive (as there are few active management costs), and relatively painless in a managerial and political sense (as it often involves no major shift in management direction). It also avoids the risks of unintended consequences of active management.

This approach can be criticized as being one of "unmanaged retreat." In vulnerable forest systems, it would open the way to sweeping, sudden change, perhaps by major fire or insect disturbance, followed by an ecosystem shift that cannot be predicted in detail, but which would almost certainly result in a landscape with greatly reduced tree cover. Over the long run, this approach may fail to protect some aspects of biodiversity.

FROZEN LANDSCAPE MANAGEMENT MODEL

In this model we assume that the ecosystem structure present on a particular landscape at some given point in time is the correct one. There are at least two model variants. The more modest variant seeks to maintain the landscape as it is now by resisting (further) anthropogenic change. The more radical, purist or ambitious variant seeks to restore the landscape as of some point in the past by reversing anthropogenic changes. Most ambitious is the attempt to recreate the landscape as of a point prior to any (European) anthropogenic impacts.

The Montana state lands forestry policy (MDNRC, 1996), which aims to replicate forests that were historically present on the Montana landscape, and the Shilo vegetation management plan (Dillon Ltd., 1996), which aims to maintain Manitoba's Spruce Woods vegetation cover types and distributions as of 1988, are two operative examples of this model. Similarly, the operative vegetation management plan for Duck Mountain Provincial Park in Saskatchewan targets the restoration of the grassland component of that park to its pre-European areal extent (Wright et al., 1995).

This model has several advantages. First, it is easily understood in principle by professionals and the public. Second, when we target the preservation of the existing landscape we have the advantage of knowing exactly what we are trying to do and can measure success or failure relatively clearly. Even if targeting the re-creation of a past landscape, we are sometimes able to establish with reasonable certainty the nature and proportions of the basic historical ecological landscape constituents. Third, the fact these landscapes exist, or once existed, provides empirical credibility to the targeted landscape. They are not "pie in the sky" visualisations.

Nonetheless, the model has weaknesses. Our landscapes, in the absence of human intervention, naturally change radically on many time scales, be it from year to year or from century to century. It is hard to see what ecological basis there could be to justify fixing a Prairie Provinces landscape as of an arbitrary point in time. Most crucially, from a practical standpoint, it becomes increasingly difficult, and eventually materially impossible, to maintain any fixed historic landscape as climate change advances.

MANAGED CHANGE MANAGEMENT MODEL

In this model, we accept landscape change driven by climate change as inevitable, but by active management seek to strategically delay, ameliorate and direct change. Lopoukhine (1990) suggested that active management is the only alternative for protected areas given the reality of climate change impacts. Scott and Suffling (2000) agreed that active management is warranted in response to climate change and note that intervention strategies will often be species-specific. Zoning will be useful within this model to allow differentiated management interventions and differentiated ecosystem outcomes within a given site.

A high disturbance rate changes the species composition of a forest regime and increases the rate of response to climate change (Overpeck, Rind and Goldberg, 1990). Where the objective is to delay change, interventions may include the aggressive control of wildfires, the active suppression of unwanted pathogens, and the aiding of regeneration of key extant species.

We may seek not only to delay landscape evolution, but also to direct it. Under the managed change model, if a particular species' demise is inevitable, replacement with a next-best substitute could be considered. For example, if at Cypress Hills the decline of lodgepole pine is unavoidable while ponderosa pine appears to be viable, ponderosa might be introduced as part of a managed change of the pine ecosystem.

Managed change is a conservation model employed in Britain—e.g. in response to loss of coastal conservation areas as a result of sea-level rise (a phenomenon driven there largely, though not entirely, by climate change). The strategy includes the creation of substitute ecosystems (e.g. new salt-water marshes) as offsets for those systems likely to disappear.

In the context of Prairie Provinces ecosystems, managed change could imply taking into account ecosystem changes occurring beyond any one individual forest or grassland system and considering the "big picture" of change in the wider landscape. The knowledge that a particular species or ecosystem may be valued or threatened outside our region, for example, may incline us to consider its introduction here, should new climate conditions support it.

By extending the life of existing ecological landscapes so long as reasonably possible, and by guiding their evolution, managed change can provide several public benefits. In some cases it could help preserve biodiversity. It could help maintain valued landscapes, and as the Prairies become more arid, residents' interest in the preservation of already highly-valued landscapes centred on trees or water can only be expected to increase. Managed change can reduce the risk of catastrophic change. Also, by providing a variety of landscapes, it helps keep options open. This is particularly important given that we do not know with certainty the speed or final outcomes of emissions-driven climate change impacts.

But there are disadvantages to the managed change model. One serious concern is the unpredictability of outcomes. For example, we may do the best research possible on the benefits and risks of introducing a new species as an adaptation strategy, but there can never be any certainty as to the outcome of such an experiment once implemented in the field. Some landscape experiments may turn out badly, and there is no guarantee of reversibility.

A major practical concern is that active management is necessarily more expensive than passive management. Active management requires careful species and ecosystems monitoring and study. As climate change advances, active management implies increasingly intrusive and, therefore, increasingly expensive, interventions. When we consider the scale of some regions at risk, such as the southern boreal forest, it is reasonable to wonder about the practicality of intensive management.

"ACCELERATED CAPTURE"

Carr, Weedon and Cloutis (2004: 26), commenting on the above three models, noted a fourth option: "accelerate capture before loss." In this option, resource exploitation would be accelerated if necessary to maximize the one-off economic gain from a resource not expected to endure. For example, the British Columbia government is trying to find ways to maximize the economic value of vast areas of standing dead lodgepole pine in the knowledge that this pine is deteriorating. As Alberta's mountain glaciers retreat, there may, perversely, be extra money to be made by advertising to potential tourists that the time window to see these charismatic landscape megafeatures is closing. While Carr, Weedon and Cloutis' (2004) observation says something about capturing maximum economic value, it does not advise as to how to manage the resulting post-economic-harvest landscape. As a model, it simply offers short-term advice.

LONG-RUN LANDSCAPE VISIONS

Looking beyond the issue of the immediate and necessary management of climate change impacts, it is intriguing to imagine what types of visions may compete for dominance in the post-impacts world. If, optimistically, sometime many decades into the future we succeed at stabilizing net greenhouse gas emissions, and climate change slows and environmental conditions stabilize, where will we have arrived in terms of future landscapes and conservation philosophies? Henderson et al. (2002) considered three possibilities. The future remains uncertain, and no single vision will ever completely dominate, but the climate change impacts adaptation journey may push us in the direction of the maximum diversity and garden visions described below.

THE WILDERNESS VISION

This vision is philosophically linked to the passive management model. The objective is the restoration of a low-input management style on whatever new ecosystems are in reasonably stable and self-sustaining equilibrium with the eventual end state of emissions-driven climate change. The "less human intervention required to support it, the more satisfactory the destination ecosystem," might be one measure of success under this vision. This vision is essentially a rebirth of the North American idealization of wilderness landscapes. The vision could deliver stable eco-landscapes at low management cost. However, the choice of destination ecosystems would be relatively narrow and climatically prescribed. We may lose some types of biodiversity along the way.

THE MAXIMUM DIVERSITY VISION

In this vision, the maintenance or increase of local or world biodiversity is considered the overriding objective. New genotypes of extant species, or entirely new species, could be introduced to either buffer an existing site ecosystem or to provide habitat for species threatened elsewhere. Under this vision, we would consider introducing and supporting new ecosystems, particularly to substitute for those under threat elsewhere. A high degree of intervention and management would likely be required indefinitely.

Permanently intensively managed ecosystems are common in current European conservation practice (Henderson, 1992). They are philosophically very distant from the classical wilderness preferences of many North American conservationists. Nonetheless, this approach is gaining ground in North America—we increasingly design permanently managed habitats for some species, such as managed marshes for waterfowl. To gain the necessary financial and stakeholder support for managed systems, the focus is often species specific, although a range of other species may collaterally benefit. This vision could deliver a high degree of biodiversity in response to climate change, but it would be expensive, potentially arbitrary in its specific species or ecosystem focus (economically valuable or charismatic species and ecosystems could benefit disproportionately), and would not appear "natural" to some conservationists.

THE GARDEN VISION

In this variegated and shifting vision, ecosystem management is entirely driven by personal and societal preferences for particular species and ecosystems—preferences which derive from a complicated tangle of cultural, aesthetic, historical or utilitarian predispositions. No account is taken of local or global biodiversity considerations. Flora and fauna are introduced, maintained or eradicated in a public parallel to the kind of individual decision-making that generates the landscape of a private garden.

A bit of the gardening approach already exists in modern North American conservation practice. We all have our favoured species we are inclined to support by habitat manipulation or breeding site provision, be it orchids, trout, wood ducks, bluebirds, or elk. We even have preferences amongst ages or types of individuals within a given species. For example, younger, vigorous aspen is perceived by some as more desirable than a "decadent" or insect-infested stand, however equally natural both states of aspen may be.

The gardening vision of nature conservation is widely applied at conservation sites in Britain and is also widespread in countries such as France and Japan. Resulting landscapes tend to be ecologically diverse, aesthetically attractive and high maintenance. In North America, the gardening approach would likely favour the introduction or support of economically valuable fish and game animals within a landscape, habitat manipulation to favour species such as hunted waterfowl or big game, and the retention or planting of aesthetically pleasing vegetation, such as exotic trees. Hunting, fishing and recreation values would be maximized. Although this vision does not target biodiversity, it often delivers reasonable diversity in practice. However, it can at times damage ecological integrity via species introductions, while unfavoured species or ecosystems can be lost through disinterest and neglect. It is arbitrary in nature, expensive in practice, and often popular with particular stakeholders.

CONCLUSIONS

The Prairie Provinces face major climate change impacts on ecosystems and landscapes. These climate change impacts combine and interact with other ongoing human impacts. The speed of impacts and change is accelerating.

Key climate change impacts on Prairie Provinces' ecosystems include:

- An increased rate and intensity of forest disturbances, such as fire and pathogens;
- Possible loss of forest cover in grassland-forest ecotone regions, such as the southern boreal forest, the Alberta foothills forest, and in the island forests of the grasslands;
- Grassland invasion in all the prairie-parkland national parks;
- Increased stress on aquatic ecosystems from warmer and drier conditions—many prairie aquatic species are at risk of extirpation;
- Declines in migratory waterfowl populations with loss of wetlands;
- An increase in plant productivity with a longer and warmer growing season and increasing atmospheric CO_2—at many sites this impact may be limited or overwhelmed by moisture limitations or other constraints; and
- The evolution of new landscape ecosystems; e.g. a drier climate in southern Alberta and Saskatchewan would potentially support shortgrass prairie currently found farther south.

We have many adaptation options, and some alternative choices about future ecosystems, but it will not be possible to maintain Prairie Provinces' ecosystems as they were or as we know them now. Antecedent landscapes can no longer be effectively targeted. Passivity in the face of impacts may shrink our ecosystem options, particularly in prairie forests. However, active management entails some risk and expense.

Williams, Jackson and Kutzbach (2007) noted that the world will experience the loss of some climate types entirely over the next century, while novel climates will also emerge. These climate developments will drive the disappearance of some ecosystems and the emergence of new ones. We are living in a new age of creation ecology. We have options, but the past is not one of them. The future ecosystems that result from climate change in the Prairie Provinces will be unprecedented.

REFERENCES

Alberta Reforestation Standards Science Council. 2001. Linking Regeneration Standards to Growth and Yield and Forest Management Objectives [online]. Alberta Sustainable Resource Development Department. Available from World Wide Web: www3.gov.ab.ca/srd/forests/fmd/arssc/pdf/ARSSC_Report.pdf

Alberta Sustainable Resource Development. 2005. Standards for Tree Improvement in Alberta [online]. Available from World Wide Web: www3.gov.ab.ca/srd/forests/fmd/manuals/index.html. Date accessed: April 13, 2008

Anderson, J.C., I. Craine, A.W. Diamond and R. Hansell. 1998. "Impacts of Climate Change and Variability on Unmanaged Ecosystems, Biodiversity, and Wildlife." Pp. 121–88 in G. Koshida and W. Avis (eds.). *The Canada Country Study: Climate Impacts and Adaptation* (Volume 7: National Sectoral Volume). Environment Canada.

Aspen FACE Experiment. 2009. Available from World Wide Web: *http://aspenface.mtu.edu/index.html.* Date accessed: April 13, 2008

Baker, B.B., J.D. Hanson, R.M. Bourdon and J.B. Eckert. 1993. "The Potential Effects of Climate Change on Ecosystem Processes and Cattle Production on U.S. Rangelands." *Climate Change* 25: 97–117.

Baulch, H., D. Schindler, M. Turner, D. Finlay, M. Paterson and R. Vinebrooke. 2005. "Effects of Warming on Benthic Communities in a Boreal Lake: Implications of Climate Change." *Limnology and Oceanography* 50: 1377–92.

Beaubien, E.G. 1997. "Plantwatch: Tracking the Biotic Effects of Climate Change Using Students and Volunteers. Is Spring Arriving Earlier on the Prairies?" Pp. 66–68 in *The Ecological Monitoring and Assessment Network Report (Environment Canada) on the 3rd National Science meeting, January 1997: Saskatoon, Saskatchewan.*

Beaubien, E.G. and H.J. Freeland. 2000. "Spring Phenology Trends in Alberta, Canada: Links to Ocean Temperature." *International Journal of Biometeorology* 44, no. 2: 53–59.

Bendzsak, M. 2006. Evaluation of Conifer Tree Species Alternatives for Island Forest Renewal. Saskatchewan Forest Centre: Prince Albert, 37 pgs.

Bethke, R.W. and T.D. Nudds. 1995. "Effects of Climate Change and Land Use on Duck Abundance in Canadian Prairie-Parklands." *Ecological Applications* 5, no. 3: 588–600.

Boisvenue, C. and S. Running. 2006. "Impacts of Climate Change on Natural Forest Productivity—Evidence Since the Middle of the 20th Century." *Global Change Biology* 12: 862–82.

Campbell, B.D. and D.M. Stafford Smith. 2000. "A Synthesis of Recent Global Change Research on Pasture and Rangeland Production: Reduced Uncertainties and Their Management Implications." *Agriculture, Ecosystems and Environment* 82: 39–55.

Carr, A., P. Weedon and E. Cloutis. 2004. *Climate Change Implications in Saskatchewan's Boreal Forest Fringe and Surrounding Agricultural Areas.* Geospatial Consulting: Prince Albert, Saskatchewan.

Clair, T., B. Warner, R. Robarts, H. Murkin, J. Lilley, L. Mortsch and C. Rubec. 1998. "Canadian Inland Wetlands and Climate Change." Pp. 189–218 in G. Koshida and W. Avis (eds.). *The Canada Country Study: Climate Impacts and Adaptation* (Volume 7: National Sectoral Volume). Environment Canada.

Coffin, D.P. and W.K. Lauenroth. 1996. "Transient Responses of North American Grasslands to Changes in Climate." *Climatic Change* 34: 269–78.

Collatz, G.J., J.A. Berry and J.S. Clark. 1998. "Effects of Climate and Atmospheric CO_2 Partial Pressure on the Global Distribution of C_4 Grasses: Present, Past, and Future." *Oecologica* 114: 441–54.

Davis, M.B. and R.G. Shaw. 2001. "Range Shifts and Adaptive Responses to Quarternary Climate Change." *Science* 292: 673–79.

Davis, M.B. and C. Zabinski. 1992. "Changes in Geographical Range Resulting from Greenhouse Warming Effects on Biodiversity in Forests." Pp. 297–308 in R.L. Peters and T.E. Lovejoy (eds.). *Global Warming and Biological Diversity.* Yale University Press: New Haven, Connecticut.

de Groot, W., P. Bothwell, D. Carlsson, K. Logan, R. Wein and C. Li. 2002. *Forest Fire Management Adaptation to Climate Change in the Prairie Provinces.* Canadian Forestry Service; University of Alberta; Prairie Adaptation Research Collaborative: Regina.

Dillon Ltd. 1996. *Natural Resources Management Plan.* Dillon Ltd.: Winnipeg.

EPA (United States Environmental Protection Agency). 2002. *U.S. Climate Action Report—2002—Third National Communication of the United States of America under the United Nations Framework Convention on Climate Change.* Environmental Protection Agency: Washington, DC.

Epstein, H.E., R.A. Gill, J.M. Paruelo, W.K. Lauenroth, G.J. Jia and I.C. Burke. 2002. "The Relative Abundance of Three Plant Functional Types in Temperate Grasslands and Shrublands of North and South America: Effects of Projected Climate Change." *Journal of Biogeography* 29: 875–88.

Ermine, Willie, Ralph Nilson, David Sauchyn, Ernest Sauve and Robin Smith. 2005. "ISI ASKIWAN—The State of the Land: Prince Albert Grand Council Elders' Forum on Climate Change." *Journal of Aboriginal Health* 2, no. 1: 62–75.

Ferguson, S., I. Stirling and P. McLoughlin. 2005. "Climate Change and Ringed Seal (Phoca hispida) Recruitment in Western Hudson Bay." *Marine Mammal Science* 21: 121–35.

Ficke, A., C. Myrick and L. Hansen. 2007. "Potential Impacts of Global Climate Change on Freshwater Fisheries." *Reviews in Fish Biology and Fisheries* 17: 581–613.

Frelich, L. and K. Puettmann. 1999. "Restoration Ecology." Pp. 499–524 in M. Hunter (ed.). *Maintaining Biodiversity in Forest Ecosystems.* Cambridge University Press: Cambridge.

Gitay, H., S. Brown, W. Easterling and B. Jallow. 2001. "Ecosystems and Their Goods and Services" Pp. 235–342 in J. McCarthy, O. Canziani, N. Leary, D. Dokken and K. White (eds.). *Climate Change 2001: Impacts, Adaptation, and Vulnerability, IPCC.* Cambridge University Press: Cambridge.

Gracia, C., S. Sabaté, and A. Sánchez, A. 2002. "El Cambio Climático y la Reducción de la Reserva de Agua en el Bosque Mediterráneo." *Ecosistemas* 11, no. 2. Available from World Wide Web: www.aeet.org/ecosistemas/022/investigacion4.htm. Date accessed: April 13, 2008

Grissom, P., M. Alexander, B. Cella, F. Cole, J. Kurth, N. Malotte, D. Martell, W. Mawdsley, J. Roessler, R. Quillin and P. Ward. 2000. "Effects of Climate Change on Management and Policy: Mitigation Options in the North American Boreal Forest." Pp. 85–101 in E. Kasischke and B. Stocks (eds.). *Fire, Climate Change, and Carbon Cycling in the Boreal Forest.* Springer-Verlag: New York.

Halpin, P. 1997. "Global Climate Change and Natural-Area Protection: Management Responses and Research Directions." *Ecological Applications* 7: 828–43.

Hamburg, S.P. and C.V. Cogbill. 1988. "Historical Decline of Red Spruce Populations and Climatic Warming." *Nature* 331: 428–31.

Harvell, C., C. Mitchell, J. Ward, S. Altizer, A. Dobson, R. Ostfeld and M. Samuel. 2002. "Climate Warming and Disease Risks for Terrestrial and Marine Biota." *Science* 296: 2158–62.

Haysom, K.A. and S.T. Murphy. 2003. The Status of Invasiveness of Forest Tree Species Outside Their Natural Habitat: A Global Review and Discussion Paper. Forestry Department, Food and Agriculture Organization of the United Nations, Working Paper FBS/3E.

Henderson, N. 1992. "Wilderness and the Nature Conservation Ideal: Britain, Canada, and the United States Contrasted." *Ambio* 21: 394–99.

——. 2006. "Climate Change Impacts in Saskatchewan." Pp. 23–25 in *Ecoliving: Working Together for a Sustainable World.* Regina Ecoliving Inc.: Regina.

Henderson, N., T. Hogg, E. Barrow and B. Dolter. 2002. *Climate Change Impacts on the Island Forests of the Great Plains and the Implications for Nature Conservation Policy.* Prairie Adaptation Research Collaborative: Regina.

Herrington, R., B. Johnson and F. Hunter. 1997. "Responding to Global Climate Change in the Prairies" in *Canada Country Study: Climate Impacts and Adaptation* (Volume III). Environment Canada: Ottawa.

Hoekman, S.T., L. S. Mills, D.W. Howerter, J.H. Devries and I.J. Ball. 2002. "Sensitivity Analyses of the Life Cycle of Mid-Continent Mallards." *Journal of Wildlife Management* 66, no. 3: 883–900.

Hogg, E.H. 1994. "Climate and the Southern Limit of the Western Canadian Boreal Forest." *Canadian Journal of Forest Research* 24: 1835–45.

——. 1997. "Temporal Scaling of Moisture and the Forest-Grassland Boundary in Western Canada." *Agriculture and Forest Meteorology* 84: 115–22.

Hogg, E.H. and P.Y. Bernier. 2005. "Climate Change Impacts on Drought-Prone Forests in Western Canada." *Forestry Chronicle* 81: 675–82.

Hogg, E., J.P. Brandt and B. Kochtubajda. 2001. "Responses of Western Canadian Aspen Forests to Climate Variation and Insect Defoliation During the Period 1950–2000." Draft: project A039 CCAF.

Hogg, E.H. and P.A. Hurdle. 1995. "The Aspen Parkland in Western Canada: A Dry-Climate Analogue for the Future Boreal Forest?" *Water, Air and Soil Pollution* 82: 391–400.

Hogg, E.H. and A.G. Schwarz. 1997. "Regeneration of Planted Conifers across Climatic Moisture Gradients on the Canadian Prairies: Implications for Distribution and Climate Change." *Journal of Biogeography* 24: 527–34.

Hunter, M. 2007. "Climate Change and Moving Species: Furthering the Debate on Assisted Colonization." *Conservation Biology* 21: 1356–58.

Inkley, D.B., M.G. Anderson, A.R. Blaustein, V.R. Burkett, B. Felzer, B. Griffith, J. Price and T.L. Root. 2004. "Global Climate Change and Wildlife in North America." *The Wildlife Society Technical Review 04–2.* The Wildlife Society: Bethesda, Maryland, 26 pgs.

IPCC (Intergovernmental Panel on Climate Change). 2007. *Climate Change 2007: The Physical Science Basis: Contribution of Working Group 1 to the Fourth Assessment Report of the Intergovernmental Panel on Climate Change* (S. Solomon, D. Qin, M. Manning, Z. Chen, M. Marquis, K. Averyt, M. Tignor and H. Miller, eds.). Cambridge University Press: Cambridge and New York. 996 pgs.

Jacobs, K. 2002. *Connecting Science, Policy and Decision-Making: A Handbook for Researchers and Science Agencies.* National Oceanic and Atmospheric Administration: Silver Spring, Maryland.

James, P., K Murphy, R. Espie, D. Gauthier and R. Anderson. 2001. *Predicting the Impact of Climate Change on Fragmented Prairie Biodiversity: A Pilot Landscape Model.* Saskatchewan Environment and Resource Management/Canadian Plains Research Center: Regina.

Johnson, W., B. Millett, T. Gilmanov, R. Voldseth, G. Guntenspergen and D. Naugle. 2005. "Vulnerability of Northern Prairie Wetlands to Climate Change." *BioScience* 55: 863–72.

Johnston, M.H. 1996. "The Role of Disturbance in Boreal Mixedwood Forests of Ontario." Pp. 33–40 in C.R. Smith and G.W. Crook (compilers). *Advancing Boreal Mixedwood Management in Ontario: Proceedings of a Workshop.* Canadian Forest Service, Great Lakes Forestry Centre: Sault Ste. Marie, Ontario.

Johnston, M. and T. Williamson. 2005. "Climate Change Implications for Stand Yields and Soil Expectation Values: A Northern Saskatchewan Case Study." *Forestry Chronicle* 81: 683–90.

Joyce, L.A., D. Ojima, G.A. Seielstad, R. Harriss and J. Lackett. 2001. "Potential Consequences of Climate Variability and Change for the Great Plains." Pp. 191–217 in *National Assessment Synthesis Team, The Potential Consequences of Climate Variability and Change* (Report for the US Global Change Research Programme). Cambridge University Press: Cambridge.

Kelly, A.E. and M.L. Goulden. 2008. "Rapid Shifts in Plant Distribution with Recent Climate Change. *PNAS* 105: 11823–26.

Kharin, V.V. and F.W. Zwiers. 2000. "Changes in the Extremes in an Ensemble of Transient Climate Simulations with a Coupled Atmosphere-Ocean GCM." *Journal of Climate* 13: 3760–88.

Kochy, M. and S. Wilson. 2001. "Nitrogen Deposition and Forest Expansion in the Northern Great Plains." *Journal of Ecology* 89: 807–17.

Kurz, W. and M. Apps. 1993. "Contribution of Northern Forests to the Global C Cycle: Canada as a Case Study." *Water, Air and Soil Pollution* 70: 163–76.

Laporte, M., L. Duchesne and S. Wetzel. 2002. "Effect of Rainfall Patterns on Soil Surface CO_2 Efflux, Soil Moisture, Soil Temperature and Plant Growth in a Grassland Ecosystem of Northern Ontario, Canada: Implications for Climate Change." *BMC Ecology* 2: 1–6.

LaRoe, E.T. and D.H. Rusch. 1995. "Changes in Nesting Behavior of Arctic Geese." Pp. 388–89 in E.T. LaRoe, G.S. Farris, c.e. Puckett, P.D. Doran and M.J. Mac (eds.). *Our Living Resources: A Report to the Nation on the Distribution, Abundance, and Health of U.S. Plants, Animals, and Ecosystems.* US Department of the Interior, National Biological Service: Washington, DC.

Ledig, F. and J. Kitzmiller. 1992. "Genetic Strategies for Reforestation in the Face of Global Climate Change." *Forest Ecology and Management* 50: 153–69.

Lemon, E. (ed.). 1983. *CO_2 and Plants—The Response of Plants to Rising Levels of Atmospheric Carbon Dioxide.* Westview Press: Boulder.

Lenten, T., H. Held, E. Kriegler, J. Hall, W. Lucht, S. Rahmstorf and H. Schellnhuber. 2008. "Tipping Elements in the Earth's Climate System." *Proceedings of the National Academy of Sciences 105:* 1786–93.

Leopold, Aldo. 1949. *A Sand County Almanac.* Oxford University Press: Oxford.

Lloyd, A.H. 2005. "Ecological Histories from Alaskan Tree Lines Provide Insight into Future Change." *Ecology* 86, no. 7: pp. 1687–95.

Long, S.P., E.A. Ainsworth, A. Rogers and D.R. Ort. 2004. "Rising Atmospheric Carbon Dioxide: Plants FACE the Future." *Annual Review of Plant Biology* 55: 591–628.

Long, S.P., and P.R. Hutchin. 1991. "Primary Production in Grasslands and Coniferous Forests with Climate Change: An Overview." *Ecological Applications* 1: 139–56.

Lopoukhine, N. 1990. "National Parks, Ecological Integrity and Climate Change." Pp. 317–28 in G. Wall and M. Sanderson (eds.). *Climatic Change: Implications for Water and Ecological Resources* (occasional paper no. 11), Dept. of Geography publication series, University of Waterloo: Waterloo.

Macdonald, G., J. Szeicz, J. Claricoates and K. Dale. 1998. "Response of the Central Canadian Treeline to Recent Climatic Changes." *Annals of the Association of American Geographers* 88: 183–208.

Malcolm, J. and A. Markham. 2000. *Global Warming and Terrestrial Biodiversity Decline.* World Wildlife Fund: Gland, Switzerland.

Malcolm, J.R. and L.F. Pitelka. 2000. *Ecosystems and Global Climate Change: A Review of Potential Impacts on U.S. Terrestrial Ecosystems and Biodiversity.* Pew Center on Global Change.

Manitoba Conservation Department. 2000. *An Action Plan for Manitoba's Network of Protected Areas* (3rd edition). Department of Conservation: Winnipeg.

——. 2005. Forest Renewal in Manitoba web site. Available from World Wide Web: *www.gov.mb.ca/conservation/forestry/forest-renewal/fr1-intro.html.* Date accessed: April 13, 2008

McCarty, J.P. 2001. "Ecological Consequences of Recent Climate Change." *Conservation Biology* 15: 320–31.

McLachlan, J., J. Hellmann and M. Schwartz. 2007. "A Framework for Debate of Assisted Migration in an Era of Climate Change." *Conservation Biology* 21: 297–302.

MDNRC (Montana Department of Natural Resources and Conservation). 1996. State Forest Land Management Plan Final Environmental Impact Statement Record of Decision May 30, 1996, Helena.

Melville, G.E. 2001. *Climate Change and an Ecosystem-Resource Adaptation Approach for Vulnerable Lakes in the Boreal Plain Ecozone.* Saskatchewan Research Council Publication 11374–1E01: Saskatoon.

Morgan, J., D. Milchunas, D. LeCain, M. West and A. Mosier. 2007. "Carbon Dioxide Enrichment Alters Community Structure and Accelerates Shrub Growth in the Shortgrass Steppe." *Proceedings of the National Academy of Sciences* 104: 14724–29.

Myneni, R., C. Keeling, C. Tucker, G. Asrar and R. Nemani. 1997. "Increased Plant Growth in the Northern High Latitudes from 1981–1991." *Nature* 386: 698–702.

Nash, R. 1982. *Wilderness and the American Mind* (3rd edition). Yale University Press: New Haven.

Natural Resources Canada. 2006. *The State of Canada's Forests 2005–2006.* Natural Resources Canada: Ottawa.

Nie, D., M.B. Kirkham, L.K. Ballou, D.J. Lawlor, and E.T. Kanemasu. 1992. "Changes in Prairie Vegetation Under Elevated Carbon Dioxide Levels and Two Soil Moisture Regimes." *Journal of Vegetation Science* 3: 673–78.

Norby, R.J., E.H. DeLucia, B. Gielen, C. Calfapietra, C.P. Giardina, J.S. King, J. Ledford, H.R. McCarthy, D.J.P. Moore, R. Ceulemans, P. De Angelis, A.C. Finzi, D.F. Karnosky, M.E. Kubiske, M. Lukac, K.S. Pregitzer, G.E. Scarascia-Mugnozza, W.H. Schlesinger and R. Oren. 2005. "Forest Response to Elevated CO_2 is Conserved Across a Broad Range of Productivity." *Proceedings of the National Academy of Sciences* 102: 18052–56.

Obbard, M. 2006. *Temporal Trends in Body Condition of Southern Hudson Bay Polar Bears* (climate change research information note number 3). Ontario Ministry of Natural Resources, 8 pgs.

Ojima, D.S., W.J. Parton, M.B. Coughenour, J.M.O. Scurlock, T.B. Kirchner, T.G.F. Kittel, D.O. Hall, D.S. Schimel, E. Garcia Moya, T.G. Gilmanov, T.R. Seastedt, Apinan Kamnalrut, J.I. Kinyamario, S.P. Long, J-C. Menaut, O.E. Sala, R.J. Scholes, and J.A. van Veen. 1996. "Impact of Climate and Atmospheric Carbon Dioxide Changes on Grasslands of the World." Pp. 271–311 in A.I. Breymeyer, D.O. Hall, J.M. Melillo and G.I. Agren (eds.). *Global Change: Effects on Coniferous Forests and Grasslands.* John Wiley and Sons: Chichester.

O'Riordan, T., C. Wood and A. Shadrake. 1993. "Landscapes for Tomorrow." *Journal of Environmental Planning and Management* 36: 123–47.

Overpeck, J.T., D. Rind and R. Goldberg. 1990. "Climate-Induced Changes in Forest Disturbance and Vegetation." *Nature* 343: 51–53.

Parisien, M., V. Kafka, N. Flynn, K. Hirsch, B. Todd and M. Flannigan. 2005. *Fire Behaviour Potential in Central Saskatchewan under Predicted Climate Change* (PARC Summary Document). Prairie Adaptation Research Collaborative: Regina.

Parmesan, C. and G. Yohe. 2003. "A Globally Coherent Fingerprint of Climate Change Impacts Across Natural Systems." *Nature* 421: 37–42.

Parton, W.J., M.B. Coughenour, J.M.O. Scurlock, D.S. Ojima, T.G. Gilmanov, R.J. Scholes, D.S. Schimel, T.B. Kirchner, J.-C. Menaut, T.R. Seastedt, E. Garcia Moya, Apinan Kamnalrut, J.I. Kinyamario and D.O. Hall. 1996. "Grassland Ecosystem Modeling: Development and Test of Ecosystems Models for Grassland Systems." Pp. 229–67 in A.I. Breymeyer, D.O. Hall, J.M. Melillo and G.I. Agren (eds.). *Global Change: Effects on Coniferous Forests and Grasslands.* John Wiley and Sons: Chichester.

Parton, W.J., D.S. Ojima and D.S. Schimel. 1994. "Environmental Change in Grasslands: Assessment Using Models." *Climatic Change* 28: 111–41.

Paruelo, J.M. and W.K. Lauenroth. 1996. "Relative Abundance of Plant Functional Types in Grasslands and Shrublands of North America." *Ecological Applications* 6: 212–24.

Pernetta, J. 1994. "Editorial Preface." Pp. vi-viii in J.C. Pernetta, R. Leemans, D. Elder and S. Humphrey (eds.). *Impacts of Climate Change on Ecosystems and Species: Implications for Protected Areas, International Union for Conservation of Nature and Natural Resources.* Gland, Switzerland.

Peters, R.L. 1992. "Conservation of Biological Diversity in the Face of Climate Change." Pp. 15–30 in R.L. Peters and T.E. Lovejoy (eds.). *Global Warming and Biological Diversity.* Yale University Press.

Peterson, A. 2003. "Projected Climate Change Effects on Rocky Mountain and Great Plains Birds: Generalities of Biodiversity Consequences." *Global Change Biology* 9: 647–55.

Pielou, E. 1991. *After the Ice Age: The Return of Life to Glaciated North America.* University of Chicago Press: Chicago.

Pitt, J., S. Larivière and F. Messier. 2008. "Survival and Body Condition of Raccoons at the Edge of the Range." *Journal of Wildlife Management* 72: 389–95.

Poiani, K. and W. Johnson. 1993. "Potential Effects of Climate Change on a Semi-Permanent Prairie Wetland." *Climate Change* 24: 213–32.

Rehfeldt, G.E., C.C. Ying, D.L. Spittlehouse and D.A. Hamilton. 1999. "Genetic Responses to Climate in Pinus contorta: Niche Breadth, Climate Change and Reforestation." *Ecological Monographs* 69: 375–407.

Reichard, S. and C. Hamilton. 1997. "Predicting Invasions of Woody Plants Introduced into North America." *Conservation Biology* 11: 193–203.

Sala, O.E., F.S. Chapin 3rd, J.J. Armesto, E. Berlow, J. Bloomfield, R. Dirzo, E. Huber-Sanwald, L.F. Huenneke, R.B. Jackson, A. Kinzig, R. Leemans, D.M. Lodge, H.A. Mooney, M. Oesterheld, N.L. Poff, M.T. Sykes, B.H. Walker, M. Walker and D.H. Wall. 2000. "Global Biodiversity Scenarios for the Year 2100." *Science* 287, no. 5459: 1770–74.

Saporta, R., J. Malcolm and D. Martell. 1998. "The Impact of Climate Change on Canadian Forests." Pp. 319–82 in G. Koshida and W. Avis (eds.). *The Canada Country Study: Climate Impacts and Adaptation* (Volume 7: National Sectoral Volume), Environment Canada.

Schindler, D.W. and W.F. Donahue. 2006. "An Impending Water Crisis in Canada's Western Prairie Provinces." *Proceedings of the National Academy of Sciences* 103: 7210–16.

Scholze, M., W. Knorr, N. Arnell and I. Prentice. 2006. "A Climate-Change Risk Analysis for World Ecosystems." *Proceedings of the National Academy of Sciences* 103: 13116–20.

Scott, D. and C. Lemieux. 2005. "Climate Change and Protected Area Policy and Planning in Canada." *Forestry Chronicle* 81: 696–703.

Scott, D. and R. Suffling (2000). *Climate Change and Canada's National Park System.* Parks Canada, Environment Canada, and the University of Waterloo.

Soja, J., N. Tchebakova, N. French, M. Flannigan, H. Shugart, B. Stocks, A. Sukhinin, E. Parfenova, F. Chapin and P. Stackhouse. 2007. "Climate-Induced Boreal Forest Change: Predictions Versus Current Observations." *Global and Planetary Change* 56: 274–96.

Solomon, A. 1994. "Management and Planning of Terrestrial Parks and Reserves During Climate Change." Pp. 1–12 in J.C. Pernetta, R. Leemans, D. Elder and S. Humphrey (eds.). *Impacts of Climate Change on Ecosystems and Species: Implications for Protected Areas, International Union for Conservation of Nature and Natural Resources.* Gland, Switzerland.

Thornley, J. and M. Cannell. 2004. "Long-Term Effects of Fire Frequency on Carbon Storage and Productivity of Boreal Forests: A Modeling Study." *Tree Physiology* 24: 765–73.

Thorpe, J. 2007. *Impacts of Climate Change on Range Site Classification in the Canadian Prairies.* Saskatchewan Research Council Publication No. 11929–1E07.

Thorpe, J., N. Henderson and J. Vandall. 2006. *Ecological and Policy Implications of Introducing Exotic Trees for Adaptation to Climate Change in the Western Boreal.* Saskatchewan Research Council, SRC Publication No. 11776–1E06: Saskatoon, Saskatchewan, 111 pgs.

Thorpe, J., B. Houston and S. Wolfe. 2004. *Impacts of Climate Change on Grazing Capacity of Native Grasslands in the Canadian Prairies.* Saskatchewan Research Council Publication No. 11562–1E04.

Thorpe, J., S. Wolfe, J. Campbell, J. LeBlanc and R. Molder. 2001. *An Ecoregion Approach for Evaluating Land Use Management and Climate Change Adaptation Strategies on Sand Dune Areas in the Prairie Provinces* (Publication no. 11368-1E01). Saskatchewan Research Council, Saskatoon.

Vandall, J.P., N. Henderson and J. Thorpe. 2006. *Suitability and Adaptability of Current Protected Area Policies under Different Climate Change Scenarios: The Case of the Prairie Ecozone, Saskatchewan.* Saskatchewan Research Council Publication No. 11755–1E06.

Vilkka, L. 1997. *The Intrinsic Value of Nature.* Rodopi: Amsterdam.

Walther, G-R., S. Berger and M.T. Sykes. 2005. "An Ecological 'Footprint' of Climate Change." *Proceedings of the Royal Society* B272: 1427–2.

Wand, S.J.E., G.F. Midgley, M.H. Jones and P.S. Curtis. 1999. "Responses of Wild C_4 and C_3 Grass (Poaceae) Species to Elevated Atmospheric CO_2 Concentration: A Meta-analytic Test of Current Theories and Perceptions." *Global Change Biology* 5: 723–41.

Wang, G., S. Chin and W. Bauerle. 2006. "Effect of Natural Atmospheric CO_2 Fertilization Suggested by Open-Grown White Spruce in a Dry Environment." *Global Change Biology* 12: 601–10.

Watts, W. 1983. "Vegetational History of the Eastern United States." Pp. 294–310 in S. Porter (ed.). *Late Quaternary Environments of the United States, Volume 1, the Late Pleistocene.* University of Minnesota Press: Minneapolis.

Welch, D. 2005. "What Should Protected Areas Managers Do in the Face of Climate Change?" *George Wright Forum* 22: 75–93.

Wheaton, Elaine. 1997. "Forest Ecosystems and Climate." Appendix B: pp. 1–31 in R. Herrington, B. Johnson and F. Hunter (eds.). *Responding to Climate Change in the Prairies: Volume Three of the Canada Country Study: Climate Change and Adaptation.* Environment Canada.

Wilcove, D. 2008. *No Way Home: The Decline of the World's Great Animal Migrations.* Island Press: Washington DC.

Wilcove, D.S., D. Rothstein, J. Dubow, A. Phillips, and E. Losos. 1998. "Quantifying Threats to Imperilled Species in the United States." *BioScience* 48: 607–15.

Williams, C. E. 1997. "Potential Valuable Ecological Functions of Non-Indigenous Plants." Pp. 26–36 in J. Luken and J. Theriet (eds.). *Assessment and Management of Plant Invasions.* Springer Verlag: New York.

Williams, J., S. Jackson and J. Kutzbach. 2007. "Projected Distributions of Novel and Disappearing Climates By 2100 A.D. ." *Proceedings of the National Academy of Sciences* 104: 5738–42.

Winslow, J.C., E.R. Hunt Jr. and S.C. Piper. 2003. "The Influence of Seasonal Water Availability on Global C_3 Versus C_4 Grassland Biomass and its Implications for Climate Change Research." *Ecological Modelling* 3233: 1–21.

Wolfe, S. and J. Thorpe. 2005. "Shifting Sands: Climate Change Impacts on Sand Hills in the Canadian Prairies and Implications for Land Use Management." *Prairie Forum* 30, no. 2: 123–41.

Wright, R., J. Vandall, C. Lockerbie, G. Gray and K. Giannetta. 1995. *Duck Mountain Provincial Park Vegetation Management Plan.* Saskatchewan Environment and Resource Management: Regina.

CHAPTER 7

AGRICULTURE

Elaine Wheaton and Suren Kulshreshtha

"Agriculture is both extremely important to the Canadian economy and inherently sensitive to climate." —LEMMEN AND WARREN, 2004: XI

"Agricultural production, more so than any other form of production, is impacted the most by the weather."—STROH CONSULTING, 2005: 4

INTRODUCTION

The above quotes are especially true for the Canadian Prairies. The Prairies have one of the world's most important agricultural resource bases, yet the area is vulnerable because of often scarce and variable water resources and the variability of climate. Some of the largest year-to-year differences in agricultural production of both crops and livestock are weather and climate related. Agriculture has adapted to considerable agroclimatic changes and is adapting to present climate change, and incurring adaptation costs and benefits.

This section builds upon the highlights of recent literature and emphasizes the prairie and decision-making relevant aspects, presenting findings of the main recent estimated impacts of future climate change on agriculture in the Canadian Prairies, as well as adaptations and vulnerabilities. Key recommendations for further research are included.

CONCEPTUAL IMPACTS ON AGRICULTURE AND ADAPTATIONS

Knowledge of the current and possible future impacts of climate change is critical to suitable adaptation. Both positive and negative effects of climate change are expected for agriculture, although the nature of the impacts depends upon the effectiveness of adaptation. Conceptually, under climate change, the agriculture industry in the Prairie Provinces could be impacted by two types of processes: first, climate change affecting prairie agriculture directly; and, second, prairie agriculture affected through trade linkages with rest of the world. The first type of impact is a result of bio-physical changes,[1] perhaps modified by adaptation measures, whereas the second type of impacts depends on how climate change affects other parts of the world, which in turn would affect demand for prairie agricultural products. A further distinction needs to be made between direct economic impact of climate change from changes in average temperature and precipitation and that through extreme events (e.g. drought and floods). An overview of this approach is shown in Figure 1.

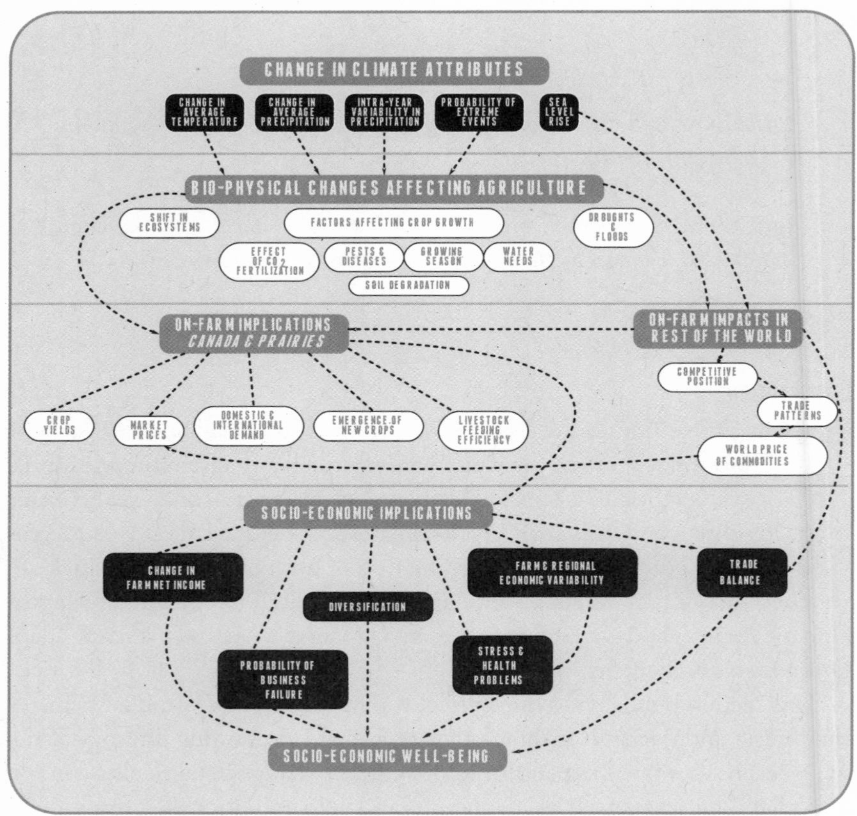

Figure 1. Interrelationships between Climate Change and Socio-Impacts related to Agriculture

The top section of Figure 1 illustrates various bio-physical impacts associated with a changing climate. Agriculture could benefit from several aspects of the warming climate, depending on the rate and amount of climate change and ability to adapt. Benefits could result from such factors as longer growing seasons, increasing growing degree-days and a warmer winter. Some changes, such as temperature increase, could be positive for growth and yield to certain thresholds.

Some of the above bio-physical changes will be translated into socio-economic impacts not only on the prairie region, but also on farms in other parts of Canada and in the rest of the world. Some of these changes may result in more/less income in the hands of producers. A higher frequency of extreme weather events may result in more variable income, which may lead to higher stress and other health-related problems. Higher/lower income in the hands of producers would affect their purchasing power for farm supplies and personal expenditures, which would have implications for rural communities and regional and national industries. The higher economic position of producers would reflect the new opportunities presented by the climate change and the ability of the producers to take advantage of these opportunities. Depending on how climate change affects other parts of the world, it is conceivable that prairie agriculture would be in a more competitive position, which could result in higher trade to other parts of the world. This will bring new economic opportunities to the region.

BIO-PHYSICAL EFFECTS OF CLIMATE CHANGE ON AGRICULTURE

Potential changes in agro-climatic and other relevant variables, along with their likely advantages and disadvantages, are summarized in Table 1 (see Table 1 at end of chapter). Agricultural producers are fairly adaptive and may be able to take advantage of the positive changes, although several aspects of the changing climate could be adverse and challenging. These include changed timing of precipitation, and increased risk of droughts, hot spells, and excessive moisture. An increased speed and large magnitude of change would make adaptation more difficult.

Climatic change information has been used to drive crop production models for many years now (e.g. Williams et al., 1988). However, results of assessments are still wide ranging, depending on several factors including the climate scenarios and impact models used, scale of application, and assumptions (e.g. Wall, Smit, and Wandel, 2004). The possible net effects of climate change on Canadian agriculture are uncertain and depend largely on adaptation measures assumed (Lemmen and Warren, 2004), as well as the type of climate scenarios used.

No estimates of the direct impacts of climate change on livestock in the Prairies were found. Much more information exists for changes in livestock feed. Changes in relevant weather elements (e.g. hot and cold spells, flooding, blizzards, ice storms) are summarized in Table 1. Deaths of livestock associated with hot spells and poor quality water, for example, have occurred in the Prairies. Alternatively, milder winters, for example, can be beneficial for livestock. Indirect effects of climate change through changing insects and diseases are knowledge gaps and should also be considered.

What climate change means for agricultural land use in the Prairies is still uncertain. Annual crop production may be negatively affected, with further warming beyond moderate levels, implying an increase in the importance of grazing operations. Projected changes for grass production and land capability for rangeland grazing are relatively modest in the long term (Thorpe, Houston and Wolfe, 2004). Thorpe's (2007) analysis of the impact of climate change on land capability for rangeland grazing in the Canadian Prairies indicates a gradual decrease in grazing capacity, reduced economic return from grazing, and a need for more aggressive adaptation. Sykes (2008) examined the likely responses of seeded pasture grasses and grazing cattle, and possible climate change impacts on the composition of mixed native pasture associations in Saskatchewan. Results suggest that climate change would cause significant changes in soil moisture, productivity and quality of tame pastures, live-weight of grazing cattle, and species composition of native pasture. Enhanced adaptation requires support for monitoring networks to improve the capacity to detect these changes (Thorpe, Houston and Wolfe, 2004). Although adaptation research was not a main component of Cohen et al. (2002), they demonstrated the value of adjusting to climate change, as they found that some grazing systems in Saskatchewan may benefit from climate change, if the adaptation options are viable and adopted.

Climatic change effects on water are extremely important for agriculture. Water stress during critical times for plants (e.g. flowering) is especially harmful. The use of crop varieties with greater drought and heat tolerance is a useful adaptation measure; however, some yield may still be sacrificed. Water stress is also a problem for livestock since animals require more water during times of heat stress.

Although agriculture in Canada withdraws much less water than does thermal power generation and manufacturing, it is Canada's largest net consumer of water (71%) (Coote and Gregorich, 2000; Marsalek et al., 2004). Irrigation is a major agricultural adaptation to soil water deficits.[2] Agricultural water use in Canada has shown a slow, steady growth from 1972, and this trend is likely to continue (Coote and Gregorich, 2000). Both irrigated agriculture and large-scale livestock production in Western Canada are constrained by water availability (Miller et al., 2000), especially in drought years (Wheaton et al., 2005). Irrigation is more prominent in the west, especially Alberta and British Columbia. About 60% of Canada's total irrigated cropland is in Alberta (Harker et al.,

2004). Irrigation was an important adaptation to drought in 2001–2002 in Canada (Wheaton et al., 2005), and may continue to be increasingly important in the future.

Much of the livestock industry, and its associated water demands, is also in the Prairie Provinces. In 2001, these provinces had more than 61.6% of the total beef cattle, dairy cattle, hogs, poultry, and other animals in Canada (Beaulieu and Bédard, 2003). Populations of both cattle and hogs have increased steadily in the past 10 years (Statistics Canada, 2003). The demand for water for irrigation and livestock is expected to rise with increasing temperatures and expansion in those sectors. Nagy and Kulshreshtha (2008) reported that the Prairie Provinces' agricultural water use makes up almost 80% of the total Canadian agricultural water use.

The increasing demand for water for both dryland and irrigated land may be offset to some extent by enhanced water-use efficiency with higher levels of atmospheric CO_2. However, knowledge of the effect of CO_2 on water-use efficiency and plant productivity is still uncertain and depends on the crop grown, as well as other limiting factors such as nutrients and water availability (Van de Geijin and Goudriaan, 1996; Ziska and Bunce, 2007). Easterling et al. (2007) estimated that C_3 crop yields increase by 10–25% under unstressed conditions at 550 ppm CO_2 concentrations and C_4 crop yields increase by 0–10%. Model projections of future food supply, however, may have over-estimated the positive effect of increased CO_2. Soil moisture and nutrient availability are the two most important edaphic factors for crop yield. Pollution, such as ozone exposure, and other stresses also limit the CO_2 response. The genotypic response to CO_2 could be substantial. For all cultivars, high temperature results in pollen sterility, regardless of the level of CO_2. The efficacy of existing models requires improvements, including more biotic and abiotic factors.

Shifts of agro-climate zones are occurring now, and these changes would accelerate with climate change. New threats and/or opportunities would emerge. In addition, agriculture on the Prairies would be affected by many international changes, as discussed in the next section.

ECONOMIC ASSESSMENT

Although there appears to be a consensus that the sector of the Prairies most affected by climate change would be agriculture, the economic impacts of climate change on this sector have not been frequently studied. The few studies undertaken in this area are summarized in this section. In addition to direct economic impacts of climate change, extreme events (e.g. the 2001–2002 droughts, floods) would also accentuate their magnitude. Rosenzweig et al. (2002) reported very high damage to crop production from flooding and heavy precipitation in the United States. Social impacts of floods include changes in quality of life, community population, tourism and recreational activities, and uncertainty of public policies (Kulkarni, 2002). Flooding and excess soil moisture cause serious damage to communities and agricultural production in the Prairies, especially in wetter areas, such as in northeast Saskatchewan and northern Manitoba. However, even the driest areas are vulnerable to heavy precipitation events. For example, the largest eight-hour precipitation event on the Canadian Prairies occurred in the Vanguard area, in southwest Saskatchewan, on July 3, 2000. Damage included flooded buildings, washed-out roads and rail lines, compromised drinking water supplies, and decreased agricultural production (Hunter et al., 2002).

Conceptually, total economic impacts could be a sum of two types of impacts: First, direct impacts on prairie agriculture; and, second, indirect impacts through trade with other regions of the world. These are described below.

DIRECT ECONOMIC CHANGES RESULTING FROM SHIFTS IN CLIMATE CHARACTERISTICS

Results for the economic analysis of climate change for the prairie region have been highly variable from region to region, as well as from study to study. Two types of methodologies have been followed: use of integrated tools including simulation models and a Ricardian Analysis.

Arthur (1988) estimated impacts of climate change using an integrated assessment model.[3] This study involved climate-related variables along with a soil moisture model, crop yield simulation, and farm-level impacts using a linear programming model. Results at the farm level were linked to an input-output model to estimate regional economic impacts. According to this study, agriculture in the region would experience a small economic impact of climate change, perhaps in the magnitude of 1–8%, which in fact did not even approach the magnitude of the 1961-type drought. Manitoba, being the least moisture deficient of the three provinces, was predicted to gain under climate change. It should be noted that this study did not consider the impacts of a higher frequency of droughts.

Mooney and Arthur (1990) estimated that for Manitoba, under a changed climate, yield of conventional crops will decline. However, producers would shift to higher-value crops, which may result in an increased gross margin of 53% from the business-as-usual situation. For Saskatchewan, more adverse impacts, including large economic losses, were initially predicted (Williams et al., 1988; van Kooten, 1992). More recent estimates suggest that the northern production areas in the United States (neighbouring the Canadian Prairies) are expected to gain from increased length of growing seasons and warmer temperatures (Bloomfield and Tubiello, 2000). Although some uncertainty remains, the carbon dioxide fertilization effect may also benefit agriculture, thereby reducing the economic losses. While the agricultural systems of the Midwest United States would not alter drastically, farmers will need to adjust their production practices such as shifting dates of planting, changing crop mix, and finding better-suited cultivars (Pfeifer and Habeck, 2002).

Adams and McCarl (2001) used an optimization model called "ASM–Agricultural Sector Model" to estimate the impact of climate change on United States agriculture. This model considers both crop and livestock production and processing. Welfare change is measured in terms of change in producer and consumer surplus under a climate change scenario versus a scenario of no climate change. Results for the Northern Plains[4] of the United States indicate that the region would gain under moderate climate change (temperature increase of 1.5–2.5°C), but could be a loser as temperature increases reach 5°C, on average. These changes in welfare were directly a result of change in value of crop production.

The second type of modelling, a Ricardian Analysis, is a cross-sectional analysis where differences in climatic patterns are explicitly used to predict the level of some economic indicator. In the context of climate change, the economic variable of interest is land values. Mendelsohn (2001) applied this method[5] to various counties of the United States to explain the impact on land values caused by climatic variables, seasonal temperature and rainfall, and socio-economic variables, as well as altitude, soil type, wetlands, and other biophysical variables. However, countries in lower latitudes are expected to be damaged by warming. Results for the Northern Plains were similar to those of Adam and McCarl (2001).

The Ricardian approach was applied to the Prairie Provinces by Weber and Hauer (2003). This study estimated a positive impact on agriculture with average gains of $1,551/ha, an increase of approximately 200% over the 1995 values. In relative terms, gains in the Prairies are significant—ranging from 17% in Manitoba to 38% in Saskatchewan (Weber and Hauer, 2003: 175). These land values are driven by the increase in length of growing season and production of more valuable crops.

Direct changes in agriculture are bound to have impacts on the rest of the prairie economy (Arthur and van Kooten, 1992; Wheaton et al., 1992; Williams et al., 1988; Wheaton et al., 2005, 2008). According to Arthur (1988), since on-farm expenditures remain fairly constant across climate scenarios, discretionary expenditures of households and enterprises—e.g. farm machinery and equipment as well as other farm investments—would change in response to changing cash flows. Arthur and Freshwater (1986) also suggested that the greatest losses will be felt in the trade and service sectors which are affected by a decline in farmers' discretionary expenditures.

Trade-Induced Economic Impacts of Climate Change

Climate change effects on crop yields are likely to vary greatly from region to region throughout the world. Using the Basic Linked System (BLS), Fischer et al. (1996) estimated that impacts appear to be less adverse, or even positive, for mid- to high-latitude regions relative to low-latitude ones. Relative productivity of agriculture changes in favour of developed countries (see Table 2 at end of chapter). However, yields of other crops are expected to improve more favourably than those for the cereal crops. For all crops, North America may see an increase in yields of 5.9%. The world price is also predicted to be higher under climate change—a 306% increase without physiological (CO_2 fertilization effect) change, and a 24% increase with it. Economic impacts of climate change are highly sensitive to assumption of price level (Abler et al., 2002).

Mendelsohn (2003) also indicated that results of various country assessments imply that countries farther north than the United States[6] will tend to benefit from warming as they will enjoy benefits from warming in virtually every sector. However, countries in lower latitudes are expected to be damaged by warming.

These changes may result in changes in trade flows, as demand for agricultural products in many developing countries grows and their respective food supply dwindles. However, no study was found that has reported implications of these changes in production and demand on trade flows among various parts of the world. Neither has there been an assessment of the impacts of these changes on the export of agricultural products from the Canadian Prairies. It is our considered opinion that future agricultural trade flows from the Canadian Prairies would face a high degree of uncertainty under climate change.

LIMITATIONS OF ECONOMIC ANALYSIS STUDIES FOR THE PRAIRIE REGION

There at least four major limitations to the available evidence on the economic effects of climate change on agriculture:

1. Very few studies have addressed livestock production. Losses will occur through heat effects on livestock, and the incidence of new pests and diseases, but this may be compensated by improved feed efficiency under a warmer climate. Understanding the effect on livestock enterprises during drought periods is even more complex. During such periods, facing feed and forage scarcity, a common practice of ranchers and other cow-calf operators is to cull the cow herd. This has negative implications for production levels in future time periods.
2. Most economic impact studies have ignored the frequency of extreme events, such as droughts and floods. Droughts, particularly sustained droughts, can have a devastating effect on the regional economy (Wheaton et al., 2005; 2008).[7] Sauchyn (2007) states that the most challenging future climate will include long-duration droughts and climatic variability beyond historical experience.
3. Most studies have not followed integrated approaches, such as bio-economic models, to estimate the economic impacts of climate change on prairie agriculture.

4. Studies rarely consider indirect effects of climate changes on other aspects that affect crop production (e.g. pests, diseases, weeds, and soils).

The prairie region would be affected, however, not only by the changes within the region, but also by those happening outside the region, given trade links within Canada as well as with rest of the world. For example, Canada's position in the production and trade of wheat and grain corn may improve relative to rest of the world (Smit, 1989). Reilly (2002) has predicted that wheat production in Canada, which occurs mostly in the Prairies, will rise 4.5–20%, whereas coarse grains and other production of crops may increase or decrease depending upon the GCM selected.

ADAPTATION

Adaptation is required to minimize the adverse effect of climate change, which contributes to the vulnerability of the industry and the people associated with it. Vulnerability depends on the effectiveness and timing of adaptation and the distribution of coping capacity, which can be very uneven. Understanding of the current sensitivity of agriculture to climate variability and its coping capacity remains limited. The most recent IPCC reports address both global and regional (i.e. North America) agriculture. Easterling et al. (2007) report that in mid- to high-latitude regions, moderate to medium warming (1–3°C), along with associated CO_2 increase and precipitation changes, can benefit crop and pasture yields. However, further warming has increasingly negative impacts in all regions. Also, adaptation leads to rising relative benefits with low to moderate warming, but such adaptations could be strained by a lack of financial resources required for adaptation, change in lifestyle resulting from change in enterprise mixes or cultural practices, or learning skills for better management of agricultural production processes after adaptation.

Adaptation could take place at the level of the farm, as well as at larger levels such as agri-business and municipal to national governments (e.g. Smit and Skinner, 2002; Wheaton and McIver, 1999). In the context of globalization, adaptation has become a central issue for agri-business, national policy makers, and at the international level.[8] Information regarding impacts and adaptations at these levels is rare, but work describing government policies during drought situations, for example, may be relevant. Wittrock and Koshida (2005) describe the government responses and programs related to the 2001 and 2002 droughts in Canada, showing that several government response and safety net programs partially reduced some of the negative socio-economic impacts of these record droughts.

Mortsch (2006) suggested a number of adaptive measures that could be undertaken to alleviate some of the adverse impacts of climate change on agriculture. The suggested list included:

- For crop production:
  - switch timing of planting and harvesting to take advantage of longer growing seasons;
  - switch from spring wheat to higher yielding winter wheat; and
  - change irrigation and tillage practices.
- For livestock:
  - reduce heat stress by proving shade, using sprinklers, or by lessening crowding;
  - shift to heat-resistant breeds or replace cattle with sheep;
  - increase stocking density (often double normal level) during first half of the growing season followed by no grazing through the remainder of the season. Cattle can be sent to feedlots early or finished on-farm with pastures with supplements;
  - lengthen grazing period with fewer number of cattle stocked; and
  - move herds to where pasture and feed are more plentiful.
- Under drought situations:
  - enhance water conservation and water use efficiency (e.g. improve irrigation efficiency);
  - change current agricultural practices (use drought-resistant cultivars,[9] change time of planting, switch to dryland farming);
  - develop drought monitoring and seasonal climate predictions;
  - develop drought risk management plans that enhance the capacity to cope outside current ranges; and
  - purchase crop insurance.
- For agricultural water supply:
  - enhance water conservation (e.g. snow management to increase storage);
  - improve irrigation efficiency; and
  - improve on-farm water infrastructure and management (e.g. wells, pasture pipelines, constructed ponds, and small reservoirs).
- For reducing soil erosion:
  - undertake more conservation tillage practices (no-till or reduced till) that provide benefits of storing more soil water, enhancing soil carbon.

Some of the large suite of agricultural adaptations, especially at the farm level, is summarized by Wheaton (2004). Options for producers include:

- find out what climate changes are occurring now and estimate the effects. Effects at the local to international scales have implications for achieving agricultural objectives;
- find out more about future climate change trends (as presented in Chapter 4 of this book) and their possible effects;

- estimate how sensitive planning and operations are to these changes;
- evaluate your options to determine which farm management practices work the best in a changing climate. Examples include timing of seeding, crop types, diversification (e.g. crops, field, animal locations), inputs (e.g. fertilizer, herbicides, pesticides), and financial options (e.g. insurance, off-farm work);
- use water, energy, and soil conservation practices and other sustainable agricultural practices;
- use improved safety procedures to prepare for and deal with extreme weather events (e.g. windstorms, heat waves, tornados, dust storms, intense rain storms and floods; and
- work towards building a flexible and resilient agriculture management system that will be able to cope with change and surprise.

In the context of adaptation to climate change, two other issues need to be addressed. First, although a number of adaptation measures have been suggested—and could be available—their relative efficiency in a given region under climate change has yet to be assessed. For example, Pfeifer and Habeck (2002) have noted that a country's climate sensitivity is influenced not only by its climate, but also by social factors associated with the agro-climatic zone in which it is located. Second, the process of adaptation could be hampered by barriers that may be present in a given socio-economic-political context. Wheaton et al. (2007) documented barriers to adaptation to a severe drought. These included lack of knowledge of water supplies and water use, lack of funds and research, and difficulty (such as reluctance) in making changes. Thus, adaptations are most effective if they are implemented properly, facilitated, coordinated and have few, if any, barriers.

Swanson, Hiley, and Venema (2007) identified regions within the Prairie Provinces with higher level of agricultural adaptive capacity. Their methodology was based on six sets of indicators of capacity: (1) economic resources, (2) technology, (3) infrastructure, (4) information, skills and management, (5) institutions and networks, and (6) equity. Analysis was conducted using Census Division level data from the 2001 Census of Agriculture (Statistics Canada, 2001) and Alasia (2004). Results are summarized in Figure 2. In the whole prairie region, the highest adaptive capacity for agriculture was found around the towns of Morden and Winkler. In Alberta, the south-central part of the province, and, in Saskatchewan, the Census Districts surrounding Saskatoon and Regina, were determined to have the highest adaptive capacity.

The shifting of agro-climatic zones occurring now, and expected to accelerate, may be among the most important impacts. New or changed threats and opportunities are becoming more likely—e.g. the increased probability of drought in

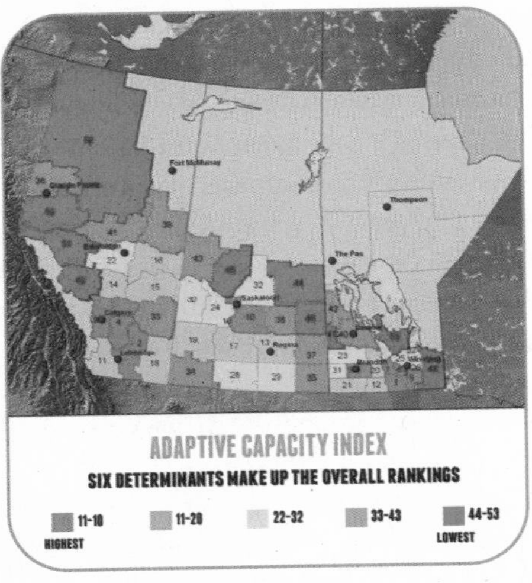

Figure 2. Adaptive Capacity Index for the Prairie Provinces (Source: Swanson, Hiley, and Venema (2007)).

regions where frost or excess moisture are currently greater threats. Nyirfa and Harron (2001) examined shifts in land suitability for agriculture with climate change and also found significantly higher moisture limitations. Adaptations to these types of changes are more challenging because people have had less experience in dealing with them.

The availability of options to adapt is not the only prerequisite to adaptation. The options must have other characteristics such as suitability (e.g. easy to use, reasonable cost, producer willingness to use), and must be implemented. Bradshaw, Dolan and Smit (2004) tested the likely uptake of adaptation by documenting the adoption of crop diversification in Canadian prairie agriculture (15,000 operations) for the period 1994–2002. They found that individual farms have become more specialized in their cropping patterns. Many options may seem suitable, but their application may not be ideal, or it may be slow or may not occur. Reactive policies and practices will be insufficient to meet the climate change challenge, and planned proactive adaptation is needed (Sauchyn, 2007).

Knowledge of the current physical and social vulnerabilities in the agricultural sector is essential for the anticipation of problems, so that risks can be managed appropriately (Diaz and Gauthier, 2007). Wheaton et al. (2007) and Wittrock et al. (2006, 2007) used the droughts of 2001 and 2002 to improve the understanding of the current adaptation processes in Canadian agriculture and in rural communities.

CONCLUSIONS

Positive impacts and opportunities for agriculture may result from continued expansion of the growing season, increasing heat units, and milder, shorter winters. Negative impacts of climate change include the extremes such as drought, intense storms, increased variability, and changes that are rapid or above certain thresholds. The net impacts on agriculture are not clear and depend heavily on assumptions such as the coping range and effectiveness of adaptation measures. Several aspects of adaptation, however, are not well understood, including both current and future adaptation processes and effectiveness.

Although the current adaptive and production capacity of prairie agriculture appears relatively high (e.g. Sauchyn, Kenney and Stroich, 2005; Burton and Lim, 2005), coping thresholds are periodically exceeded and will be exceeded, perhaps more often, in the future. For example, the recent droughts of 2001 and 2002 caused severe crop and livestock production losses, as further discussed in a case study presented in this book.

Some institutional adaptations will also be required, since changes brought about by weather and related climate change will be one of the many factors affecting the economic status of agriculture. Existing policies and programs may not be enough to cope with these changes and may need to be totally revamped, or at least modified to meet the challenge.

AREAS FOR FURTHER CONSIDERATION

A Canadian framework for advancing research on vulnerability, impacts and adaptation research in agriculture and other sectors is needed. The federal and provincial ministers of Environment and Energy agreed to support the development and implementation of a *National Adaptation Framework* (Charlottetown, PEI, May 2002). However, Wall, Smit and Wandel (2004) report that only modest progress has occurred in the agriculture sector (SSCAF, 2003). The ministers further recommend that the following themes would strengthen climate/agriculture research:

- broaden the context to indicate other conditions that shape agriculture as well as climate;
- consider multiple spatial and temporal scales;
- identify ways to enhance the adaptive capacity of agricultural systems; and
- enhance science/policy linkages.

Burton and Lim (2005) also point to an adaptation policy framework as a vehicle to facilitate and improve the understanding of adaptation.

In general, research needs described in Lemmen and Warren (2004) have been met with insufficient action, so most of these remain outstanding. Examples of substantial and critical research needs include:

- Estimates of changes in crop production are wide ranging with both positive and negative changes expected. Improved modelling and estimates are needed to narrow this range and better assess the direction of change and reasons for the differences;
- Estimates of effects on livestock production are rare and are needed because of the importance of the sector;
- Improved understanding and modelling of the role of pests and diseases in climate impact and adaptation assessment;

- Enhanced understanding of the adaptation process, including current vulnerability. It is improving, but further modelling is required. Research is needed regarding improved means of increasing adaptive capacity in the agriculture sector;
- Estimates of the impacts (adaptations to and vulnerabilities) of extreme climatic events;
- Estimates of impacts on and adaptations of agri-business;
- Water supply is one of the most important challenges facing the agricultural sector. Interacting impacts of the agriculture and water sector must be assessed, adaptations tested and vulnerabilities determined;
- Estimates and understanding of current as well as future vulnerabilities; and
- Improved tools for coupled biophysical and economic impact-adaptation-vulnerability assessment.

Field et al. (2007) point out several areas where the understanding of climate change impacts for agriculture can be improved, and these gaps are relevant to the Prairies:

- sensitivity of agriculture to extremes, such as droughts and excessive moisture;
- lack of tools and estimates of the conditions that lead to tipping points, where a system changes rapidly;
- interacting responses of diverse sectors impacted by climate change;
- impacts of climate change in the context of other trends that are likely to complicate the development of strategies for dealing with climate change;
- understanding of the options for proactive adaptation to conditions outside the range of historical experience.

There is also a need for an integrated bio-physical/socio-economic assessment of climate change impacts, similar to that shown in Figure 1. The scope of the investigation needs to be national, regional, and international in nature since all these factors would shape the nature of impacts under the changed climate and the adaptation measures that would be effective in the future.

ENDNOTES

1. It is recognized that managerial decisions, either involving adaptation to these changes or a lack thereof, may also be major factors in determining the magnitude of these impacts.
2. A detailed discussion of irrigation in the Prairies is provided in Chapter 16 of this book.
3. Details on this model are presented in Arthur (1988).
4. This region is similar to the Canadian Prairies. It includes North Dakota, South Dakota, and Montana.
5. A similar analysis for Zimbawbe has also been reported by Mano and Nhemachena (2007).
6. These countries include Canada as well as some European countries located in higher latitudes.
7. More details on the economic impacts of the 2001 and 2002 droughts are provided in Chapter 15 of this book.
8. This issue is further discussed by Burton and Lim (2005).
9. A number of drought-induced genes have been identified in a wide range of wheat varieties (Deng et al., 2005).

REFERENCES

Abler, D., L. Shortle, J. Carmichael and R. Horan. 2002. "Climate Change, Agriculture, and Water Quality in the Chesapeake Bay." *Climatic Change* 55, no. 3: 339–59.

Adams, R. and B. McCarl. 2001. "Agriculture: Agronomic—An Economic Analysis." In R. Mendelsohn (ed.). *Global Warming and the American Economy—A Regional Assessment of Climate Change Impacts.* Edward Elgar: Northampton, MA.

Alasia, A. 2004. "Mapping the Socio-Economic Diversity of Rural Canada." *Rural and Small Town Analysis Bulletin* 5, no. 2. Catalogue No. 21–006-XIE. Statistics Canada: Ottawa.

Arthur, L. and D. Freshwater. 1986. *Analysis of the Economic Effects of a Prolonged Agricultural Drought in Manitoba.* Research Bulletin 86–2. Department of Agricultural Economics, University of Manitoba: Winnipeg.

Arthur, L.M. 1988. "Socio-Economics Impact of Climate Change on Prairie Agriculture: The Greenhouse Effect." Pp. 187–202 in B.L. Magill and F. Geddes (eds.). *The Impact of Climate Variability and Change on the Canadian Prairies—Proceedings of the Symposium/Workshop.* Alberta Environment: Edmonton.

Arthur, L.M. and G.C. van Kooten. 1992. "Climate Impacts on the Agribusiness Sectors of a Prairie Economy." *Prairie Forum* 17: 97–109.

Barrow, E., B. Maxwell and P. Gachon (eds.). 2004. *Climate Variability and Change in Canada.* Environment Canada: Toronto, Ontario, 114 pgs.

Beaulieu, M.S. and F. Bédard. 2003. *A Geographic Profile of Canadian Livestock, 1991–2001.* Agriculture and Rural Working Paper Series, Working Paper No. 62. Statistics Canada: Ottawa, ON. 32 pgs.

Bloomfield, J. and F. Tubiello. 2000. *Impacts of Climate Change in the United States: Agriculture*[online]. [accessed September 12, 2005]. Available from World Wide Web: *http://www.climatehotmap.org/impacts/agriculture.html*

Bradshaw, B., H. Dolan and B. Smit. 2004. "Farm-Level Adaptation to Climatic Variability and Change: Crop Diversification in the Canadian Prairies. *Climatic Change* 67: 119–41.

Brown, R. 2006. *Snow Cover Response to Climate Warming* [online]. [accessed September 12, 2005]. Available from World Wide Web: *http://www.socc.ca/snow/variability/index.cfm*

Burton, I. and B. Lim. 2005. "Achieving Adequate Adaptation in Agriculture." *Climatic Change* 70: 191–200.

CCIS (Canadian Climate Impact Scenarios Project). 2005. *Bioclimate Profiles* [online]. Prepared for Canadian Climate Adaptation Fund. Canadian Institute for Climate Studies, Victoria, B.C.. [accessed September 12, 2005]. Available from World Wide Web: *http://www.cics.uvic.ca/scenarios/bcp/select.cgi?&sn=55*

——. 2006. *Introduction* [online]. This project was operated by the Canadian Institute for Climate Studies at the University of Victoria, Victoria, British Columbia. [accessed 20, 2006]. Available from World Wide Web: *http://www.cics.uvic.ca/scenarios*

Cohen, R.D.H., C.D. Sykes, E.E. Wheaton and J.P. Stevens. 2002. *Evaluation of the Effects of Climate Change on Forage and Livestock Production and Assessment of Adaptation Strategies on the Canadian Prairies.* University of Saskatchewan: Saskatoon, SK. 50 pgs.

Coote, D.R. and L.J. Gregorich (eds.). 2000. *The Health of Our Water—Toward Sustainable Agriculture in Canada.* Research Planning and Coordination Directorate. Research Branch, Agriculture and Agri-Food Canada: Ottawa, ON. 173 pgs.

Deng, X., L. Shan, S. Inanaga and M. Inoue. 2005. "Water-Saving Approaches for Improving Wheat Production." *Journal of the Science of Food and Agriculture* 85, no. 8: 1379–99.

Diaz, H. and D. Gauthier. 2007. "Institutional Capacity for Agriculture in the South Saskatchewan River Basin." In E. Wall, B. Smit and J. Wandel (eds.). *Farming in a Changing Climate: Agricultural Adaptation in Canada.* UBC Press: Vancouver.

Easterling, W.E., P.K. Aggarwal, P. Batima, K.M. Brander, L. Erda, S.M. Howden, A. Kirilenko, J. Morton, J-F. Soussana, J. Schmidhuber and F.N. Tubiello. 2007. " Food, Fibre and Forest Products." Pp. 273–313 in M.L. Parry, O.F. Canziani, J.P. Palutikof, P.J. van der Linden and c.e. Hanson (eds.). *Climate Change 2007: Impacts, Adaptation and Vulnerability. Contribution of Working Group II to the Fourth Assessment Report of the Intergovernmental Panel on Climate Change.* Cambridge University Press: Cambridge, UK.

Field, C.B., L.D. Mortsch, M. Brklacich, D.L. Forbes, P. Kovacs, J.A. Patz, S.W. Running and M.J. Scott. 2007. "North America." Pp. 617–52 in M.L. Parry, O.F. Canziani, J.P. Palutikof, P.J. van der Linden and c.e. Hanson (eds.). *Climate Change 2007: Impacts, Adaptation and Vulnerability. Contribution of Working Group II to the Fourth Assessment Report of the Intergovernmental Panel on Climate Change.* Cambridge University Press: Cambridge, UK.

Fischer, G., K. Frohberg, M. Perry, and C. Rosenzweig. 1996. "The Potential Effects of Climate Change on World Food Production and Security." In F. Bazzaz and W. Sombroek (eds.). *Global Climate Change and Agricultural Production—Direct and Indirect Effects of Changing Hydrological, Pedological and Plant Physiological Processes.* John Wiley and Sons: New York.

Gameda, S., B. Qian and A. Bootsma. 2005. "Climate Change Scenarios for Agriculture." Presentation to: *Adapting to Climate Change in Canada 2005,* Montreal, Quebec. Agriculture and Agri-Food Canada: Ottawa, ON.

Harker, B., J. Lebedin, M.J. Goss, C. Madramootoo, D. Neilsen, B. Paterson and T. van der Gulik. 2004. "Land-Use Practices and Changes—Agriculture." Pp. 49–55 in Environment Canada, *Threats to Water Availability in Canada.* NWRI Scientific Assessment Report Series No. 3 and ACSD Science Assessment Series No. 1. National Water Research Institute: Burlington, ON.

Hunter, F.G., D.B. Donald, B.N. Johnson, W.D. Hyde, J.M. Hanesiak, M.O.B. Kellerhals, R.F. Hopkinson and B.W. Oegema. 2002. "The Vanguard Torrential Storm (Meteorology and Hydrology)." *Canadian Water Resources Journal* 27, no. 2: 213–27.

Kharin, V.V. and F.W. Zwiers. 2000. "Changes in the Extremes in an Ensemble of Transient Climate Simulations with a Coupled Atmosphere-Ocean GCM." *Journal of Climate* 13: 3760–88.

——. 2005. "Estimating Extremes in Transient Climate Change Simulations." *Journal of Climate* 18: 1156–73.

Kulkarni, T. 2002. "Vulnerability of Landscape Hazards to Climate Change." C-CIARN Landscape Hazards Workshop, Gatineau, Quebec. October 3–5. Natural Resources Canada: Ottawa.

Leggett, J., W.J. Pepper and R.J. Swart. 1992. "Emissions Scenarios for the IPCC: An Update." Pp. 69–96 in J.T. Houghton, B.A. Callander and S.K. Varney (eds.). *Climate Change 1992: The Supplementary Report to the IPCC Scientific Assessment.* Cambridge University Press: Cambridge, UK.

Lemmen, D.S. and F.J. Warren (eds.). 2004. *Climate Change Impacts and Adaptation: A Canadian Perspective.* Natural Resources Canada: Ottawa, ON. 174 pgs.

Mano, R. and C. Nhemachena. 2007. *Assessment of the Economic Impacts of Climate Change on Agriculture in Zimbabwe: A Ricardian Analysis.* Policy Research Working Paper Series: 4292. World Bank: Washington, DC.

Marsalek, J., W.E. Watt, L. Lefrancois, B.F. Boots and S. Woods. 2004. "Municipal Water Supply and Urban Development." Pp. 35–40 in Environment Canada, *Threats to Water Availability in Canada.* NWRI Scientific Assessment Report Series No. 3 and ACSD Science Assessment Series No. 1. National Water Research Institute: Burlington, ON.

Mendelsohn, R. 2001. "Agriculture: A Ricardian Analysis." In R. Mendelsohn (ed.). *Global Warming and the American Economy—A Regional Assessment of Climate Change Impacts.* Edward Elgar: Northampton, MA.

———. 2003. "Assessing the Market Damages from Climate Change." In J. Griffin (ed.). *Global Climate Change—The Science, Economics and Politics.* Edward Elgar: Northampton, MA.

Miller, J.J., K.F.S.L. Bolton, R.C. de Loë, G.L. Fairchild, L.J. Gregorich, R.D. Kreutzwiser, N.D. MacAlpine, L. Ring and T.S. Veeman. 2000. "Limits on Rural Growth Related to Water." Pp. 131–39 in D.R. Coote and L.J. Gregorich (eds.). *The Health of Our Water—Toward Sustainable Agriculture in Canada.* Research Planning and Coordination Directorate, Research Branch. Agriculture and Agri-Food Canada: Ottawa, ON.

Mooney, S. and L.M. Arthur. 1990. "The Impacts of Climate Change on Agriculture in Manitoba." *Can. J. Agric. Econ.* 38, no. 4: 685–94.

Mortsch, L. 2006. "Impact of Climate Change on Agriculture, Forestry, and Wetlands." Pp. 45–68 in J. Bhatti, R. Lal, M. Apps and M. Price (eds.). *Climate Change and Managed Ecosystems.* Taylor and Francis: New York.

Nagy, C. and S. N. Kulshreshtha. 2008. *Canadian Provincial Agricultural Waster Use Module (C-PAWUM).* Department of Bioresource Policy, Business and Economics. University of Saskatchewan: Saskatoon, SK. 51 pgs.

Nyirfa, W. and W. Harron. 2001. *Assessment of Climate Change on the Agricultural Resources of the Canadian Prairies.* Report to PARC QS-3. Agriculture and Agri-Food Canada: Regina, SK. 27 pgs.

Pfeifer, R.A. and M. Habeck. 2002. "Farm Level Economic Impacts of Climate Change." Pp. 159–78 in O. Doering, J. Randolph, J. Southworth and R. Pfeifer (eds.), *Effects of Climate Change and Variability on Agricultural Production Systems.* Kluwer Academic Publishers: Boston, MA.

Reilly, J.M. 2002. "Impacts of Variability on Agriculture." In J.M. Reilly (ed.). *Agriculture: The Potential Consequences of Climate Variability and Change.* Cambridge University Press: Cambridge, MA.

Rosenzweig, C., F. Tubiello, R. Goldberg, E. Miller and J. Bloomfield. 2002. "Increased Crop Damage in the US from Excess Precipitation under Climate Change." *Global Environmental Change* 12, no. 3: 197–202.

Sauchyn, D. 2007. "Climate Change Impacts on Agriculture in the Prairies." Pp. 80–93 in E. Wall, B. Smit and J. Wandel (eds.). *Farming in a Changing Climate: Agricultural Adaptation in Canada.* University of British Columbia Press: Vancouver, B.C..

Sauchyn, D., S. Kenney and J. Stroich. 2005. "Drought, Climate Change, and the Risk of Desertification on the Canadian Plains." *Prairie Forum* 30, no. 1: 143–56.

Smit, B. 1989. "Climate Warming and Canada's Comparative Position in Agriculture." *Climate Change Digest* CCD 89–01.

Smit, B. and M. Skinner. 2002. "Adaptation Options in Agriculture to Climate Change: a Typology." *Mitigation and Adaptation Strategies for Global Change* 7: 85–114.

SSCAF (Standing Committee on Agriculture, Fisheries and Forestry). 2003. *Climate Change: We are at Risk.* Standing Senate Committee on Agriculture and Forestry: Ottawa, ON.

Statistics Canada. 2001. *Total Area of Farms, Land Tenure and Land in Crops, by Province (1981—2001 census of Agriculture)* [online]. [accessed September 12, 2005]. Available from World Wide Web: *http://www.statcan.ca/english/freepub/95F0355XIE/quality.htm*

——. 2003. *Livestock Estimates.* Statistics Canada: Ottawa, ON.

Stroh Consulting. 2005. *Agriculture Adaptation to Climate Change in Alberta: Focus Group Results.* Prepared for Alberta Agriculture, Food and Rural Development: Edmonton, AB.

Swanson, D., J. Hiley and H. Venema. 2007. *Indicators of Adaptive Capacity to Climate Change for Agriculture in the Prairie Regions of Canada.* International Institute for Sustainable Development and Agriculture and Agri-Food Canada: Winnipeg.

Sykes, C. 2008. "Simulation of the Effects of Climate Change on Forage and Cattle Production in Saskatchewan." PhD dissertation, University of Saskatchewan: Saskatoon, SK.

Thorpe J. 2007. *Impacts of Climate Change on Range Site Classification in the Canadian Prairies.* SRC Publication No. 11929–1E07. Saskatchewan Research Council: Saskatoon, SK.

Thorpe, J., B. Houston and S. Wolfe. 2004. *Impact of Climate Change on Grazing Capacity of Native Grasslands in the Canadian Prairies.* SRC Publication No. 11561–1E04. Saskatchewan Research Council: Saskatoon, SK.

Van de Geijin, S.C. and J. Goudriaan. 1996. "The Effects of Elevated CO_2 and Temperature Change on Transpiration and Crop Water Use." In F. Barraz and W. Sombroek (eds.). *Global Climate Change and Agricultural Production. Direct and Indirect Effects of Changing Hydrological, Pedological, and Plant Physiological Processes.* FAO: Rome, Italy.

van Kooten, G.C. 1992. "Economic Effects of Global Warming on Agriculture with some Impact Analyses for Southwestern Saskatchewan." In E. Wheaton, V. Wittrock and G.D.V. Williams (eds.). *Saskatchewan in a Drier World.* SRC Publication No. E-2900–17-E-92. Saskatchewan Research Council: Saskatoon, SK.

Wall, E., B. Smit and J. Wandel. 2004. *Canadian Agri-Food Sector Adaptation to Risks and Opportunities from Climate Change: Position Paper on Climate Change, Impacts, and Adaptation in Canadian Agriculture.* C-CIARN Agriculture: Guelph, ON. 56 pgs.

Weber, M. and G. Hauer. 2003. "A Regional Analysis of Climate Change Impacts on Canadian Agriculture." *Canadian Public Policy* 29, no. 2: 163–80.

Wheaton, E. 2004. *Climate Change: Past and Future Climate Trends, Impacts, Adaptations and Vulnerabilities.* Website content prepared for Climate Change Saskatchewan. SRC Publication No. 11905–1E04. Saskatchewan Research Council: Saskatoon, SK.

Wheaton, E.E., L.M. Arthur, B. Chorney, S. Shewchuk, J. Thorpe, J. Whiting and V. Wittrock. 1992. "The Prairie Drought of 1988." *Climatological Bulletin* 26, no. 3: 188–204.

Wheaton, E., G. Koshida, B. Bonsal, T. Johnston, W. Richards and V. Wittrock. 2007. *Agricultural Adaptation to Drought (ADA) in Canada: The Case of 2001 to 2002. Synthesis Report.* Prepared for Government of Canada's Climate Change Impacts and Adaptation Program. SRC Publication No. 11927–1E07. Saskatchewan Research Council: Saskatoon, SK. 35 pgs.

Wheaton, E., S. Kulshreshtha, V. Wittrock and G. Koshida. 2008. "Dry Times: Lessons from the Canadian Drought of 2001 and 2002." *The Canadian Geographer* 45: 391–411.

Wheaton, E.E. and D.C. MacIver. 1999. "A Framework and Key Questions for Adapting to Climate Variability and Change." *Mitigation and Adaptation Strategies for Global Change* 4: 215–25.

Wheaton, E., V. Wittrock, S. Kulshreshtha, G. Koshida, C. Grant, A. Chipanshi and B. Bonsal. 2005. *Lessons Learned from the Canadian Droughts Years of 2001 and 2002: Synthesis Report.* SRC Publication No. 11602–46E03. Saskatchewan Research Council: Saskatoon, SK.

Williams, G.D.V., R.A. Fautley, K.H. Jones, R.B. Stewart and E.E. Wheaton. 1988. "Estimating Effects of Climatic Change on Agriculture in Saskatchewan, Canada." Pp. 221–371 in M.L. Parry, T.R. Carter and N.T. Konjin (eds.). *The Impact of Climatic Variations on Agriculture,* Vol. 1. Kluwer Academic Publishers: Dordrecht/Boston/London.

Williams, G.D.V. and E.E. Wheaton. 1998. "Estimating Biomass and Wind Erosion Impacts for Several Climatic Scenarios: A Saskatchewan Case Study." *Prairie Forum* 23, no. 1: 49–66.

Wittrock, V., D. Dery, S. Kulshreshtha and E. Wheaton. 2006. *Vulnerability of Prairie Communities' Water Supply during the 2001 and 2002 Droughts: A Case Study of Cabri and Stewart Valley, Saskatchewan.* Prepared for Institutional Adaptation to Climate Change (IACC): A Project of SSRHC—MCRI Program. SRC Publication No. 11899–2E06. Saskatchewan Research Council: Saskatoon, SK. 107 pgs.

Wittrock, V. and G. Koshida. 2005. *Canadian Droughts of 2001 and 2002: Government Response and Safety Net Programs—Agriculture Sector.* SRC Publication No. 11602–2E03. Saskatchewan Research Council: Saskatoon, SK.

Wittrock, V. E., Wheaton, M. Khakapour and S. Kulshreshtha. 2007. *Vulnerability of Prairie Communities During the 2001 and 2002 Droughts: Case Studies of Taber and Hanna, Alberta and Outlook Saskatchewan.* SRC Publication No. 11899–8E07. Saskatchewan Research Council: Saskatoon, SK. 124 pgs.

Ziska, L. and J. Bunce. 2007. "Predicting the Impact of Changing CO_2 on Crop Yields: Some Thoughts on Food." *New Phytologist* 175: 607–18.

Indices	Changes (with respect to 1961–1990, unless noted)	Climate models and emission scenarios	Period and spatial pattern	Reference	Possible advantages for agriculture?	Possible disadvantages for agriculture?
Thermal Indices						
Growing degree-days	25 to 40% 42 to 45%	CSIROMk2b Bll, greater changes with the other models CGCM GA1	2050s Greater changes in the north 2050s for Lethbridge and Yorkton	Thorpe, Houston and Wolfe, 2004 CCIS, 2005	More crop options; more crops per year; improved crop quality; shifts to earlier spring and later fall growth	Accelerated maturation rates and lower yields; increased insect activity; changed herbicide and pesticide efficacy; changing insects and diseases could be negative for human, plant, and livestock health
Heating degree-days	-23%	CGCM GA1	2050s for Lethbridge and Yorkton	CCIS, 2005	Decreased heating costs	
Cooling degree-days	146 to 218%	CGCM GA1	2050s for Lethbridge and Yorkton	CCIS 2005		Increased ventilation needed for barns, increased cooling shelters, and air conditioning
Hot spells: 20 y return period of maximum temperature	1 to 2°C increase from 2000	CGCM2 A2	2050	Kharin and Zwiers, 2005	Better vacation weather?	Heat stress to plants and animals; increased transpiration rates can reduce yields; increased need for water for cooling and drinking
Cold spells: 20 y return period of minimum temperature	2 to >4°C increase from 2000	CGCM2 A2	2050	Kharin and Zwiers, 2005	Decreased heat stress to animals	Increased pest and diseases Winterkill potential increases

Table 1. Future possible changes in agroclimates for the agricultural region of the Canadian Prairies, and examples of possible advantages and disadvantages for agriculture (*table continued next page*).

Indices	Changes (with respect to 1961–1990, unless noted)	Climate models and emission scenarios	Period and spatial pattern	Reference	Possible advantages for agriculture?	Possible disadvantages for agriculture?
Moisture Indices						
Soil moisture capacity (fraction), annual	>0 to <-0.2 Mostly drying	CGCM2 A2 ensemble mean	2050s Greatest decreases in south to south-east	Barrow, Maxwell and Gachon, 2004		Increased moisture stress to crops and decreased water availability
Palmer Drought Severity Index	Severe droughts twice as frequent	Goddard Institute for Space Studies	Doubled CO_2	Williams et al., 1988 Southern SK	No flooding damages	Increased damages and losses from droughts; increased costs of adaptation, etc.
Moisture deficit; annual precipitation minus potential evapotranspiration (P-PE)	-60 to -140 mm, i.e., increased deficit	GCM1 and HadCM3	2050s	Gameda, Qian, and Bootsma 2005	As for droughts	As for droughts
	0 to -75 mm	CGCM1 GA	2050s		As above	As above
	-11 to 4 mm	CGCM1 GA1	2050s for Lethbridge and Yorkton	Nyirfa and Harron 2001 CCIS 2006	Direction of change not clear	Direction not clear
Aridity Index; ratio of annual precipitation and potential evapotranspiration (P/PE)	Area of AI<0.65 increases by 50%	CGCM2 B2	2050s	Sauchyn et al. 2005	As above	As above
Number of dry days: time between 2 consecutive rain days (=>1 mm)	Modest & insignificant changes	CGCM2 A2	2080–2100	Kharin and Zwiers 2000		
Number of rain days	Modest & insignificant changes	CGCM2 A2	2080–2100	Kharin and Zwiers 2000		
Precipitation extremes: 20 y return period of annual extremes	Increase of 5 to 10 mm and return period decreases by about a factor of 2	CGCM2 A2	2050	Kharin and Zwiers 2005		More flooding and erosion concerns; more difficult planning for extremes
Snowcover	Widespread reductions	CGCM2 IS92a	Next 50–100 years	Brown 2006	Decreased snow plowing; Increased grazing season	Decreased quantity and quality of water supplies

Table 1. (*table continued next page*).

Indices	Changes (with respect to 1961–1990, unless noted)	Climate models and emission scenarios	Period and spatial pattern	Reference	Possible advantages for agriculture?	Possible disadvantages for agriculture?
Other Indices						
Wind speed, annual	<5 to >10%	CGCM2 A2 ensemble mean	2050s	Barrow, Maxwell and Gachon, 2004	Greater dispersion of air pollution	Greater soil erosion of exposed soils; damage to plants and animals
Wind erosion of soil	16% -15%	Manabe and Stouffer Goddard Institute for Space Studies	Doubled CO_2 Doubled CO_2	Williams and Wheaton, 1998	Direction of change is not clear	Direction of change is not clear
Incident solar radiation	<-2 to <-6 Wm^{-2}	CGCM2 A2 ensemble mean	2050s Greatest decreases in central north	Barrow, Maxwell and Gachon, 2004	Decreased radiation may partially offset heat stress	Reduced plant growth if thresholds are exceeded
Climate severity index	-3 to -9	CGCM1 IS92a	2050s Greatest improvements in AB and MB	Barrow, Maxwell and Gachon, 2004		Less severe climates for outside work; more suitable for animals
Carbon Dioxide	Various emission scenarios used, e.g., $1\%y^{-1}$	IS92a		Leggett, Pepper, and Swart, 1992	Increased plant productivity, depending on other limits	Possible effects on quality of yield

Table 1 (*concluded*).

Notes: Most of the advantages and disadvantages are summarized from Wheaton (2004).
Data for additional variables (e.g. timing of maximum precipitation, freezing rain events, water body ice formation and break-up days, snow cover season length and snow depth) are not readily available. Climate Severity Index is an annual measure of the impact of climate on human comfort and well-being and of the risk of certain climatic hazards to human health and life, with a scale ranging from 0 to 100 (Barrow, Maxwell, and Gachon, 2004). Higher CSI indicates more severe climates. Severity is weighted equally between winter and summer discomfort factors and psychological, hazards, and outdoor mobility factors.

Region of the World	Percent Change over Base (No Climate Change) Level Production		
	Cereals	Other Crops	All Crops
	Developed Regions		
North America	+2.7	+12.6	+5.9
Western Europe and Other Developed Market Economies	+6.2	+13.5	+10.9
Eastern Europe plus USSR	+17.7	+25.9	+22.8
Pacific OECD Countries	+8.3	+13.9	+11.0
	Developing Regions		
Africa	-20.6	+0.8	-3.0
Latin America	-16.7	-6.1	-8.7
West Asia	-12.2	-4.6	-6.5
South Asia	-9.8	-4.0	-6.7
Centrally Planned Asia	+3.3	+9.4	+6.9
Pacific Asian Countries	-14.9	-11.1	-11.4

Table 2. Climate Change* Impacts on Different Regions of the World, 2060, Based on Goddard Institute for Space Studies (GISS) Model

* Scenario assumed direct physiological effects of increased atmospheric CO_2 concentrations (at 555 ppm), and where applicable some farm level adaptations. Source: Compiled from Fischer et al. (1996)

CHAPTER 8

VULNERABILITY OF FOREST MANAGEMENT TO CLIMATE AND CLIMATE CHANGE IN THE CANADIAN PRAIRIES

Mark Johnston and Tim Williamson

INTRODUCTION

In the Prairie Provinces, managed forest Crown land totals approximately 36 million ha, of which about 24 million ha is in Alberta and 6 million ha each in Saskatchewan and Manitoba (Natural Resources Canada, 2008). Most of this land is allocated to forest companies through Forest Management Agreements (FMA). The FMAs are contractual agreements between the company and the provincial government that grant rights to harvesting and also require certain forest management activities to be undertaken by the company: forest regeneration, protection of water quality and fish habitat, submission of forest management plans, etc. Generally, FMAs are based on a 20-year planning cycle and are usually renewed every 10 years. The provincial governments maintain ownership of the land while the company has the right to annually harvest a given volume of wood, usually based on requirements of the wood processing facilities associated with the FMA (pulp mills, saw mills, etc.). The forest products industry in the Prairie Provinces generated revenues of approximately $8 billion and employed about 30,000 people in 2007 (Natural Resources Canada, 2008).

Forest management in the prairie region is strongly affected by climate and climatic variability. Climate determines tree growth and the distribution of species, and strongly influences disturbances that structure forest ecosystems (Kimmins, 2004) Vulnerabilities range from immediate effects of individual events such as intense rainfall preventing woods operations, to the long-term impacts of forest fires on wood supply. Forests are vulnerable at the species level, e.g. the loss of spruce to the spruce budworm, and at the landscape level where large fire events change the nature of the forest for decades. In addition, forest management is affected by other environmental factors (e.g. acid rain, nitrogen deposition) and socio-economic factors (e.g. national and global market forces, land use change). The impacts of climate and climate variability must be considered in the larger context of these other factors.

CURRENT VULNERABILITY OF FOREST MANAGEMENT TO CLIMATE AND CLIMATE VARIABILITY

SHORT-TERM VULNERABILITIES

Short-term climatic events can affect forest operations and access to harvestable wood supplies. Impacts include flooding that leads to the loss of roads, bridges and culverts; temperatures that affect the ability to construct and maintain ice roads; water-logged soils in cut blocks preventing equipment operations; and temperatures and moisture conditions that affect the length of frozen ground conditions for winter operations (Sauchyn and Kulshreshtha, 2008; Archibald et al., 1997).

The interaction of climate and forest management determines vulnerability that arises from climate change impacts on soil and water. The primary impacts to soil result from the effects of harvest activities in cut blocks on soil physical properties. In wet areas or periods of high precipitation, soils are disturbed by equipment operations and may be deeply rutted, affecting long-term site productivity, the ability to regenerate the site, and the potential for soil erosion (Archibald et al., 1997; Grigal, 2000). Steep topography can exacerbate these conditions and may lead to landslides (Grigal, 2000). Other soil impacts are due to improper maintenance of roads causing erosion and water control structures (waterlogged soils, flooding). Forest operations are often carried out in winter in the absence of wet conditions that prevent operations and because the frozen soil is relatively impervious to impacts by heavy equipment (Grigal, 2000). Forest management also includes the responsibility for rehabilitating forest roads and returning them to productivity through regenerating forest cover. Flooding or severe erosion caused by extreme precipitation events can reduce or eliminate the opportunities for rehabilitation and regeneration (Van Rees and Jackson, 2002).

The main impacts to water are associated with road crossings over creeks and rivers (Gunn and Sein, 2000). Climatic events, especially extreme precipitation, can result in sedimentation of watercourses at road crossings if they are not properly constructed or maintained, affecting water quality and fisheries habitat (Steedman, 2000).

LONG-TERM VULNERABILITIES

In the long-term, climate affects the growth and continued productivity of forest stands. Temperature, moisture and nutrient availability, and atmospheric CO_2

concentrations all affect tree growth directly (Kimmins, 2004). Climatic factors, mediated by soils and topography, also affect the species composition of forest stands and landscapes by determining which species occur on various sites (Rowe, 1996). In addition, forest productivity and species composition at the landscape level are affected by large-scale disturbance, which is strongly controlled by climatic conditions. For Canadian forests, the most important disturbance agents are forest fires (Weber and Flannigan, 1997) and insect outbreaks (Volney and Fleming, 2000). For example, insect pests in the Prairie Provinces affected on average 3.1 million ha between 1975 and 2003, with a maximum extent of 10–12 million ha in the mid-1970s (NFDP, 2008). Forest fires in the Prairie Provinces burned, on average, somewhat less than 1 million ha per year between 1975 and 2006, but reached 3–4 million ha during some years in the 1980s (NFDP, 2008).

Carbon Sequestration

There is increasing interest in carbon sequestration as an additional benefit of forests and forest management. Climate directly influences carbon sequestration by affecting tree growth as described above. Carbon is lost as a result of large-scale disturbance, especially forest fires. Amiro et al. (2001) estimated CO_2 emissions from burned forest for an area roughly similar to that of the prairie boreal forest (Boreal Plains, Boreal Shield West, see Figure 1 in Chapter 6 of this book). They found that annual emissions averaged 39 megatonnes (Mt) CO_2 during the period 1959–99, but reached 59 Mt CO_2 in 1981, a particularly severe fire year. This is an amount similar to the annual CO_2 emissions of Saskatchewan in the early 1990s (Environment Canada, 2008). Forest management activities can increase sequestration through activities such as thinning or fertilization, but this potential can be reduced by climatic events that limit tree growth. Kurz et al. (2008a) suggest that the managed forests of Canada could be a source of between 30 and 245 Mt CO_2e/yr^{-1} during the first Kyoto Protocol commitment period (2008–12). They attribute this large range of projected emissions to wide variations in expected levels of forest fires and insect outbreaks.

KEY VULNERABILITIES TO FUTURE CLIMATE CHANGE

FOREST ECOSYSTEM VULNERABILITIES

As indicated in Chapter 2, climate scenarios for the Prairie Provinces suggest that the future will bring warmer winters with greater precipitation and earlier springs. Summers may be somewhat warmer, but will be dryer due to increased evapotranspiration. In addition, extreme precipitation and drought events may become more frequent. Under these conditions, excess spring moisture from earlier and more rapid snowmelt, earlier and perhaps longer spring road weight restrictions, and waterlogged conditions in operating areas can be expected. This could affect both woods operations and the construction and use of forest roads. Sites prone to erosion (e.g. road crossings) could become more vulnerable to higher and more intense precipitation (Spittlehouse and Stewart, 2003). Flooding would be a concern and will require closer attention to proper sizing of culverts and other water control structures (Spittlehouse and Stewart, 2003). In areas where winter operations are important, the shorter length of frozen ground conditions will limit woods operations and affect scheduling of harvesting equipment among cutting areas. Harvest access that depends on ice roads would also be vulnerable to warmer winters, with a reduced haul season (Sauchyn and Kulshreshtha, 2008; Chapter 20 in this volume).

Longer-term effects will be manifested in changes to tree growth and, hence, the volume of timber available for harvest. Forest productivity is determined by a number of environmental factors, most of which will be affected by climate change. The most important of these are temperature, moisture availability, nutrient availability (this is discussed in the following section), and atmospheric CO_2 concentration. Recent research has shown a variety of responses to changes in these factors, including long-term increases in growth, short-term increases followed by acclimation, and negative impacts on growth.

Temperature

Higher temperatures increase the rate of both carbon uptake (photosynthesis) and carbon loss (respiration), so the effect of higher temperatures will depend on the net balance between these processes (Amthor and Baldocchi, 2001). Most of the literature suggests that respiration may increase, but the increases are likely to be small; this will vary among species, season, and site conditions. Both photosynthesis and respiration have been shown to adjust to a change in environmental conditions (acclimation), so any increases may be short-lived. Finally, changes to photosynthesis have been shown to be highly dependant on nutrient availability (especially nitrogen) and water availability (Baldocchi and Amthor, 2001). Generally, net primary productivity is expected to increase under warmer temperatures if water and nutrients are not limiting (Norby et al., 2005).

Soil warming experiments have shown increases in growth and nitrogen availability. Experimental soil warming in northern Sweden (64°N) in Norway spruce stands showed increased basal area growth, and also showed that the addition of fertilizer and water dramatically increased volume growth relative to warming alone (Stromgren and Linder, 2002). In a wide-ranging review of other soil warming experiments, increased rates of nitrogen (N) availability have been found in nearly all locations and vegetation types (Rustad et al., 2001). However, this is dependant on water availability, and will also be affected by N deposition from industrial sources (Kochy and Wilson, 2001).

Higher temperatures may also result in a longer growing season. Zhou et al. (2001) found that the average period of vegetation greenness (i.e. growing season) increased by 12 days in North America and 18 days in northern Eurasia between 1981 and 1999. In addition, flowering and fruiting may occur earlier than at present, with unknown consequences for tree regeneration and interaction with pollinators. Similarly, McDonald et al. (2004) found that the mean date of spring thaw in the North American boreal forest advanced by 13 days between 1988 and 2001. Goetz et al. (2005) reported similar patterns in tundra regions of Canada and Alaska. However, they also found that photosynthesis in unburned boreal forest areas varied by up to ±15% between 1982 and 2003, and showed no systematic pattern in growing season length. Much of the variability in photosynthesis was attributed to the impacts of large forest fires, but could also have been affected by drought, nutrient availability and insect outbreaks (Goetz et al., 2005). These authors emphasize the importance of interactions between ecophysiological processes and large-scale disturbance.

Available Soil Moisture

Soil available water-holding capacity (AWC) is a critical factor in determining water availability for uptake by the tree's root system. Work in northern Saskatchewan has shown that potential biomass production is highly sensitive to climate change, and that differences in AWC strongly affect how productivity will change (Johnston, 2001). On sites with low AWC, productivity declines under all future climate scenarios as projected by the Canadian Global Climate Model (CGCM1) (Johnston, 2001). On sites with moderate AWC, productivity goes up initially in response to warmer temperatures, but then declines as water availability declines in later decades. On sites with high AWC, productivity continued to increase through the 21st century since available soil water is sufficient to support the increased growth (Johnston, 2001). Similarly, Johnston and Williamson (2005) found that simulated future drought reduced productivity of white spruce in Saskatchewan by about 20% on sites with low AWC. Much of the southern boundary of the boreal forest in the Prairie Provinces is currently vulnerable to drought impacts, and this is expected to increase in the future (Hogg and Bernier, 2005).

The net effect of water availability will be determined by its seasonal distribution (spring versus summer) relative to the demand for water from the vegetation. Higher availability may not benefit the trees if uptake is limited due to frozen soils in spring with dormant root systems. Lower soil water availability in summer will also cause decreases in growth. Alternatively, if the trees are able to take advantage of the early spring melt, growth could be enhanced (Cohen and Miller, 2001). In managed forests, planted seedlings are vulnerable to climatic conditions in the first few years of growth, particularly drought, and natural regeneration following disturbance is also highly vulnerable to climate in the early stages of establishment (Parker et al., 2000; Spittlehouse and Stewart, 2003).

Carbon Dioxide (CO_2)

Increased CO_2 concentration affects a number of productivity-related factors. Plants take up CO_2 through stomata in the leaves, but lose water at the same time through transpiration. Under higher levels of atmospheric CO_2, less water is lost for a given unit of CO_2 uptake (Long et al., 2004). This increase in water-use efficiency (WUE) could be particularly important on water-limited sites, such that tree growth might continue where it would be severely limited under current CO_2 levels. For example, Johnston and Williamson (2005) used a forest ecosystem model to explore responses of white spruce productivity under a range of future climate conditions in Saskatchewan. They found that even under severe drought conditions, increased WUE due to increased CO_2 concentrations resulted in an increase in productivity relative to current conditions. Productivity declined by about 20% when the WUE effect was not included in the model.

Several ongoing experiments are exposing forest stands to increased levels of CO_2 in a natural environment. These Free Air Carbon dioxide Enrichment (FACE) experiments have been established in an aspen forest in northern Wisconsin and in a loblolly pine plantation in North Carolina. These stands are exposed to levels of CO_2 roughly twice that of the pre-industrial period. Results at the pine site have shown that the initial increase in net primary productivity (NPP) is relatively short-lived (3–4 years), and only occurs when soil nutrient and water levels are relatively high (DeLucia et al., 1999; Oren et al., 2001). In the aspen FACE study, trees were exposed to CO_2, CO_2 +ozone (O_3) and O_3 alone. The NPP increased under the CO_2 -only treatment, but did not differ from untreated trees when CO_2 was combined with O_3. Ozone apparently compensated for the increase in NPP caused by CO_2 alone (Isebrands et al., 2001). Long et al. (2004) carried out a meta-analysis of plant growth at a variety of FACE sites around the world. They found a greater response to increased CO_2 concentrations in trees than other vegetation, with biomass production increasing an average of about 20%. Norby et al. (2005) found that productivity in forest stands across a range of FACE sites (ca. 550 ppm CO_2) was remarkably consistent, showing an increase in NPP of $23\pm2\%$.

Disturbance

Future disturbance regimes are also expected to be considerably different from those of today. For the prairie region, forest fires are expected to be more frequent (Bergeron et al., 2004), of higher intensity (Parisien et al., 2004) and burn over larger areas (Flannigan et al., 2005), although the magnitude of these changes is difficult to predict. Insect outbreaks are also expected to be more frequent and severe (Volney and Fleming, 2000). Of particular concern is the mountain pine beetle (MPB), currently in a major outbreak phase in the B.C. interior. The beetle is limited by winter temperatures less than -39°C, and occurs throughout the B.C. interior and in outlying populations of lodgepole pine as far east as Slave Lake in central Alberta. Under a warming climate, this limiting temperature is likely to occur further to the north and west, allowing the beetle to spread into jack pine in the Prairie Provinces. The MPB has been shown experimentally to develop successful populations in jack pine, so there is every expectation that jack pine will prove to be a suitable host (Langor et al., 2006). The overlap in range between jack pine and lodgepole pine in north-central Alberta will provide the physical connection between the two host species, and recent experience has shown that strong winds can move large numbers of beetles several hundred kilometres. Jack pine's distribution is nearly continuous from Alberta to New Brunswick, so the spread of the beetle across the Canadian boreal forest is a possible future scenario. Further detail is available in Logan, Réginière and Powell (2003), Carroll et al. (2004), Moore et al. (2005), and Taylor et al. (2006). The long-term effect of insect outbreaks on forest management is difficult to predict, but recent research suggests increased tree mortality resulting from the interaction of insects, drought and fire in the southern margin of the boreal forest in the Prairie Provinces (Hogg and Bernier, 2005; Volney and Hirsch, 2005).

Increased rates of fire disturbance will differentially affect tree species due to differences in flammability and their ability to regenerate following fire (Johnston, 1996). Some coniferous species are inherently more flammable than hardwood species (Parisien et al., 2004); so increased forest fire activity will likely favour hardwood species (e.g. aspen) over some conifers such as white spruce. However, other conifers such as jack pine are well adapted to reproduce following fire, so a long-term increase in forest frequency may lead to an increase in aspen and jack pine at the expense of spruce and other species less resistant to fire (Williamson et al., 2009). Impacts on forest management will be determined by the relative importance of hardwoods and softwoods to the local forestry economy. For example, oriented strand board (OSB) mills in Canada generally use 90–100% hardwood (mainly aspen) as feedstock, so an increase in disturbance regime that favours aspen will result in an increase in wood supply for OSB in the long term. In contrast, saw mills that depend on fire-susceptible softwood species for lumber production may experience a decline in wood supply under increased forest fire activity. This may result in significant relocation of wood-processing plants within the Prairie Provinces, which may have an effect on their competitive position both domestically and internationally.

Carbon Sequestration

Carbon (C) sequestration will be strongly affected by the small- and large-scale processes described above, with losses or gains in biomass C occurring with decreases or increases in tree growth. Increased fire and insect activity will likely lead to larger losses of carbon. Kurz et al. (2008b) use the Canadian Carbon Budget Model of the Canadian Forest Sector to explore the carbon implications of the mountain pine beetle outbreak in British Columbia. They estimate that the cumulative impact of the beetle outbreak during 2000–20 will be 270 Mt carbon. This impact converted the forest from a small net carbon sink to a large net carbon source, both during and immediately after the outbreak. In the worst year, the impacts resulting from the beetle outbreak in British Columbia were equivalent to 75% of the average annual direct forest fire emissions from all of Canada during 1959–99 given by Amiro et al. (2001). Increased fire activity will also release more carbon than currently, but the increase is difficult to predict.

In addition, changes in soil C stocks are sensitive to environmental conditions. Recent research has shown the high sensitivity of soil organic matter decomposition to temperature, and suggests that future warming will likely reduce soil C stocks in addition to affecting C in forest biomass (Knorr et al. 2005; Fang et al. 2005)

ECONOMIC AND SOCIAL BENEFITS

Climate change may have both positive and negative impacts on the forest sector. Given the importance of forest products to the building sector, increases in natural hazards will likely stimulate the forest products sector. For example, the price of OSB panels went up by over 50% in the weeks following Hurricane Katrina in the fall of 2005 (NAHB, 2005). On sites with adequate water and nutrients, increased tree growth may result in an increased wood supply. This could depress the market price but provides a benefit to consumers (Sohngen and Sedjo, 2005). Alternatively, large-scale disturbance and mortality could reduce the wood supply, increasing prices and leading to local or regional wood shortages (Sohngen and Sedjo, 2005). Associated impacts could include mill closures and the attendant economic effects on small forest-dependent communities (Williams, Parkins and McFarlane, 2005). Changes in species composition due to disturbance and changed growth conditions in the future may require mills to modify processing capacity and produce new products. Salvage harvesting following large-scale disturbances may provide additional woody biomass for use in bioenergy production, currently being considered in British Columbia in forests affected by the mountain pine beetle (Kumar, Flynn and Sokhansanj, 2005). However, impacts of intensive salvage harvesting on ecosystem function and biodiversity may be adverse (Lindenmayer et al., 2004).

ADAPTATION IN THE FORESTRY SECTOR

Adaptation and adaptive capacity in the forestry sector have not been examined thoroughly. O'Shaughnessy (2001) made a preliminary attempt to determine adaptation options for the forest sector in the Prairie Provinces, and identified the need for stronger interactions between the forest management and research communities. A general conclusion was that forest management is well placed to develop adaptive strategies because of its reliance on the criteria for Sustainable Forest Management (SFM) as set out by the Canadian Council of Forest Ministers (CCFM, 2003). These criteria and the associated indicators define SFM and provide guidance on how to maintain sustainability in the face of changing environmental and socio-economic conditions. However, these criteria and indicators have not been tested specifically for their application to climate change adaptation, and this should be a high priority for the forest management community.

Consistent with Criteria and Indicators is the provincial government requirement that forest companies carry out long-term forest management plans (e.g. ASRD, 1998; MC, 1999; SE, 2005). These plans provide the biophysical and socio-economic context for a company's forest operations and describe generally how their activities will be carried out over the planning horizon. These plans provide the ideal vehicle for incorporating climate change considerations into forest management, given their long-term perspective and general level of detail (Spittlehouse, 2005).

Most forest companies in Canada are undergoing one of several certification procedures that indicate that their products are produced from a sustainably managed forest land base. As of late 2005, over 113 million ha had been certified in Canada (FPAC, 2005). To the extent that sustainable forest management is consistent with climate change adaptation, certification should support adaptive capacity in the forestry sector. However, these certification systems have not been rigorously tested specifically for their relevance to climate change adaptation.

Spittlehouse (2005) identified four areas of activity to ensure appropriate adaptive responses in forest management: research and education; understanding the vulnerability of forest resources; policies to facilitate implementation of adaptation in forest management; and monitoring systems to identify problems induced by climate change. Spittlehouse and Stewart (2003) suggest several adaptation options, including:

- Genetic Management: breeding for pest resistance and climate stresses and extremes;
- Forest Protection: altering forest structure and developing "fire-smart" landscapes, i.e., creating areas of reduced flammability through fuel modification;

- Forest Regeneration: assisting the migration of commercial tree species from their present to future ranges through artificial regeneration;
- Silvicultural Management: pre-commercial thinning to enhance growth and insect/disease resistance;
- Forest Operation: mitigating climate change impacts on infrastructure, fish, and potable water supplies, and on the timing of peak flow and volume in streams resulting from increased winter precipitation and earlier snow melt;
- Non-timber Resources: minimizing fragmentation of habitat and maintaining connectivity; and
- Park and Wilderness Area Management: manage these areas to delay, ameliorate and direct change.

Potential adaptations such as these may provide "no-regrets" options that provide immediate benefits while positioning forest managers to better cope with future climate change.

KEY UNCERTAINTIES, RESEARCH GAPS AND PRIORITIES

FUTURE CLIMATE AND SOCIO-ECONOMIC SCENARIOS

Regional climate change assessments in the UK (Holman et al., 2005a, 2005b) and the European Union assessment known as the ATEAM project (Schröter et al., 2005) have shown that socio-economic scenarios are often more important than climate scenarios in impacts assessments, particularly in determining economic impacts and adaptive capacity. The approach these authors have taken is to "downscale" the scenarios developed in the Special Report on Emission Scenarios (Nakicenovic and Swart, 2000). For example, Abildtrup et al. (2006) describe the development of agricultural scenarios used in the ATEAM assessment (Schröter et al., 2005). They found that assumptions about commodity prices, for example, were more important in determining potential adaptation actions than were the projected biophysical impacts of climate change. There is a critical need to develop these scenarios for Canada, ideally at the regional level.

Global-scale climate scenarios are increasingly available especially with the release of the IPCC Fourth Assessment Report (2007). However, for regional impacts assessments, the global products are not of sufficient resolution. Recent regional-scale climate scenarios (e.g. Canadian Regional Climate Model, Laprise et al., 2003; Sushama et al., 2006) are addressing this need, as are recent high-resolution downscaled products (e.g. Price et al., 2004; McKenney et al., 2004). There is a need for linked climate-ecosystem models that capture the complex relationships between the atmosphere and the land surface including non-linearities and threshold effects, multiple trophic levels, disturbance dynamics, and changes in land use (Feddema et al., 2005). An example of this approach being developed in Canada is the coupling of the Canadian Global Circulation Model (CGCM2, Flato et al., 2000), the Canadian Land Surface Scheme (Verseghy et al., 1993) and the Canadian Terrestrial Ecosystem Model (Arora, 2003).

ECOSYSTEM MODELS

To represent ecosystem impacts, comprehensive ecosystem models are required that include both local-scale ecosystem processes (e.g. productivity) and landscape-scale processes (e.g. seed dispersal, disturbance). These dynamic global vegetation models can be used as stand-alone simulators or coupled directly to GCMs. Examples include the Integrated Biosphere Simulator (Foley at al., 1996), Lund-Potsdam-Jena model (Gerber, Joos and Prentice, 2004) and MC1 (Bachelet et al., 2001). There is a strong need for further regional-scale application of these models, including detailed parameterization and validation of results. This approach has been applied in several recent European forestry assessments (Kellomäki and Leinonen, 2005; Schröter et al., 2005; Koca, Smith and Sykes, 2006).

ADAPTATION AND ADAPTIVE CAPACITY

Relative to other sectors, the adaptive capacity of the forest sector is not well understood. Currently the assumption is that adaptive capacity should be adequate to the extent that sustainable forest management is broadly consistent with climate change adaptive capacity. However, the degree of support for climate change adaptation in the Criteria and Indicators and forest certification systems needs to be rigorously tested. There is also a gap in understanding the extent to which existing government forest management policy either encourages or discourages the implementation of adaptive strategies. There may exist government policies that would discourage or even prevent forest managers from implementing adaptive options. These policies need to be identified and modified. Similarly, there may be need for new policies that would enable adaptation. There is also a need for determining the relative importance of adaptive responses vs. other priorities for forest managers, and to develop approaches that incorporate climate change considerations into existing policy instruments, e.g. long-term forest management plans. Finally, investment in innovation has recently been low in the Canadian forest products industry. Lack of innovation will prevent adaptation options from being implemented under future climate and socio-economic change. Investment in innovative technology and forest management practices will help ensure an adaptive forest sector in Canada, both now and in the future (Ohlson, McKinnon and Hirsch, 2005).

CONCLUSIONS

Forests in Canada's Prairie Region are likely to be highly vulnerable to the impacts of climate change, particularly along the southern fringe. Forest dieback in response to recent drought has underscored the potential for widespread impacts. In particular, forests occupying marginal sites and separated from the main forest area (e.g. the "Island Forests") are particularly at risk and are unlikely to maintain tree cover in the future. Forests along the southern edge are also likely to experience increased insect attack and more frequent and severe fires. These multiple sources of vulnerability will combine to make the impacts of climate change among the most severe in Canada. The forest sector (industry, government and NGOs) will need to jointly develop adaptation options that will increase the likelihood of maintaining forest health in the face of this vulnerability. In addition, adaptive capacity will need to be increased in order for forest management to be sustainable in the future. Finally, climate change must be viewed as one of several agents of change to which the forest sector will have to adapt in the coming decades.

REFERENCES

Abildtrup, J., E. Audsley, M. Fekete-Farkas, C. Giupponi, M. Gylling, P. Rosato and M. Rounsevell. 2006. "Socio-Economic Scenario Development for the Assessment of Climate Change Impacts on Agricultural Land Use: A Pairwise Comparison Approach." *Environmental Science and Policy* 9: 101–15.

ASRD (Alberta Sustainable Resource Development). 1998. *Interim Forest Management Planning Manual. Edmonton* [online]. [accessed October 5, 2009]. Available from World Wide Web: *http://www3.gov.ab.ca/srd/forests/fmd/manuals*

Amiro, B.D., J.B. Todd, B.M. Wotton, K.A. Logan, M.D. Flannigan, B.J. Stocks, J.A. Mason, D.L. Martell and K.G. Hirsch. 2001. "Direct Carbon Emissions from Canadian Forest Fires, 1959–1999." *Canadian Journal of Forest Research* 31: 512–25.

Amthor, J.S. and D.D. Baldocchi. 2001. "Terrestrial Higher Plant Respiration and Net Primary Production." Pp. 33–59 in J. Roy, B. Saugier and H. Mooney (eds.). *Terrestrial Global Productivity*. Academic Press: San Diego.

Archibald, D.J., W.B. Wiltshire, D.M. Morris and B.D. Batchelor. 1997. *Forest Management Guidelines for the Protection of the Physical Environment, Version 1.0*. Report no. 51032, Ontario Ministry of Natural Resources: Toronto.

Arora, V.K. 2003. "Simulating Energy and Carbon Fluxes over Winter Wheat using Coupled Land Surface and Terrestrial Ecosystem Models." *Agricultural and Forest Meteorology* 118: 21–47.

Bachelet, D., J.M. Lenihan, C. Daly, R.P. Neilson, D.S. Ojima and W.J. Parton. 2001. *A Dynamic Vegetation Model for Estimating the Distribution of Vegetation and Associated Ecosystem Fluxes of Carbon, Nutrients, and Water Technical Documentation, Version 1.0*. General Technical Report PNW-GTR-508. USDA Forest Service: Corvallis, OR.

Baldocchi, D.D. and J.S. Amthor. 2001. "Canopy Photosynthesis: History, Measurements and Models." Pp. 9–31 in J. Roy, B. Saugier and H. Mooney (eds.). *Terrestrial Global Productivity*. Academic Press: San Diego.

Bergeron, Y., M. Flannigan, S. Gauthier, A. Leduc and P. Lefort. 2004. "Past, Current and Future Fire Frequency in the Canadian Boreal Forest: Implications for Sustainable Forest Management." *Ambio* 33: 356–60.

CCFM (Canadian Council of Forest Ministers). 2003. *Defining Sustainable Forest Management in Canada: Criteria and Indicators* [online]. Canadian Council of Forest Ministers: Ottawa, ON. [accessed October 5, 2009]. Available from World Wide Web: *http://www.ccfm.org/review_e.html*

Carroll, A.L., S.W. Taylor, J. Régnière and L. Safranyik. 2004. "Effects of Climate Change on Range Expansion by the Mountain Pine Beetle in British Columbia." Pp. 223–32 in T.L. Shore, J.E. Brooks and J.E. Stone (eds.). *Mountain Pine Beetle Symposium: Challenges and Solutions, October 30–31, 2003*, Kelowna, British Columbia, Canada. Information Report B.C.-X-399. Natural Resources Canada, Canadian Forest Service, Pacific Forestry Centre: Victoria, B.C..

Cohen, S. and K. Miller (eds). 2001. "North America." Pp. 735–800 in J. McCarthy, O. Canziani, N. Leary, D. Dokken and K. White (eds.). *Climate Change 2001: Impacts, Adaptation, and Vulnerability*, IPCC. Cambridge University Press: Cambridge.

DeLucia, E., J. Hamilton, S. Naidu, R. Thomas, J. Andrews, A. Finzi, M. Lavine, R. Matamala, J. Mohan, G. Hendrey and W. Schlesinger. 1999. "Net Primary Production of a Forest Ecosystem with Experimental CO_2 Enrichment." *Science* 284: 1177–79.

Environment Canada. 2008, *National Inventory Report: Greenhouse Gas Sources and Sinks in Canada, 1990–2006* [online]. [accessed October 5, 2009]. Available from World Wide Web: *http://www.ec.gc.ca/pdb/ghg/inventory_report/2006_report/tdm-toc_eng.cfm*

Fang, C., P. Smith, J.B. Moncrieff and J.U. Smith, 2005, "Similar Response of Labile and Resistant Soil Organic Matter Pools to Changes in Temperature." *Nature* 433: 57–59.

Feddema, J., K. Oleson, G. Bonan, L. Mearns, W. Washington, G. Meehl and D. Nychka. 2005. "A Comparison of a GCM Response to Historical Anthropogenic Land Cover Change and Model Sensitivity to Uncertainty in Present-Day Land Cover Representations." *Climate Dynamics* 25: 581–609.

Flannigan, M.D., K.A. Logan, B.D. Amiro, W.R. Skinner and B.J. Stocks. 2005. "Future Area Burned in Canada." *Climatic Change* 72: 1–16.

Flato, G.M., G.J. Boer, W.G. Lee, N.A. McFarlane, D. Ramsden, M.C. Reader and A.J. Weaver. 2000. "The Canadian Centre for Climate Modelling and Analysis Global Coupled Model and its Climate." *Climate Dynamics* 16: 451–67.

Foley, J.A., I.C. Prentice, N. Ramankutty, S. Levis, D. Pollard, S. Sitch and A. Haxeltine. 1996. "An Integrated Biosphere Model of Land Surface Processes, Terrestrial Carbon Balance and Vegetation Dynamics." *Global Biogeochemical Cycles* 10: 603–28.

FPAC (Forest Products Association of Canada). 2005. "Forestry Certification" [online]. FPAC Market Acceptance Customer Briefing Note, September 2005. Forest Products Association of Canada, Ottawa, ON. [accessed October 5, 2009]. Available from World Wide Web: *http://www.certificationcanada.org/pdfs/certification_2005.pdf*

Gerber, S., F. Joos and I.C. Prentice. 2004. "Sensitivity of a Dynamic Global Vegetation Model to Climate and Atmospheric CO_2." *Global Change Biology* 10: 1–17.

Goetz, S.J., A.G. Bunn, G.J. Fiske and R. A. Houghton. 2005. "Satellite-Observed Photosynthetic Trends Across Boreal North America Associated with Climate and Fire Disturbance." *Proceedings of the National Academy of Sciences* 102: 13521–25.

Grigal, D.F. 2000. "Effects of Extensive Forest Management on Soil Productivity." *Forest Ecology and Management* 138: 167–85.

Gunn, J.M. and R. Sein. 2000. "Effects of Forestry Roads on Reproductive Habitat and Exploitation of Lake Trout in Three Experimental Lakes." *Canadian Journal of Fisheries and Aquatic Sciences* 57 (Suppl. 2): 97–104.

Holman, I.P., M.D.A. Rounsevell, S. Shackley, P.A. Harrison, R.J. Nicholls, p.m. Berry and E. Audsley. 2005a. "A Regional, Multi-sectoral and Integrated Assessment of the Impacts of Climate and Socio-Economic Change in the UK. Part I, Methodology." *Climatic Change* 71: 9–41.

Holman, I.P., R.J. Nicholls, P.M. Berry, P.A. Harrison, E. Audsley, S. Shackley and M.D.A. Rounsevell. 2005b. "A Regional, Multi-sectoral and Integrated Assessment of the Impacts of Climate and Socio-Economic Change in the UK. Part II, Results." *Climatic Change* 71: 43–73.

Hogg, E.H. and P.Y. Bernier. 2005. "Climate Change Impacts on Drought-Prone Forests in Western Canada." *The Forestry Chronicle* 81: 675–82.

IPCC (Intergovernmental Panel on Climate Change). 2007. "Summary for Policymakers." In S. Solomon, D. Qin, M. Manning, Z. Chen, M. Marquis, K.B. Averyt, M. Tignor and H.L. Miller (eds.). *Climate Change 2007: The Physical Science Basis. Contribution of Working Group I to the Fourth Assessment Report of the Intergovernmental Panel on Climate Change.* Cambridge University Press: Cambridge, New York.

Isebrands, J., E. McDonald, E. Kruger, G. Hendrey, K. Percy, K. Pregitzer, J. Sober and D. Karnosky. 2001. "Growth Responses of *Populus tremuloides* Clones to Interacting Elevated Carbon Dioxide and Tropospheric Ozone." *Environmental Pollution* 115: 359–71.

Johnston, M.H. 1996. "The Role of Disturbance in Boreal Mixedwood Forests of Ontario." Pp. 33–40 in C.R. Smith and G.W. Crook (compilers). *Advancing Boreal Mixedwood Management in Ontario: Proceedings of a Workshop.* Canadian Forest Service, Great Lakes Forestry Centre: Sault Ste. Marie, ON.

——. 2001. *Sensitivity of Boreal Forest Landscapes to Climate Change.* Final report submitted to Natural Resources Canada, Climate Change Action Fund. Saskatchewan Research Council Publication No. 11341–7E0, Saskatchewan Research Council: Saskatoon, SK. 34 pgs.

Johnston, M. and T. Williamson. 2005. "Climate Change Implications for Stand Yields and Soil Expectation Values: A Northern Saskatchewan Case Study." *The Forestry Chronicle* 81: 683–90.

Kellomäki, S. and S. Leinonen (eds.). 2005. *Management of European Forests Under Changing Climatic Conditions.* Research Report 163, Faculty of Forestry, University of Joensuu: Joensuu, Finland. 21 pgs.

Kimmins, J.P. 2004. *Forest Ecology: A Foundation for Sustainable Forest Management and Environmental Ethics in Forestry* (3rd Edition). Prentice Hall: Upper Saddle River, NJ.

Knorr, W., I.C. Prentice, J.I. House and E.A. Holland. 2005. "Long-Term Sensitivity of Soil Carbon Turnover to Warming." *Nature* 433: 298–301.

Koca, D., B. Smith and M.T. Sykes. 2006. "Modelling Regional Climate Change Effects on Potential Natural Ecosystems in Sweden." *Climatic Change* 78: 381–406.

Kochy, M. and S.D. Wilson. 2001. "Nitrogen Deposition and Forest Expansion in the Northern Great Plains." *Journal of Ecology* 89: 807–17.

Kumar, A., P.C. Flynn and S. Sokhansanj. 2005. *Feedstock Availability and Power Costs Associated with Using B.C.'s Beetle-Infested Pine* [online]. Final Report prepared for Biocap Canada Foundation and B.C. Ministry of Forests and Range. Biocap Canada Foundation, Queen's University: Kingston, ON. [accessed October 5, 2009]. Available from World Wide Web: *http://www.biocap.ca*

Kurz, W., G. Stinson, G. Rampley, C. Dymond and E. Neilson. 2008a. "Risk of Natural Disturbances Makes Future Contribution of Canada's Forests to the Global Carbon Cycle Highly Uncertain." *Proceedings of the National Academy of Science* 105: 1551–55.

Kurz, W., C. Dymond, G. Stinson, G. Rampley, E. Neilson, A. Carroll, T. Ebata and L. Safranyik. 2008b. "Mountain Pine Beetle and Forest Carbon Feedback to Climate Change." *Nature* 452: 987–90.

Langor, D., A. Rice, D. Williams and M. Thormann. 2006. "Mountain Pine Beetle Invasion of Jack Pine: A Prognosis" [online]. Presentation to Alberta Sustainable Resource Development. [accessed October 5, 2009]. Available from World Wide Web: *http://www.srd.gov.ab.ca/forests/pdf/Dave_Langor_2006.pdf*

Laprise, R., D. Caya, A. Frigon and D. Paquin. 2003. "Current and Perturbed Climate as Simulated By the Second-Generation Canadian Regional Climate Model (CRCM-II) Over Northwestern North America." *Climate Dynamics* 21: 405–21.

Lindenmayer, D.B., D.R. Foster, J.F. Franklin, M.L. Hunter, R.F. Noss, F.A. Schmiegelow and D. Perry. 2004. "Salvage Harvesting Policies after Natural Disturbance." *Science* 303: 1303.

Logan, J.A., J. Régnière and J.A. Powell. 2003. "Assessing the Impacts of Global Warming on Forest Pest Dynamics." *Frontiers in Ecology and Environment* 1: 130–37.

Long, S.P., E.A. Ainsworth, A. Rogers and D.R. Ort. 2004. "Rising Atmospheric Carbon Dioxide: Plants FACE the Future," *Annual Review of Plant Biology* 55: 591–628.

MC (Manitoba Conservation). 1999. *Ten-year Forest Management Plan Submission Guidelines* [online]. [accessed October 5, 2009]. Available from World Wide Web: *http://www.gov.mb.ca/conservation/forestry/forest-practices/practices/fpp-guideline-pdfs.html*

McDonald, K.C., J.S. Kimball, E. Njoku, R. Zimmermann and M. Zhao. 2004. "Variability in Springtime Thaw in the Terrestrial High Latitudes: Monitoring a Major Control on the Biospheric Assimilation of Atmospheric CO_2 with Spaceborne Microwave Remote Sensing" [online]. *Earth Interactions* 8 (paper no. 20). [accessed October 5, 2009]. Available from World Wide Web: *http://EarthInteractions.org*

McKenney, D.W., M.F. Hutchinson, P. Papadopol and D.T. Price. 2004. "Evaluation of Alternative Spatial Models of Vapour Pressure in Canada." Paper 6.2 in Proceedings, *26th Agricultural and Forest Meteorology Conference*. American Meteorology Society: Vancouver, B.C., August 22–26, 2004.

Moore, R.D., I.G. McKendry, K. Stahl, H.P. Kimmins and Y.H. Lo. 2005. *Mountain Pine Beetle Outbreaks in Western Canada: Coupled Influences of Climate Variability and Stand Development.* Final Report for the Climate Change Action Fund, Project A676. Climate Change Impacts and Adaptation Directorate, Natural Resources Canada: Ottawa, ON.

Nakicenovic, N. and R. Swart (eds.). 2000. *Special Report on Emissions Scenarios: A Special Report of Working Group III of the Intergovernmental Panel on Climate Change* [online]. Cambridge University Press: Cambridge, U.K., 599 pgs. [accessed October 5, 2009]. Available from World Wide Web: *http://www.grida.no/climate/ipcc/emission/index.htm*

NAHB (National Association of Home Builders). 2005. "Rebuilding Katrina-Destroyed Homes at Least a Year Away." *Nation's Building News*, October 10, 2005.

NFDP (National Forestry Database Program). 2008. *A Compendium of Canadian Forestry Statistics* [online]. [accessed October 5, 2009]. Available from World Wide Web: *http://nfdp.ccfm.org/*

Natural Resources Canada. 2008. *Canada's Forests, Statistical Data* [online]. Natural Resources Canada: Ottawa, ON. [accessed October 5, 2009]. Available from World Wide Web: *http://canadaforests.nrcan.gc.ca/*

Norby, R.J., E.H. DeLucia, B. Gielen, C. Calfapietra, C.P. Giardina, J.S. King, J. Ledford, H.R. McCarthy, D.J.P. Moore, R. Ceulemans, P. De Angelis, A.C. Finzi, D.F. Karnosky, M.E. Kubiske, M. Lukac, K.S. Pregitzer, G.E. Scarascia-Mugnozza, W.H. Schlesinger and R. Oren. 2005. "Forest Response to Elevated CO_2 is Conserved Across a Broad Range of Productivity." *Proceedings of the National Academy of Sciences* 102: 18052–56.

Ohlson, D.W., G.A. McKinnon and K.G. Hirsch. 2005. "A Structured Decision-Making Approach to Climate Change Adaptation in the Forest Sector." *The Forestry Chronicle* 81: 97–103.

Oren, R., D. Ellsworth, K. Johnsen, N. Phillips, B. Ewers, C. Maier, K. Schafer, H. McCarthy, G. Hendrey, S. McNulty and G. Katul. 2001. "Soil Fertility Limits Carbon Sequestration by Forest Ecosystems in a CO_2-enriched Atmosphere." *Nature* 411: 469–72.

O'Shaughnessy, S. 2001. "A Framework for Determining the Ability of the Forest Sector to Adapt to Climate Change." M.Sc. Thesis, Department of Geography, University of Saskatchewan: Saskatoon, SK.

Parisien, M.A., K.G. Hirsch, S.G. Lavoie, J.B. Todd and V.G. Kafka. 2004. *Saskatchewan Fire Regime Analysis*. Information Report NOR-X-394, Northern Forestry Centre, Canadian Forest Service, Natural Resources Canada: Edmonton, AB. 61 pgs.

Parker, W.C., S.J. Colombo, M.L. Cherry, M.D. Flannigan, S. Greifenhagen, R.S. McAlpine, C. Papadopol and T. Scarr. 2000. "Third Millennium Forestry: What Climate Change Might Mean to Forests and Forest Management in Ontario." *The Forestry Chronicle* 76: 445–63.

Price, D.T., D.W. McKenney, P. Papadopol, T. Logan and M. F. Hutchinson. 2004. "High Resolution Future Scenario Climate Data for North America." Paper 7.7 in Proceedings, *26th Agricultural and Forest Meteorology Conference*, American Meteorology Society: Vancouver, B.C., August 22–26, 2004.

Rowe, J.S. 1996. "Land Classification and Ecosystem Classification." *Environmental Monitoring and Assessment* 39: 11–20.

Rustad, L., J. Campbell, G. Marion, R. J. Norby, M.J. Mitchell, A. E, Hartley, J.H.C. Cornelissen and J. Gurevitch. 2001. "A Meta-analysis of the Response of Soil Respiration, Net Nitrogen Mineralization, and Aboveground Plant Growth to Experimental Ecosystem Warming." *Oecologia* 126: 543–62.

Saskatchewan Environment. 2005. *Forest Planning Manual* [online]. [accessed October 5, 2009]. Available from World Wide Web: *http://www.se.gov.sk.ca/forests/forestmanagement/Sask_Leg_Manuals.htm#ForestPlanningManual*

Sauchyn, D. and S. Kulshreshtha. 2008. "Prairies." Pp. 275–328 in D. Lemmen, F. Warren, J. Lacroix, and E. Bush (eds.). *From Impacts to Adaptation: Canada in a Changing Climate 2007*. Government of Canada: Ottawa.

Schröter, D., W. Cramer, R. Leemans, I.C. Prentice, M.B. Araujo, N.W. Arnell, A. Bondeau, H. Bugmann, T.R. Carter, C.A. Gracia, A.C. de la Vega-Leinert, M. Erhard, F. Ewert, M. Glendining, J.I. House, S. Kankaanpaa, R.J.T. Klein, S. Lavorel, M. Lindner, M.J. Metzger, J. Meyer, T.D. Mitchell, I. Reginster, M. Rounsevell, S. Sabate, S. Sitch, B. Smith, J. Smith, P. Smith, M.T. Sykes, K. Thonicke, W. Thuiller, G. Tuck, S. Zaehle and B. Zierl. 2005. "Ecosystem Service Supply and Vulnerability to Global Change in Europe." *Science* 310: 1333–37.

Sohngen, B. and R. Sedjo. 2005. "Impacts of Climate Change on Forest Product Markets: Implications for North American Producers." *The Forestry Chronicle* 81: 669–74.

Spittlehouse, D. 2005. "Integrating Climate Change Adaptation into Forest Management." *The Forestry Chronicle* 81: 691–95.

Spittlehouse, D.L. and R.B. Stewart. 2003. "Adaptation to Climate Change in Forest Management" [online]. B.C. *Journal of Ecosystems and Management* 14: 1–11. [accessed October 5, 2009]. Available from World Wide Web: Available at: *http://www.forrex.org/jem/2003/vo14/no1/art1.pdf*

Steedman, R.J. 2000. "Effects of Experimental Clearcut Logging on Water Quality in Three Small Boreal Forest Lake Trout Lakes." *Canadian Journal of Fisheries and Aquatic Sciences* 57 (Suppl. 2): 92–96.

Stromgren, M. and S. Linder. 2002. "Effects of Nutrition and Soil Warming on Stemwood Production in a Boreal Norway Spruce Stand." *Global Change Biology* 8: 1195–1204.

Sushama, L., R. Laprise, D. Caya, A. Frigon and M. Slivitzky. 2006. "Canadian RCM Projected Climate-Change Signal and its Sensitivity to Model Errors." *International Journal of Climatology* 26: 2141–59.

Taylor, S.W., A. L. Carroll, R.I. Alfaro and L. Safranyik. 2006. "Forest, Climate and Mountain Pine Beetle Outbreak Dynamics in Western Canada." Pp. 67–94 in L. Safranyik and B. Wilson (eds.). *The Mountain Pine Beetle: A Synthesis of Biology, Management, and Impacts on Lodgepole Pine.* Mountain Pine Beetle Initiative, Natural Resources Canada, Canadian Forest Service, Pacific Forestry Centre: Victoria, B.C..

Van Rees, K.C.J. and D. Jackson. 2002. *Response of Three Boreal Tree Species to Ripping and Rollback of Roadways. Prince Albert Model Forest, Final Report.* Prince Albert, SK. 26 pgs.

Verseghy, D.L., N.A. McFarlane and M. Lazare. 1993. "A Canadian Land Surface Scheme for GCMS: II. Vegetation Model and Coupled Runs." *International Journal of Climatology* 13: 347–70.

Volney, W.J.A. and R.A. Fleming. 2000. "Climate Change and Impacts of Boreal Forest Insects." *Agriculture, Ecosystems and Environment* 82: 283–94.

Volney, W.J.A. and K.G. Hirsch. 2005. "Disturbing Forest Disturbances." *The Forestry Chronicle* 81: 662–68.

Weber, M.G. and M.D Flannigan. 1997. "Canadian Boreal Forest Ecosystem Structure and Function in a Changing Climate: Impacts on Fire Regimes." *Environmental Reviews* 5: 45–166.

Williamson, T., S. Colombo, P. Duinker, P. Gray, R. Hennessey, D. Houle, M. Johnston, A. Ogden and D. Spittlehouse. 2009. *Climate Change and Canada's Forests: From Impacts to Adaptation.* Sustainable Forest Management Network and Government of Canada: Edmonton, AB.

Williamson, T.B., J.R. Parkins and B.L. McFarlane. 2005. "Perceptions of Climate Change Risk to Forest Ecosystems and Forest-based Communities." *The Forestry Chronicle* 81: 710–16.

Zhou, L., C. Tucker, R. Kaufmann, D. Slayback, N. Shabanov and R. Myneni. 2001. "Variations in Northern Vegetation Activity Inferred from Satellite Data of Vegetation Index During 1981 to 1999." *Journal of Geophysical Research—Atmospheres* 106: 20,069–83.

CHAPTER 9

TRANSPORTATION: THE PRAIRIE LIFELINE

Danny Blair

BACKGROUND

The transportation network in the Prairies is remarkably extensive and diverse, including a very large network of roads and rails, a substantial number of winter roads, dozens of airports, ferries and boat landings, and a single ocean port. The impressive and perhaps surprising diversity is indicative of the diversity of the region's geography, and the network's extent is directly related to the region's large area.

The public road network consists of more than 540,000 km of two-lane-equivalent roads, accounting for 52% of the national total (Table 1) (Transport Canada, 2008). About 20% of the region's network is paved. Alberta, Saskatchewan and Manitoba account for 69% of the unpaved roads and 27% of the paved roads in Canada. Furthermore, the Prairies account for 26% of all the roads classified as Core or Feeder Routes in the National Highway System (Transport Canada, 2008), with 79% of these primary prairie routes located in Alberta and Saskatchewan (Table 2). Clearly, Alberta and Saskatchewan have very large road networks serving communities distributed across their large expanses. Manitoba has a much smaller road network, concentrated in the southern part of the province (Figure 1)

In contrast, Manitoba has the largest network of winter (ice) roads serving remote and northern communities not serviced by permanent roads. Every year, some 2,200 km of winter roads are constructed in Manitoba, over land and water, at a cost of over $8 million per annum (Manitoba Transportation and Government Services, 2006; Manitoba Infrastructure and Transportation, 2007; 2008)

Length (two-lane equivalent thousand km)			
	Paved	Unpaved	Total
Newfoundland and Labrador	10.6	8.6	19.3
Prince Edward Island	4.3	1.8	6.0
Nova Scotia	18.1	9.0	27.1
New Brunswick	19.5	12.0	31.5
Quebec	81.5	63.2	144.7
Ontario	119.8	71.1	191.0
Manitoba	19.3	67.3	86.6
Saskatchewan	29.5	198.7	228.2
Alberta	61.7	164.6	226.3
British Columbia	48.2	22.9	71.1
Yukon	2.2	3.5	5.8
Northwest Territories	0.9	3.6	4.5
Nunavut	—	0.3	0.3
Prairie Provinces	110.5	430.6	541.1
Total	415.6	626.7	1,042.3

Table 1: Length of Public Road Network in Canada (Transport Canada, 2008)

	Kilometres			
	Core routes	Feeder routes	Northern and remote routes	Total
Newfoundland and Labrador	1,008	298	1,163	2,469
Prince Edward Island	208	188	—	396
Nova Scotia	903	296	—	1,199
New Brunswick	987	835	—	1,822
Quebec	3,447	767	1,436	5,651
Ontario	6,131	706	—	6,836
Manitoba	982	742	368	2,093
Saskatchewan	2,450	—	238	2,688
Alberta	3,992	217	197	4,406
British Columbia	5,869	447	724	7,040
Yukon	1,079	—	948	2,027
Northwest Territories	576	—	847	1,423
Nunavut	—	—	—	—
Prairie Provinces	7,424	959	803	9,186
Total	27,631	4,495	5,921	38,047

Table 2: National Highway System (Transport Canada, 2008)

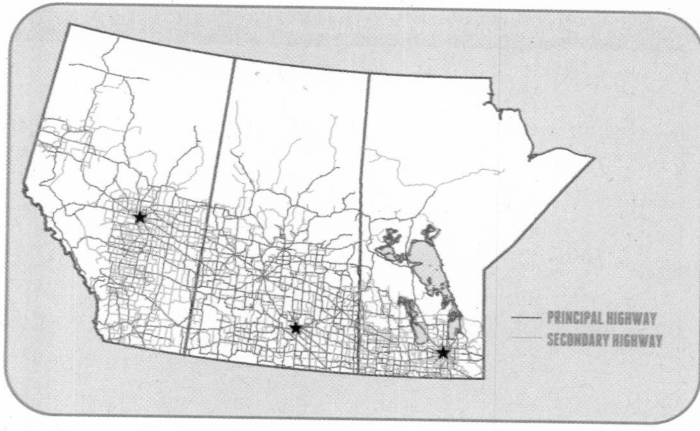

Figure 1. The principal and secondary highway networks in the Prairie Provinces (Note: The data used to make this map were provided by DMTI Spatial Inc.)

(Figure 2). In northern Saskatchewan, almost $2 million is spent annually to operate and maintain less than 300 km of winter roads on land and less than 200 km of ice roads in northern Saskatchewan (Saskatchewan Highways and Transportation, 2007). In Alberta, a 280-km winter road between Fort McMurray and Fort Chipewyan is built annually by the Regional Municipality of Wood Buffalo (RMWB, 2008); a 228-km extension of this winter road is built from Fort Chipewyan to Fort Smith in the Northwest Territories (Parks Canada, 2008). In addition, the Regional Municipality of Wood Buffalo builds a 65-km winter road from Fort McMurray to the Saskatchewan border, where it connects to a winter road built by the Province of Saskatchewan, connecting to the town of La Loche, Saskatchewan (Severs, 2005). Furthermore, countless winter service roads are built by forestry, mining and oil exploration interests in each of the provinces.

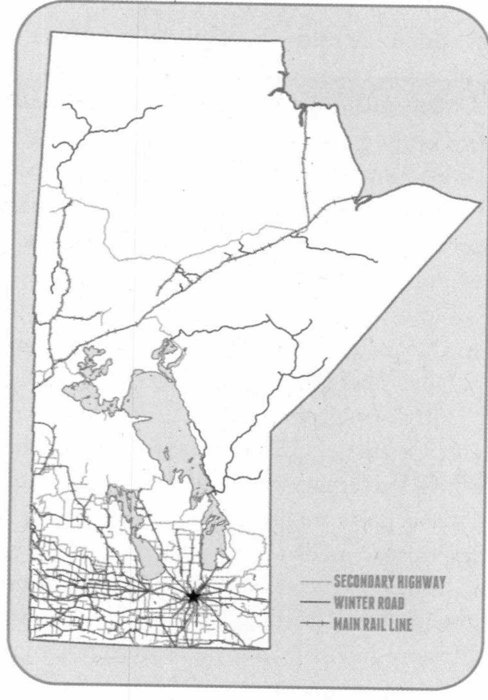

Figure 2. The 2008–09 winter road, secondary highway and main rail line networks in Manitoba (After: Manitoba Infrastructure and Transportation, 2009; DMTI Spatial Inc.)

The Prairies are also home to several well-used ferry oper-

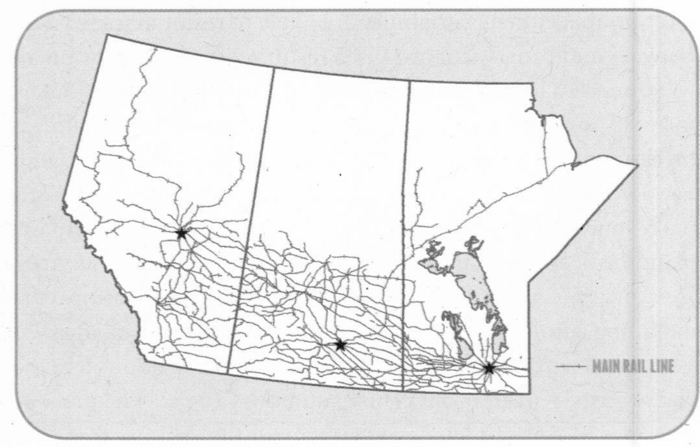

Figure 3. The main rail line network in the Prairie Provinces (The data used to make this map were provided by DMTI Spatial Inc.)

ations. Manitoba provides passenger and vehicle ferry service to five northern communities (Manitoba Infrastructure and Transportation, 2008), and Saskatchewan operates 12 ferries on the Saskatchewan River system and a barge on Wollaston Lake in the north (Saskatchewan Highways and Transportation, 2007). Alberta operates ferries at 12 locations, two of which remain in operation as winter ice bridges over the Peace River (Alberta Infrastructure and Transportation, 2007; 2008).

The railway network in the region is significant, even though many kilometres of rail have been abandoned over the last several decades (Figure 3). In 2006, more than 28,000 km of track was in operation for freight and passenger traffic in the Prairies, comprising 39% of the track in Canada (Statistics Canada, 2008). Importantly, the railway network is connected to Canada's only Arctic seaport, at Churchill, Manitoba (Newton, Fast and Henley, 2002). The port has four deep-sea berths for the loading and unloading of grain, general cargo, and tanker vessels, and is currently in operation from mid-July to the beginning of November (Tivy et al., 2007).

Fifty-one airports with scheduled service were in operation in the region in 2006 (Statistics Canada, 2006). However, there are many small airports without scheduled service scattered throughout the region, especially in the north. These local airports are important links for many remote communities, especially in response to medical emergencies, when they provide air ambulance services.

Transportation is a vital component of almost all economic and social activities, and transportation systems are very sensitive to extreme weather events (Andrey and Mills, 2003a; Field et al., 2007). With climate change resulting in warmer winter temperatures, it is likely that more of the cold-season precipitation will come in the form of rain or freezing rain. Increased frequencies of extreme precipitation events (Kharin and Zwiers, 2000) and increased inter-annual climate variability are likely to result in increased damage to roads, railways and other structures as a result of flooding, erosion and landslides.

Some climate changes may result in economic savings, such as reduced need for road snow clearing, whereas other changes may require significant capital investments, such as improvements to storm-water management (IBI Group, 1990; Marbek Resource Consultants, 2003). In any case, it is clear that substantial resources may need to be committed to the anticipation of changes and the planning of appropriate adaptive strategies (Peterson et al., 2006). Furthermore, since weather is a key component of many transportation-related safety issues, including automobile and aircraft accidents, climate change will affect the risks associated with the transportation of people and goods and, perhaps, the associated costs of insurance (White and Etkin, 1997). The demand for transportation may also be affected, since many of the region's transportation-sensitive sectors, including agriculture, energy and tourism, will also be impacted by climate change, as will those of its trading partners (Wilbanks et al., 2007).

IMPACTS ON INFRASTRUCTURE

Each component of the extensive and diverse transportation infrastructure of the Prairies requires proper design, construction and maintenance to operate as safely and reliably as possible over its design lifetime. There are significant costs associated with transportation infrastructure, most incurred by local, municipal, provincial and federal governments. Private companies and corporations also have major investments in transportation infrastructure, such as in the railway industry.

Arguably, the most significant negative impact of climate change on transportation infrastructure in the Prairies is related to winter roads (Chapter 20). Winter roads are a vital social, cultural and economic lifeline to remote communities (Foster, 1996; Kuryk, 2003; Centre for Indigenous Environmental Resources, 2006). Shipping bulk goods by air is prohibitively costly; thus, the threats that the changing climate pose to winter road operations is a concern to each of the provincial governments mandated to provide surface transport, as well as to the communities currently serviced by these roads. During the warm El Niño winter of 1997–98, a total of $15–18 million was spent airlifting supplies into remote communities in Manitoba and northern Ontario because winter roads could not be built or maintained for sufficient periods of time (Paul and Saunders, 2002; Kuryk, 2003).

Warmer winters may result in substantial reductions in the costs associated with non-ice road infrastructure. Cold temperatures and frequent freeze-thaw cycles cause much of the deterioration of paved and non-paved surfaces (Haas, Li and Tighe, 1999). In the southern parts of the Prairies, where the vast majority of the permanent road surfaces are found, reductions in the length and severity of the frost-affected season could result in long-term cost savings associated with repairs and maintenance. However, winter warm spells or increased inter-daily temperature variability may cause the frequency of freeze-thaw cycles to increase (Canadian Council of Ministers of the Environment, 2003), if only during a period of a few decades, while winters become warmer. In the southern Prairies, it is clear that many of these changes to the frost-related climate are ongoing (Vincent and Mekis, 2006). Additionally, in some northern areas, paved roads are stabilized by frozen substrates during winter, and may therefore be compromised by warmer winter temperatures.

Increases in mean temperature and the frequency of hot days during summer are expected to lead to increased road-related infrastructure costs. Asphalt-covered surfaces, particularly those with large amounts of heavy truck traffic, are especially susceptible to damage during heat waves. Potential problems include the flushing and bleeding of liquid asphalt from poorly constructed surfaces, and rutting of softened surfaces (Lemmen and Warren, 2004)—the most serious and costly type of damage to repair. Each is largely preventable with proper design and construction practices. To date, it is not clear which is likely to be greater in the region: savings associated with less frost-related damage to roads or the costs of increased damage to roads due to warmer summer temperatures (Infrastructure Canada, 2006).

Extreme cold temperatures cause broken railway ties, failure of switches, and physical stress to railway cars; thus, maintenance of railway infrastructure is likely to cost less as a result of warmer winter temperatures. In summer, however, rail damage caused by thermal expansion (Grenci, 1995; Smoyer-Tomic, Kuhn and Hudson, 2003) will likely increase as heat waves become more frequent. More significant, perhaps, is the likelihood that northern railways with lines passing through areas of permafrost, such as the one serving the Port of Churchill in northern Manitoba, will require frequent and significant repair, if not replacement, as a result of continued permafrost degradation (Nelson, Anisimov and Shiklomanov, 2002).

Sea-level rise will impact the shores of Hudson Bay (Overpeck et al., 2006; Bindoff et al., 2007), even though the land is rising due to a high rate of glacioisostatic rebound. On the one hand, the Port of Churchill and its associated facilities may experience more frequent and severe erosion by water and ice, which would affect port infrastructure. On the other hand, the significantly longer ice-free season in Hudson Bay and northern channels resulting from continued climate warming (Arctic Climate Impact Assessment, 2005; Lemke et al., 2007) will increase opportunities for ocean-going vessels to use the Port of Churchill as a point of departure and arrival for grain and other bulk commodities (Tivy et al., 2007).

OPERATIONAL ISSUES

Climate change will potentially affect transportation service availability, scheduling, efficiency, and safety (e.g. Andrey and Mills, 2003b). All modes of transport in the Prairies and elsewhere (Committee on Climate Change and U.S. Transportation, 2008) are at least occasionally unavailable, or schedules are disrupted, due to weather-related events. The majority of weather-related delays and cancellations occur in the winter, usually as a result of heavy snowfalls, blizzards and freezing rain, but also during extreme cold snaps. With warmer winters, there should be fewer and shorter delays. There is some evidence that a warmer climate will be associated with fewer and less intense blizzards (Lawson, 2003). If this is correct, there may be substantial savings to the transportation industry, particularly in the airline and trucking sectors. For trucking, there are often significant costs and penalties associated with delayed shipments of, for example, perishable produce.

A warmer climate may result in fewer weather-related accidents, injuries and fatalities (Mills and Andrey, 2002), particularly in the winter, if snowfall events become less intense and frequent. Traffic accidents are strongly and positively correlated with precipitation frequency (Andrey et al., 2003). Snowfall events account for a large proportion of the reductions in road traffic efficacy and safety in Canada (Andrey et al., 2003), and are associated with large road-clearing expenditures. For example, in the winter of 2005–06, the Manitoba government conducted 1,455,193 pass kilometres of snow clearing and 220,945 pass kilometres of ice blading; almost 57,000 tonnes of de-icing chemicals were applied to provincial highways; 42% of the year's $80 million in road maintenance expenditures was spent in the winter (Manitoba Transportation and Government Services, 2006). Thus, warmer winters with fewer snow events may substantially lower costs associated with the removal of snow and ice from roads (e.g. IBI Group, 1990; Jones, 2003), and the application of less salt on icy roads may substantially reduce damage to vehicles, bridges and other steel structures (Mills and Andrey, 2002). However, these potential savings are very temperature-sensitive, and it remains possible that the number of days requiring the application of salt may, in fact, increase if there is an increase in the number of days with freezing rain. Even if total amounts of precipitation do not change substantially, it is generally acknowledged that the frequency of extreme precipitation events will increase (Groisman et al., 2005), that more of the winter precipitation will fall as rain (Akinremi, McGinn and Cutforth, 1999), and that the distribution of precipitation throughout the year will change (Hofmann et al., 1998). Increased frequency of extreme precipitation events in the summer would likely increase the frequency of road accidents and hinder transportation operations. For example, increased and more intense precipitation in the mountains would likely result in a higher incidence of flash floods and debris flows

(Chapter 23), thereby disrupting key transportation links. Associated with increased storminess would be an increase in the frequency of extreme winds, which would cause air transport delays and risks. More extreme wind events would also negatively affect road and rail transport, for which there are well-established critical thresholds for crosswind hazards (Peterson et al., 2006). The very large number of long-haul trucks crossing the Prairies would likely be the most wind-affected component of the transportation sector.

ADAPTATION: PLANNING FOR THE FUTURE

In some ways, the transportation system in the Prairies is well-prepared for climate change. Specifically, there are well-established systems in place to deal with the many different types of weather-related hazards, some of which may become more frequent in the future climate. For example, provincial governments, municipalities and cities have fleets of plows and salt-spreaders to cope with hazardous winter driving conditions, and real-time road weather information systems have been installed at critical locations; Environment Canada has an effective system for warning the public about unsafe travel conditions that may be associated with extreme weather events; and the federal government provides the aviation industry with detailed information about atmospheric conditions that may threaten safety.

However, none of these responses to hazardous atmospheric conditions have been implemented to directly address risks associated with climate change; they are in place now, and will remain in place in the future, because they address risks that arise from the extremes that characterize the region's climate. However, it seems clear that some of these responses addressing transportation-related risk will be used more frequently in the future, and other practices less often, as the frequency and intensity of weather-related hazards change. It is unlikely that any of these risk management responses will be able to be shelved, since the possibility of heavy snows, blizzards and cold snaps, and other winter-related hazards will always remain, even in a much warmer climate. Thus, it is also unlikely that any jurisdiction will choose to, or be able to, eliminate their fleets of snow plows, but the fleets may be reduced, along with the budgets allocated to these resources (Timilsina and Kralovic, 2005; Infrastructure Canada, 2006). Furthermore, it is conceivable that reduced use of road maintenance fleets may prolong their productive lives and thereby result in smaller budget demands.

Except for the problems associated with winter roads and construction in permafrost regions, little research has been done on climate change adaptation for transportation, and thus climate change scenarios have rarely been incorporated into transportation infrastructure decisions in Canada (Infrastructure Canada, 2006). This gap may reflect levels of uncertainty, or the perceived levels of uncertainty, associated with projected climate changes. Indeed, the fact that adaptation strategies are already being adopted in the north, where climate change is abundantly evident, indicates that observable and experienced change is a powerful motivator. For example, the Province of Manitoba has committed to relocating many of its northern ice roads to land-based routes (Manitoba Infrastructure and Transportation, 2007; 2008), improving safety and providing opportunities to convert at least some of the winter roads to all-season roads in the future. Also, Manitoba's response was at least partially motivated by some very short winter road seasons in the late 1990s and early 2000s, in which very costly airlifts were required to deliver supplies to remote communities (Kuryk, 2003; Paul and Saunders, 2002; Stern, 2007). There has also been a substantial amount of discussion around the possibility of using dirigibles (commonly known as blimps) to deliver bulk goods to remote communities in Manitoba (Prentice and Turiff, 2002). Dirigibles can be quite sensitive to weather conditions, especially high winds, but they are an intriguing alternative to the costly and sometimes dangerous use of snow- and ice-based winter roads.

The potential negative effects of more frequent freeze-thaw cycles on Prairie roads is likely one of the most important impacts in need of concern and investment (Andrey et al., 1999). The costs associated with more frequent repairs to frost-damaged roads could be very high, both in rural and urban settings (Infrastructure Canada, 2006). New construction designs and materials could be implemented to prevent at least some of the frost damage, but it appears that few, if any, of these options have been implemented to date, anywhere. Similarly, more intense precipitation events in the future may result in more frequent and more severe erosion to roads and railways, but it is not evident that these climate change issues are being taken into account in current planning practices. However, at least some of these issues are indirectly being taken into account with the increased prevalence of the risk management approach in planning and decision-making processes (University of Manitoba Transport Institute, 2003; Prentice et al., 2003; Prentice, Cohen and Duncan, 2005; Noble, Bruce and Egener, 2005), including those related to transportation infrastructure.

SUMMARY

Climate change presents substantial risk to the transportation sector in the Prairies. In part, this is related to the sensitivity of the sector's infrastructure and operations to weather. The projected increase in the frequency of intense precipitation events will likely result in more erosion of roads and railways, and the projected increase in the frequency of heat waves will likely result in more frequent heat-related degradation to railways and asphalt-based roads. Thawing of permafrost will present operational and maintenance risk to northern railways, and there is evidence that winter (ice) road operations are being negatively affected by warmer and shorter winters. More frequent freeze-thaw cycles, at least in the short term, may result in higher roadway maintenance and repair costs, as would the projected increase in the frequency of winter-rainfall events. Coastal erosion associated with sea-level rise may impact the port at Churchill, and the airline industry may be affected by more frequent delays and risks associated with stormy weather.

However, climate change may also reduce some risks. Fewer extreme cold events will likely reduce the cost of winter-railway maintenance, and fewer blizzards would reduce the frequency of transportation delays, both on the ground and in the air. In the long run, much warmer winters may have fewer freeze-thaw cycles, thereby reducing the cost of roadway repairs, and shorter winters may reduce the costs associated with the plowing of snow from streets and highways. Furthermore, a longer ice-free season in the north may have positive benefits to the Port of Churchill and its associated industries.

In short, other than the negative impacts on winter roads, there is a great deal of uncertainty concerning the long-term net effects of climate change on the transportation sector in the Prairies and elsewhere. Fortunately, because the transportation sector in the prairie region has always had to deal with a large amount of climate variability, it could be said that it is already quite resilient and will, therefore, be able to adapt to many of the anticipated changes. Unfortunately, the magnitude of some the projected changes will undoubtedly exceed the existing level of adaptive capacity and will necessitate substantial amounts of change to infrastructure, operations, budgeting, and planning. To date, other than with the aforementioned winter roads, there is little evidence that these long-term problems (and opportunities) are being extensively evaluated and acted upon by governments at any level.

ACKNOWLEDGEMENTS

Thanks to Weldon Hiebert and Ryan Smith for producing the maps, DMTI Spatial Inc. for providing free access to their data, and Stephen Winnemuller for research assistance.

REFERENCES

Akinremi, O.O., S.M. McGinn, and H.W. Cutforth. 1999. "Precipitation Trends on the Canadian Prairies." *Journal of Climate* 12, no. 10: 2996–3003.

Alberta Infrastructure and Transportation. 2007. *Annual report: 2006–2007* [online]. Alberta Infrastructure and Transportation, 113 pgs. [accessed May 1, 2008]. Available from World Wide Web: *http://www.infratrans.gov.ab.ca/INFTRA_Content/Publications/Production/annualreport.pdf*

——. 2008. Personal Communication. Alberta Infrastructure and Transportation, Communications Branch, May 12, 2008.

Andrey, J. and B. Mills. 2003a. "Climate Change and the Canadian Transportation System: Vulnerabilities and Adaptations." Pp. 235–79 in J. Andrey and C.K. Knapper (eds.). *Weather and Road Transportation.* University of Waterloo, Department of Geography Publication Series, Monograph 55: Waterloo, ON.

Andrey, J. and B. Mills. 2003b. *Collisions, Casualties, and Costs: Weathering the Elements on Canadian Roads* [online]. Institute of Catastrophic Loss Reduction Research, Paper Series, No. 33. 31 pgs. [accessed May 30, 2007]. Available from World Wide Web: *http://www.iclr.org/pdf/AndreyMills_Collisions-casualties-costs_ICLR-2003-report_July4–03.pdf*

Andrey, J., B. Mills, with B. Jones, R. Haasand and W. Hamlin. 1999. *Adaptation to Climate Change in the Canadian Transportation Sector.* Report submitted to Natural Resources Canada, Adaptation Liaison Office, Natural Resources Canada, Ottawa; Transport Canada; Lemmen and Warren.

Andrey, J., B. Mills, M. Leahy and J. Suggett, J. 2003. "Weather as a Chronic Hazard for Road Transportation in Canadian Cities." *Natural Hazards* 28, nos. 2–3: 319–43.

Arctic Climate Impact Assessment. 2005. *Arctic Climate Impact Assessment.* Cambridge University Press: Cambridge, United Kingdom. 1042 pgs.

Bindoff, N.L., J. Willebrand, V. Artale, A, Cazenave, J. Gregory, S. Gulev, K. Hanawa, C. Le Quéré, S. Levitus, Y. Nojiri, C.K. Shum, L.D. Talley and A. Unnikrishnan. 2007. "Observations: Oceanic Climate Change and Sea Level." In S. Solomon, D. Qin, M. Manning, Z. Chen, M. Marquis, K.B. Averyt, M. Tignor and H.L. Miller (eds.). Climate *Change 2007: The Physical Science Basis.* Contribution of Working Group I to the Fourth Assessment Report of the Intergovernmental Panel on Climate Change. Cambridge University Press: Cambridge, New York.

Canadian Council of Ministers of the Environment. 2003. *Climate, Nature, People: Indicators of Canada's Changing Climate.* Canadian Council of Ministers of the Environment: Winnipeg, Manitoba. 45 pgs.

Centre for Indigenous Environmental Resources. 2006. *Climate Change Impacts on Ice, Winter Roads, Access Trails, and Manitoba First Nations, Final Report, November 2006.* Natural Resources Canada and Indian and Northern Affairs Canada, 210 pgs.

Committee on Climate Change and U.S. Transportation. 2008. *Potential Impacts of Climate Change on U.S. Transportation.* Transportation Research Board Special Report 290, National Research Council of the National Academies: Washington, DC. 220 pgs.

Field, C.B., L.D. Mortsch, M. Brklacich, D.L. Forbes, P. Kovacs, J.A. Patz, S.W. Running and M.J. Scott. 2007. "North America." Pp. 617–52 in M.L. Parry, O.F. Canziani, J.P. Palutikof, P.J. van der Linden and c.e. Hanson (eds.). *Climate Change 2007: Impacts, Adaptation and Vulnerability. Contribution of Working Group II to the Fourth Assessment Report of the Intergovernmental Panel on Climate Change.* Cambridge University Press: Cambridge, UK.

Foster, R.H. 1996. "Winter Roads." Pp. 175–76 in J. Welsted, J. Everitt and C. Stadel (eds.). *The Geography of Manitoba: Its Land and Its People.* The University of Manitoba Press: Winnipeg, Manitoba.

Grenci, L. 1995. "Planes, Trains and Automobiles." *Weatherwise* 48: 48.

Groisman, P.Y., R.W. Knight, D.R. Easterling, T.R. Karl and V.N. Razuvaev. 2005. "Trends in Intense Precipitation in the Climate Record." *Journal of Climate* 18, no. 9: 1326–50.

Haas, R., N. Li. and S. Tighe. 1999. *Roughness Trends at C-SHRP LTPP Sites.* Final Project Report. Roads and Transportation Association of Canada: Ottawa, Ontario. 97 pgs.

Hofmann, N., L. Mortsch, S. Donner, K. Duncan, R. Kreutzwiser, S. Kulshreshtha, A. Piggott, S. Schellenberg, B. Schertzerand and M. Slivitzky. 1998. "Climate Change and Variability: Impacts on Canadian Water." Pp. 1–120 in G. Koshida and W. Avis (eds.). *The Canada Country Study: Climate Impacts and Adaptation, Volume VII: National Sectoral Issue.* Environment Canada.

IBI Group. 1990. "The Implications of Long-Term Climatic Changes on Transportation in Canada." Environment Canada, *Climatic Change Digest*, CCD90-02. 8 pgs.

Infrastructure Canada. 2006. *Adapting Infrastructure to Climate Change in Canada's Cities and Communities: A Literature Review.* Infrastructure Canada, Research and Analysis Division. 22 pgs.

Jones, B. 2003. "The Cost of Safety and Mobility in Canada: Winter Road Maintenance." In J. Andrey and C.K. Knapper (eds.). *Weather and Road Transportation.* University of Waterloo, Department of Geography Publication Series, Monograph 55: Waterloo, ON.

Kharin, V.V. and F.W. Zwiers. 2000. "Changes in the Extremes in an Ensemble of Transient Climate Simulations with a Coupled Atmosphere-Ocean GCM." *Journal of Climate* 13: 3760–88.

Kuryk, D. 2003. "Seasonal Transportation to Remote Communities—What If?" Pp. 40–49 in B.E. Prentice, J. Winograd, A. Phillips and B. Harrison (eds.). *Moving Beyond the Roads: Airships to the Arctic Symposium II Proceedings*, October 21–23, 2003, University of Manitoba Transport Institute: Winnipeg, Manitoba.

Lawson, B.D. 2003. "Trends in Blizzards at Selected Locations on the Canadian Prairies." *Natural Hazards* 29, no. 2: 123–38.

Lemke, P., J. Ren, R.B. Alley, I. Allison, J. Carrasco, G. Flato, Y. Fujii, G. Kaser, P. Mote, R.H. Thomas and T. Zhang. 2007. "Observations: Changes in Snow, Ice and Frozen Ground." In S. Solomon, D. Qin, M. Manning, Z. Chen, M. Marquis, K.B. Averyt, M. Tignor and H.L. Miller (eds.). *Climate Change 2007: The Physical Science Basis.* Contribution of Working Group I to the Fourth Assessment Report of the Intergovernmental Panel on Climate Change. Cambridge University Press: Cambridge, New York.

Lemmen, D.S. and F.J. Warren. 2004. *Climate Change Impacts and Adaptation: A Canadian Perspective* [online]. Government of Canada. 174 pgs. [accessed May 6, 2007]. Available from World Wide Web: *http://adaptation.nrcan.gc.ca/perspective/index_e.php*

Manitoba Infrastructure and Transportation. 2007. *Annual Report: 2006–2007* [online]. Manitoba Transportation and Government Services. 129 pgs. [accessed February 1, 2009]. Available from World Wide Web: *http://www.gov.mb.ca/mit/reports/annual/2006annual.pdf*

——. 2008. *Annual Report: 2007–2008* [online]. Manitoba Transportation and Government Services. 132 pgs. [accessed February 1, 2009]. Available from World Wide Web: *http://www.gov.mb.ca/mit/reports/annual/2007annual.pdf*

——. 2009. *Winter Roads Map.* [accessed October 10, 2009]. Available from World Wide Web: *http://www.gov.mb.ca/mit/winter/maps.html*

Manitoba Transportation and Government Services. 2006. *Annual Report: 2005–2006* [online]. Manitoba Transportation and Government Services. 127 pgs. [accessed June 5, 2007]. Available from World Wide Web: *http://www.gov.mb.ca/tgs/documents/tgsannua105.pdf*

Marbek Resource Consultants. 2003. *Impacts of Climate Change on Transportation in Canada* [online]. Final workshop report prepared for Transport Canada, Canmore Workshop, January 30–31, 2003: Canmore, Alberta. 35 pgs. [accessed June 19, 2007]. Available from World

Wide Web: *http://www.tc.gc.ca/programs/environment/nwicct/docs/FullWorkshopReport/Full%20Workshop%20Report.pdf*

Mills, B. and J. Andrey, J. 2002. "Climate Change and Transportation: Potential Interactions and Impacts" [online]. Pp. 77–88 in *The Potential Impacts of Climate Change on Transportation.* Federal Research Partnership Workshop, October 1–2, 2002, Summary and Discussion Papers. Department of Transportation Center for Climate Change and Environmental Forecasting. [accessed June 5, 2007]. Available from World Wide Web: *http://climate.dot.gov/workshop1002/workshop.pdf*

Nelson, F.E., O.A. Anisimov and N.I. Shiklomanov. 2002. "Climate Change and Hazard Zonation in the Circum-Arctic Permafrost Regions." *Natural Hazards* 26, no. 3: 203–25.

Newton, S.T., H. Fast and T. Henley. 2002. "Sustainable Development for Canada's Arctic and Subarctic Communites: A Backcasting Approach to Churchill, Manitoba." *Arctic* 55, no. 3: 281–90.

Noble, D., J. Bruce and M. Egener. 2005. *An Overview of the Risk Management Approach to Adaptation to Climate Change in Canada* [online]. Prepared for Natural Resources Canada, Climate Change Impacts and Adaptation Directorate. 28 pgs. [accessed May 1, 2008]. Available from World Wide Web: *http://adaptation.nrcan.gc.ca/pdf/29156ce6051f409990f872d838bcbbbb_e.pdf*

Overpeck, J.T., B.L. Otto-Bliesner, G.H. Miller, D.R. Muhs, R.B. Alley and J.T. Kiehl. 2006. "Paleoclimatic Evidence for Future Ice-sheet Instability and Rapid Sea-level Rise." *Science* 311, no. 5768: 1747–50.

Parks Canada. 2008. Wood Buffalo National Park of Canada, Visitor Information [online]. [accessed May 1, 2008]. Available from World Wide Web: *http://www.pc.gc.ca/pn-np/nt/woodbuffalo/visit/visit1a_E.asp*

Paul, A. and C. Saunders. 2002. "Melting Ice Roads Pose Manitoba Supplies Emergency." *The Edmonton Journal* (January 14), A5.

Peterson, T.C., M. McGuirk, T. Houston, A.H. Horvitz and M.F. Wehner. 2006. *Climate Variability and Change with Implications for Transportation* [online]. National Oceanic and Atmospheric Administration and Lawrence Berkeley Laboratory. [accessed May 1, 2008]. Available from World Wide Web: *http://onlinepubs.trb.org/onlinepubs/sr/sr290Many.pdf*

Prentice, B.E., S. Cohen and D.B. Duncan (eds.). 2005. *Sustainable Northern Transportation: Airships to the Arctic Symposium III,* Proceedings. University of Manitoba Transport Institute: Winnipeg, MB. 141 pgs.

Prentice, B.E. and S. Turriff (eds.). 2002. *Airships to the Arctic Symposium,* Proceedings. University of Manitoba Transport Institute: Winnipeg, MB. 180 pgs.

Prentice, B.E., J. Winograd, A. Phillips and B. Harrison (eds.). 2003. *Moving Beyond the Roads: Airships to the Arctic Symposium II,* Proceedings. University of Manitoba Transport Institute: Winnipeg, MB. 141 pgs.

RMWB. 2008. *Regional Municipality of Wood Buffalo, Getting Around* [online]. Regional Municipality of Wood Buffalo. [accessed May 1, 2008]. Available from World Wide Web: *http://www.woodbuffalo.ab.ca/residents/getting_around/index.asp*

Saskatchewan Highways and Transportation. 2007. *Annual Report: 2006–2007* [online]. Saskatchewan Highways and Transportation. 38 pgs. [accessed May 1, 2008]. Available from World Wide Web: *http://www.publications.gov.sk.ca/details.cfm?p=23632*

Severs, Laura. 2005. "Northern Road Gives Residents Hope for Future" [online]. *Business Edge* 1, no. 2 (September 25) [accessed May 1, 2008]. Available from World Wide Web: *http://www.businessedge.ca/article.cfm/newsID/10781.cfm*

Smoyer-Tomic, K.E., R. Kuhn and A. Hudson. 2003. "Heat Wave Hazards: An Overview of Heat Wave Impacts in Canada." *Natural Hazards* 28, nos. 2–3: 463–85.

Statistics Canada. 2006. *Air Carrier Traffic at Canadian Airports* [online]. Statistics Canada. 26 pgs. [accessed May 1, 2008]. Available from World Wide Web: *http://www.statcan.ca/english/freepub/51-203-XIE/51-203-XIE2006000.htm*

——. 2008. *Rail Transportation, Length of Track Operated for Freight and Passenger Transportation, by Province and Territory (kilometres)* [online]. Statistics Canada, catalogue no. 52–216-XIE. [accessed May 1, 2008]. Available from World Wide Web: *http://www40.statcan.ca/101/cst01/trad47a.htm*

Stern, N. 2007. *The Economics of Climate Change: The Stern Review*. Cambridge University Press, Cambridge, UK.

Timilsina, G.R. and P.R. Kralovic. 2005. *Potential Effects of Climate Change on the City of Calgary: Adapting to a New Environment* [online]. Canadian Energy Research Institute. 35 pgs. [accessed May 1, 2008]. Available from World Wide Web: *http://www.imaginecalgary.ca/library/imagineCALGARY_Climate_Change_Megatrend_Paper.pdf*

Tivy, A., B. Alt, S. Howell, K. Wilson and J. Yackel. 2007. "Long-range Prediction of the Shipping Season in Hudson Bay: A Statistical Approach." *Weather and Forecasting* 22, no. 5: 1063–75.

Transport Canada. 2008. *Transportation in Canada 2008: An Overview Addendum* [online]. Transport Canada, TP 14816E. 124 pgs. [accessed October 15, 2009]. Available from World Wide Web: *http://www.tc.gc.ca/policy/reports/aca/anre2008/pdf/addendum.pdf*

University of Manitoba Transport Institute. 2003. *Proceedings of the Transportation & Climate Change in Manitoba—2003 Workshop*. Prepared for Manitoba Transportation & Government Services: Winnipeg, MB, March 12, 2003. 115 pgs.

Vincent, L.A. and E. Mekis. 2006. "Changes in Daily and Extreme Temperature and Precipitation Indices for Canada over the Twentieth Century." *Atmosphere-Ocean* 44, no. 2: 177–93.

White, R. and D. Etkin. 1997. "Climate Change, Extreme Events and the Canadian Insurance Industry." *Natural Hazards* 16, nos. 2–3: 135–63.

Wilbanks, T.J., P. Romero Lankao, M. Bao, F. Berkhout, S. Cairncross, J-P. Ceron, M. Kapshe, R. Muir-Wood and R. Zapata-Marti. 2007. "Industry, Settlement and Society." Pp. 357–90 in M.L. Parry, O.F. Canziani, J.P. Palutikof, P.J. van der Linden and c.e. Hanson (eds.). *Climate Change 2007: Impacts, Adaptation and Vulnerability. Contribution of Working Group II to the Fourth Assessment Report of the Intergovernmental Panel on Climate Change*. Cambridge University Press: Cambridge, UK.

CHAPTER 10

THE SOCIAL LANDSCAPE

Debra Davidson

INTRODUCTION

All discussions about climate change, particularly discussions about mitigation and adaptation, eventually come around to the topic of people. Societies have contributed significantly to the causes of climate change, they will be—and already have been—impacted by climate change in a host of direct and indirect ways, and the nature of future impacts to social and ecological systems alike will be determined to a great extent by the sensitivity and adaptive capacity of those social systems. In this chapter, we describe the sensitivity and adaptive capacity of social systems on the Canadian Prairies, but we begin in the introduction with a brief look at what exposure means for human systems. Subsequent sections of the chapter entail an overview of the social dimensions of climate change vulnerability on the Canadian Prairies through a multi-scale analysis, beginning with the individual/household level, and then moving on to the community level. The policy process, a third and, arguably, the most important layer of the social landscape, is discussed elsewhere (Chapter 12).

The exposure of social systems to climate change is defined by the type of climate change event. In addition, because nearly all climate change events, even temperature change, manifest themselves in regionally specific ways, defining exposure also requires identifying the location of human settlements—our first layer of the social landscape. If there are no people in a particular flood-prone region, then it stands to reason that people will not be directly exposed to the floods in question. As our human settlements are located geographically in areas that we anticipate will be the sites of specific types of climate-related events, assessing vulnerability requires describing this social landscape and the anticipated climatic events to which those people will be directly exposed. As with the rest of Canada, the population of the Prairies is heavily concentrated in urban centres, which themselves are concentrated in the southern part of the region. Therefore, climatic events that are specific to this type of community and this geographic region, particularly heat waves and drought, will have a direct effect on a large proportion of the prairie population. There are also a number of small—and not-so-small—communities in the north, however, which will be exposed to forest fires, in addition to increasing uncertainty due to water availability and changes in permafrost.

However, the exposure of social systems to the effects of climate change is not quite that clear cut. The examples just described are a form of *direct* exposure. In addition to direct exposure, social systems will also face *indirectly* the repercussions of changes to the full gamut of natural resources and amenities of value to those people. The natural resources upon which our communities and economies depend, and the land bases upon which we enjoy cultural and recreational activities, are also facing multiple types of climate change events. An evaluation of vulnerability thus involves description of the geography, both of the human communities and the social values ascribed to the extended landscape of prairie ecosystems and natural resources, and the intersection of this geography with anticipated regional expression of climatic events.

Evaluation of the other two pillars of vulnerability—sensitivity and adaptive capacity—is even more complex, and tends to be the predominant focus of social studies of climate change vulnerability. As can be expected, considering the relative youth of this area of research, combined with the scale and complexity of the subject matter, there is some scholarly disagreement regarding the definitions of these concepts. The intent of this chapter is not to engage in this conceptual discussion, but rather to apply these concepts to the Canadian Prairie context. (For readers who are interested in these discussions see, among others, Adger et al., 2004; Cutter, Boruff and Shirley, 2003. See also chapters 13 and 21.) As such, we provide a brief description of the means by which these concepts are interpreted and employed in the current assessment.

The concept of sensitivity defines differences in impact from exposure to the same event. In other words, two drought events of the same scale and magnitude, but occurring either in different places or at different points in history, can have demonstrably different impacts on social systems. The severity of impact can be influenced by everything from the degree of economic dependence on water-intensive industries, to water storage capacity, to cultural predispositions toward conservation. We like to think of sensitivity factors as generally structural in nature: characteristics of social systems that are deeply embedded and relatively difficult to change, at least in the short to medium term.

Factors that the researcher may define as sensitivity and those defined as adaptive capacity may actually overlap to a certain extent in empirical analysis. Whether a given social characteristic is a source of sensitivity, or a source of adaptive capacity, in other words, is often subject to the interpretation of the individual research team. However, adaptive capacity, at least in the eyes of the theorist, is something quite different from sensitivity. Adaptive capacity describes the potential to adjust to threats, either to reduce exposure or sensitivity, or both. While some communities may respond to current or anticipated future climatic events by, for example, developing evacuation plans, protecting those most vulnerable, and improving infrastructure, other similar communities may take no such steps. In contrast to sensitivity factors, adaptive capacity factors describe characteristics of social systems that are more dynamic.

The Third Report of the IPCC, however, has identified a set of characteristics of social systems that influence their adaptive capacity, characteristics that could inform us about the potential for adaptive capacity that may exist in the Prairies. These determinants of adaptation "influence the occurrence and nature of adaptation and thereby circumscribe the vulnerability of systems and their residual impacts" (IPCC, 2001: 893). They include a range of assets such as economic resources, technology, human capital, infrastructure, equity, and institutions. The adaptive capacity of a social system depends not only on the level and quality of each one of these assets, but also on the way in which they combine with each other to facilitate the sustainability of people and natural resources.

Another important determinant of adaptive capacity identified in the literature on adaptation and vulnerability is social capital. Social capital includes all those features of social life—social relations, formal and informal networks, group membership, knowledge sharing, and values such as trust, reciprocity and civic engagement—that promote collective action and social cooperation (Glaeser, 2001; Portes, 1998; Field, Schuller and Baron, 2000). Social capital has been considered a robust determinant of adaptive capacity. Adger (2001, 2003a, 2003b) argues that society already has the capacities to adapt to climate change, and that these capacities are "bound up in the ability of societies to act collectively" (2003a: 29), which involves the existence of social capital. Social capital facilitates the establishment of collective objectives and the organization of groups in the pursuit of those objectives, contributing to the management of scarce resources and coping with the negative outcomes of climate change. Moreover, it could complement, and even substitute, a state's efforts in terms of dealing with climatic hazards (see Adger, 2000; Sygna, 2005).

A discussion of the complete universe of social characteristics that can be described as contributors to adaptive capacity is beyond the purpose of this chapter. However, the determinants identified by the IPCC and social capital can serve as a reasonably robust proxy for adaptive capacity in discussing the Prairies' adaptive capacity.

GEOGRAPHY OF THE SOCIAL LANDSCAPE

The Prairies are relatively more rural than other regions of Canada, with 35% of the population of Saskatchewan living outside of urban areas according to Statistics Canada's 2006 Census, and 29% in Manitoba. Alberta's degree of urbanization is more typical of the rest of Canada, with 18% of the population living outside urban areas (Statistics Canada, 2006). The size of prairie cities is relatively small, with only Calgary and Edmonton housing more than one million residents. A number of rural communities are "booming"—experiencing rapid population and economic growth—while others, particularly some older agricultural communities in the southern Prairies, are in decline. Communities on either end of this spectrum will be vulnerable to climate change for different reasons (Ford and Smit, 2004).

Many rural communities are located in the south, and will thus face similar forms of regional exposure to the cities. Those agricultural regions that are currently marginal from a water availability standpoint may no longer be feasibly cultivated, while we may see a northward migration of agricultural productive zones, provided the availability of cultivable soils. A smaller number of rural communities are located in the north. For these communities, the loss of permafrost poses an immediate threat to transportation infrastructure. In addition, forest ecosystems, being particularly sensitive to changes in the water cycle, are likely to suffer from drought-weakened vegetation, and will face the compounded impact of increased outbreaks of pests that will be capable of over-wintering during milder years. Both of these trends will contribute to susceptibility to forest fires, with immediate implications for health and safety of residents, and longer-term implications for those dependent on the forest industry, and/or on subsistence and recreational activities that depend on the forest ecosystem.

While the degree of regional economic diversity varies, overall, the Prairies are heavily dependent upon natural resource-based industries covering an extensive landscape that will be exposed to a host of climate change events. Much of the southern prairie land base is dominated by agriculture and, in Alberta, conventional oil and gas drilling. Moving into the north and into the foothills of the Rockies, agriculture gives way to forestry and mining. In combination, these industries are a significant component of the prairie economy. This is illustrated with the proportion of exports derived from timber, mining, energy, and agriculture. Using data from the Department of Foreign Affairs and International Trade (2003), Wellstead (2006) reports that over the 1997–2001 period, average provincial export dependency on these natural resource-based industries was 79.95% for Alberta, 88.56% for Saskatchewan, and 57.41% for

Manitoba. Many prairie residents rely on these industries as sources of employment, especially in rural communities. Over 25% of the jobs in rural communities in Canada are in resource-based industries, and a far greater proportion is employed in sectors that service these industries. In the prairie region, 78% of resource-related jobs are in agriculture (Stedman, Parkins and Beckley, 2004).

However, material dependence is not the only source of value ascribed to prairie ecosystems. Prairie residents, as do all Canadians, also value highly their ecosystems for recreational purposes, and for ecosystem benefits. According to Environment Canada's Survey on the Importance of Nature to Canadians (DuWors et al., 1996), 84.6% of the Canadian population takes part in one or more trips to engage in nature-based activities during the year prior to the time of the survey, and the average was 13.3 trips per person for the year. A small but significant number of residents also rely on their local ecosystems for the provision of subsistence, particularly Aboriginal peoples. Many of these residents rely on subsistence food sources not solely by cultural preference, but by necessity. In remote regions, where many small communities are located, the food available in local grocery stores is expensive, of limited availability, and typically of poorer nutritional value than traditional foods harvested from the local ecosystem (Desjardins and Govindaraj, 2005).

VULNERABILITY AT THE INDIVIDUAL/HOUSEHOLD LEVEL

We can refer to the 2006 and 2001 Census data files, provided by Statistics Canada (2006a), to access several indicators of sensitivity, including population change; the percentage of selected vulnerable populations; unemployment; and employment in primary industries. The Census can also provide us with several indicators related to the IPCC's determinants of adaptive capacity, including educational attainment; median family income levels; rental rates; and mobility. Each of these is described further below (See Table 1).

	Alberta	Manitoba	Saskatchewan
Total Population	3,290,350	1,148,401	968,157
Population Change (2001–06)	10.6%	2.6%	-1.1%
Recent Immigrants as a % of Provincial population	3%	3%	1%
Aboriginal Identity as a % of Provincial population	6%	15%	15%
Lone-Parent Families as a % of total Census families	14%	17%	17%
Unemployment	4.3%	5.5%	5.6%
Primary Employment as a % of total employment	6%	6%	13%
University Education	17%	15%	13%
Median Family Income (2001)	$73,823	$58,816	$58,563
Rental Rate	26%	28%	30%
Mobility	52%	63%	64%

Table 1. Select vulnerability indicators for the Prairie Provinces, based on Statistics Canada's 2006 Census

SENSITIVITY

Social systems have difficulty adjusting to rapid changes in their population, and this has implications for the individuals living in those social systems. Rapid population growth, such as is occurring in communities in northeast Alberta, can place strains on social services, making it, for example, more difficult to access a family physician. On the other end of the spectrum, population declines, currently being experienced by several older agricultural-based communities, can mean reductions in social support networks. Particular declining trajectories, such as the exodus of youth from agricultural regions, can threaten the very sustainability of those communities. In the Prairie Provinces, many parts of Alberta are feeling the strain of rapid population growth, while most rural communities in Saskatchewan, and select agricultural communities in the other two provinces as well, are attempting to adjust to declines in population. These represent pockets of sensitivity in both cases.

We can also anticipate members of certain sub-populations to have particularly high sensitivity and/or low adaptive capacity to the extreme events and long-term changes associated with climate change. We consider just a handful of those populations here, including recent immigrants (those arriving between 2001 and 2006). Recent immigrants face a number of challenges in having access to social and economic assets, including language, employment and cultural barriers, which may restrict the ability to respond to extreme events associated with climate change. The Prairie Provinces have a relatively small proportion of recent immigrant families, but those who do reside in the Prairies are almost entirely located in the urban centres, where the access to support services for immigrants is greatest. Aboriginals are also considered particularly vulnerable, due to their high level of reliance on local ecosystems for subsistence and cultural vitality; their relatively high level of dependence on primary industry employment; and the tendency for Aboriginal communities to be associated with high unemployment, etc. Individuals of Aboriginal identity are heavily represented in the Prairies overall, particularly in Saskatchewan and Manitoba. Lone-parent families, whose incomes tend to be much lower than two-parent families, are also considered a vulnerable group. The proportion of lone-parent families tends to be relatively consistent across the Prairie Provinces, and does not vary significantly from the national average of 16%.

Unemployment in the Prairies overall is not likely to pose a significant source of household vulnerability: as of the 2006 Census, both Manitoba and Saskatchewan had unemployment rates approximating the national average, while Alberta's was below the national average. Employment in primary industries is particularly high in Saskatchewan, and high in several communities in the other two provinces as well. Employment in primary industries poses a source of sensitivity for families because such industries tend to be subject to the vagaries of dynamic global raw materials markets, and are historically associated with "boom and bust" cycles, but also because at least two of these—forestry and agriculture—are themselves especially sensitive to climate change, reliant as they are on things like water supply and intact forests.

ADAPTIVE CAPACITY

Sensitivity can be addressed through adaptation, however, and the Census data do suggest certain types of adaptive capacity in the Prairie Provinces. Education is an important indicator of human capital, suggesting the resources that households have to secure gainful employment, and address emergencies. Education rates, as represented by the percentage of the population 15 years and older with a university certificate, diploma, or degree, vary modestly across the Prairie Provinces. The highest rate is in Alberta, not surprisingly, since the Prairies' two largest cities are located there. Income is also an important indicator of adaptive capacity. This indicator was derived from the 2001 Census, as this information has not yet been released for the 2006 Census. The median income level for all families as of 2001 is significantly higher in Alberta than in the other two Prairie Provinces and, as expected, it varies within each of the provinces.

Households who rent their homes may be considered to have lower adaptive capacity, because they are not in a position to make preventative upgrades to their homes, and they also have lower levels of equity to rely on in times of crisis. The percentage of renting households does not vary tremendously; the highest rate is in Saskatchewan. Renters also tend to be more mobile, which is an indication of low collective adaptive capacity for some localities, since these households may be more inclined to move away in times of crisis, rather than contribute to community recovery. The Census also collects information more directly capturing the mobility of the population, by identifying households that lived at the same address five years prior to the Census. High levels of mobility, whether reflected in in-migration, outmigration, or both, provide a potential indicator of weak social ties at the community level—social ties that are central to social capital and hence to adaptive capacity. In this instance, Alberta communities may be more vulnerable, as only 52% of individuals reported having lived at the same address five years before the Census. The rate is much higher in both Manitoba and Saskatchewan.

Taken together, the data above suggest that Alberta could be vulnerable to climate change due to high population growth, and the high mobility of its population—two trends that clearly go together. In all other respects, however, the data reported above suggest lower vulnerability for households in Alberta in comparison to the other two Prairie Provinces. Saskatchewan, with a population that was stable or in slow decline for much of the latter part of the 20th century, relatively high percentages of vulnerable populations, high dependence on employment in primary industries, and the lowest education levels, could be much more vulnerable by comparison.

VULNERABILITY AT THE COMMUNITY LEVEL: CITIES

Cities are especially susceptible to climatic stress (Bigio, 2002), because of the high concentrations of people, and because many of the physical properties of cities augment climate risks. Heat stress, for example, is likely to be exacerbated due to the "heat island effect" (e.g. Arnfield, 2003): concrete structures absorb and radiate heat, resulting in further increases in temperatures. Historically, prairie summers have been mild relative to other regions across Canada. This mild climate has also meant relatively low rates of adaptation to heat stress, such as residential air conditioning and city shelters. Extreme heat days, in other words, may be rare but significant events for prairie cities. Higher temperatures are also of concern for cities because photochemical smog forms under heat, and certain atmospheric pollutants, such as ozone and particulate matter, rise in the heat—posing an additional risk to human health, as discussed in Chapter 11.

Another characteristic of cities is the relatively high proportion of certain vulnerable populations. In the Prairies, the larger cities have higher proportions of people living in poverty, Aboriginal peoples and recent immigrants. While the population of recent immigrants in the Prairies overall is quite small in comparison to central Canada, virtually all of these individuals are located in the larger cities. The one exception to this rule is a rapidly growing recent immigrant population in northern Alberta. Similarly, the proportion of the prairie population that is Aboriginal is relatively small, but a rapidly growing segment of this population lives in prairie cities (Weber, Davidson and Sauchyn, 2008).

Cities in the prairie region also have relatively large areas committed to green space, reflecting the value prairie residents place on them. Our green spaces will be susceptible to long-term climate processes such as the migration of ecozones, and more acute events such as drought, which can place vegetation and wildlife under extreme stress. One of the greatest expenses to prairie cities during the most recent drought was the loss of ornamental trees: a recent report estimated a cost of $10 million to replace the over 16,000 trees that died in Edmonton alone (The Last Link on the Left, 2005.).

Overall, compared to other cities both in Canada and across the world, prairie cities are not densely concentrated, suggesting that damage ensuing from extreme events, though still potentially extensive, are likely to be relatively less severe in comparison to more densely packed cities like New York or Tokyo. We can also anticipate higher levels of adaptive capacity in prairie cities compared to rural areas, as is true of cities in general. Cities have well-developed communication and transportation infrastructures and emergency response capacities, and tend to have the ear of higher-level political institutions (Cross,

2001). That said, a recent study of the adaptive capacity of Canadian cities to climate change found a disturbing lack of knowledge and awareness among decision-makers of the potential impacts of climate change and the need for adaptation strategies (Wittrock, Wheaton and Beaulieu, 2001).

Focusing on Alberta's two largest cities, we can see several examples of adaptation. Edmonton has launched several initiatives to conserve water resources, including residential meters, inclining block rates for water consumption (i.e. the cost per unit of water increases as consumption increases), and a public awareness campaign to educate residents about water conservation (Roach, Huynh and Dobson, 2004; Waller and Scott, 1998). Similar to Edmonton, Calgary uses a variety of instruments to conserve municipal water resources, including financial incentives for homeowners who install water-efficient fixtures and a number of public education programs to encourage conservation. In addition, the city has undertaken a major pilot project in water conservation: the entire neighbourhood of Crestmont in North Calgary has been specifically designed to maximize water efficiency (Roach, Huynh and Dobson, 2004).

A city's capacity to adapt is also influenced by local emergency management planning. Although prairie cities have more people and property exposed to natural hazards than do its rural regions, they also have well-developed emergency management programs to cope with such events. The City of Edmonton's Office of Emergency Preparedness, for example, monitors potential threats to public safety and coordinates the city's resources in response to emergencies. Emergency officials collaborate with neighbouring municipalities to pool resources and develop common procedures. According to the City of Edmonton (2005), after extensive damage caused by flooding in the summer of 2004, Edmonton also implemented a Flood Prevention Program, which involves upgrades to drainage infrastructure in 62 at-risk areas, professional advice for homeowners regarding ways they can improve drainage on their properties, and a public education campaign.

VULNERABILITY AT THE COMMUNITY LEVEL: SMALL TOWNS

While the majority of prairie residents live in cities, this region is still characterized by a large number of small rural communities. These communities are more prevalent in Saskatchewan and Manitoba, both with lower rates of urbanization. Furthermore, many of the rural communities in Alberta are currently in a rapid growth phase as a result of energy industry development. Rural communities in the Prairies may be more sensitive to climate change than their urban counterparts in several respects. For example, the effects of drought may be more severe in dryland rural areas, since this population is more likely to be dependent on well water, or smaller reservoirs. Rural communities are also likely to have limited adaptive capacity: few rural communities have the same population density to support high levels of social capital, or have access to the same level of municipal resources, such as emergency response and health care programs. For many rural communities, simply transporting these resources into the community, or transporting residents out of the community in times of hazard, becomes a limitation due to the small number of transport routes. Residents of Fort McMurray were recently stranded temporarily, when the only road in and out was closed because of a forest fire, making evacuation impossible, medical emergencies more dangerous, and cutting off basic supplies (Soskolne et al., 2004). Moreover, in a small town, even a modest hazardous event can be disastrous, simply because it is likely to affect a greater proportion of the population—called the "proportionality impact" (Mossler, 1996). Relatively smaller response capacity in some instances can be doubly detrimental, as the very mitigation efforts of cities may exacerbate exposure in rural communities. After the storms of 1997 in Manitoba, the flood levees installed in Winnipeg in the mid-1990s led to a net increase in the level of floodwaters by over 0.6 metres in nearby rural communities (Haque, 2000). Effects are only compounded for very small or isolated communities (Harris et al., 2000). The Prairie Provinces are home to several very small communities (less than 1000 population), many of which are also located in the far north.

Furthermore, many rural communities in the Prairies are already stressed due to a number of recent non-climate events, like the softwood lumber dispute, outbreaks of bovine spongiform encephalopathy, and drought. Also, the autonomy of community level institutions to make planning decisions is reduced by the presence of large-scale market forces in export-based, homogenous economies (Epp and Whitson, 2001; Torry, 1983; Apedaile, 1992).

Many rural communities thus are simultaneously characterized by a reduced coping range—as community and household economic and social capital decline—and reduced likelihood to engage in proactive planning, due to the low degree of salience that may be placed on climate change relative to other more immediate stressors. As indicated by one empirical study of farming communities in the Prairies (Neudoerffer, 2005), the residents of rural communities may also be far more likely to treat information about climate change with a high degree of skepticism. Another study shows, however, that community attitudes towards climate change vary, and that those community residents with medium and higher levels of social capital are, on average, more optimistic and empowered when it comes to climate change than those with low levels (Diaz and Nelson, 2006).

Certain predictions associated with climate change may bode well for agriculture, including a longer frost-free season, and the northward expansion of the productive zone. Moreover, among the industries that dominate the prairie countryside, agriculture faces the greatest *potential* for resilience to climate change, based on the structural characteristics associated with agricultural production that allow for management flexibility. Alberta Agriculture, Food and Rural Development recently sponsored a set of focus groups with agricultural producers throughout the province (Stroh Consulting, 2005). In this exercise, 39% of participants expressed strong concern for climate change, particularly in regard to drought and water availability. Many indicated they were already engaging in adaptive strategies, such as changing crop types and varieties, income stabilization, stockpiling seed, reduced tillage, crop insurance, and altering the time of certain operations, although the financial squeeze experienced by many producers was seen as a limitation in the ability to mitigate risk. Many expressed confidence that with the appropriate future strategies, including education, enhancement of insurance, and technological changes to operations, the impacts of climate change to agricultural producers can be reduced.

However, the ability of agricultural communities to cope with the uncertainties associated with climate change—much less take advantage of the potential opportunities climate change may offer—are limited, first and foremost by water availability as discussed in Chapter 13. Moreover, due to the agricultural restructuring toward larger farms experienced in the last two decades, many of these communities have experienced a significant exodus of their young population, which translates into an aging of the local population and reduced levels of social capital. Also, one of the most significant constraints on the adaptive capacity of farmers is high debt loads, which have been increasing consistently since 2000. The ability of agricultural producers to implement adaptation measures will depend to a large extent on current debt, as well as other economic factors such as prices of agricultural commodities, the changing profitability of crop production, availability of crop insurance programs, etc. (see Bradshaw, Doal and Smit, 2004).

As a harbinger of future events, the 2001–2002 drought was nothing short of devastating to the agricultural sector across the Prairies. Crop production overall was the lowest it had been in 25 years, and an estimated 28,000 jobs were lost in all three Prairie Provinces combined. Net farm income was zero or negative for the first time in 25 years, with a reduction in farm cash receipts in Alberta alone of $267 million in 2001 and $920 million in 2002 (Wheaton et al., 2005). While many communities had drought adaptation strategies in place, many were found to be insufficient (ibid.). The primary adaptive mechanism employed during the 2001–2002 drought came in the form of a joint government-industry supported crop insurance program—a resource unlikely to sustain if the frequency of such events increases as expected: payments were 500% above the 10-year average in 2002 (Wheaton et al., 2005).

The forest sector does not play as large a role in the Prairie Provinces as elsewhere in Canada, with only 2% of regional employment in the forest sector overall (Stedman, Parkins and Beckley, 2004). Forest-based communities consequently make up a small proportion of rural communities in the Prairies, and the forest industry plays a minor role in several communities for which other economic sectors dominate, such as mining. Alberta's forest industry is significantly larger than is the case for either Manitoba or Saskatchewan.

The forest industry does not have the same potential for adaptation as agriculture, because of the relative inflexibility of modern industrial forestry, associated with long-term planning horizons, and specialized, capital-intensive infrastructure (Davidson, Williamson and Parkins, 2003). This is particularly true of the Prairie Provinces, whose forest industry is relatively young, and characterized by large forest management areas managed under 10-year planning horizons, and modern, high-capacity processing facilities. As with other rural communities, forest-based communities may also be severely constrained in their emergency response capacity. During the 2002 fire season, the rural municipality of Loon Lake spent $920,000—twice its tax revenue—on fire-fighting (Canada Parliament, 2003). Forest-dependent communities in the Prairies are also associated with higher unemployment and poverty, and lower median family incomes (Stedman, Parkins and Beckley, 2004). These features of life in a forest-based community are likely to translate into low levels of social capital, and a limited degree of salience being placed on climate change adaptation.

Many nature-based tourism activities may be impacted by climate change, with implications for those communities dependent on this sector. The potential impact is most acute in Alberta, whose tourism industry plays a large role in the provincial economy (ranked fourth largest provincial industry overall), due in large part to the high levels of national and international visitors to the National Parks in the Canadian Rockies. Banff alone received between 3 and 5 million visitors per year over the past decade (Service Alberta, 2005). These communities face several challenges posed by climate change, including the loss of snow pack in general and glaciers in particular; ecotype migration; lower water levels in recreational lakes and rivers; threats to game species, especially fish; and forest fires. As the amenity value declines in certain natural regions, the number of visitors to those regions is expected to decline (Scott and Jones, 2005). Many of our resort-based communities and recreation facilities today also represent enormous fixed capital expenditures, such as ski lifts and lodges (Milne, 2001) as well as expensive vacation homes, and thus the potential economic losses could be high. However, some tourism-based communities will benefit from longer tourist seasons, especially in the spring and fall (Scott and Jones, 2005). Tourism-based communities tend to be more diversified economically, and residents are likely to have a broader skill set, including many human capital skills that are important to adaptation. For some communities, particularly those that are remote, however, there are few other viable options for diversification.

Climate change is likely to affect Aboriginal communities in two ways: (1) the economic development they do have tends to be in natural resource sectors, including forestry, fishing, and mining, and the former two in particular may be vulnerable to climate change; and (2) many of these communities are also at least partly dependent on subsistence for their livelihood, with local food supplies supplementing their diets to a far greater extent than for non-Aboriginal peoples. Direct declines or annual uncertainties in the availability of moose, caribou, deer, and wild rice will increase dependence on imported foods, which is likely to be detrimental to their health.

Aboriginal communities are already under severe economic stress and are characterized by reduced levels of human and social capital, with the highest rates of unemployment all across the Prairies. While traditionally many Aboriginal communities are associated with traditional knowledge and land management systems that were quite adaptive and served as a source of resiliency in the past, the adoption of non-traditional lifestyles in recent years has led to an erosion of local knowledge and practices, and growing dependence on waged labour and external assistance, all of which have undermined adaptive capacity (Newton, 1995; Ford and Smit, 2004; Pittman, 2009).

CONCLUSION

The impact of climate change on the Prairies will be mediated to a large degree by variations in exposure, sensitivity, and adaptive capacity of prairie social systems. The people and communities of the Prairies will be exposed to a number of climate-related changes and events, the most important of which is likely to be changes in the water cycle, and associated droughts. The Prairies are urbanized and industrialized social systems, but a large segment of the prairie economy, and several individual communities, are associated with agriculture and forestry, which will be particularly sensitive to climate change and variability. Alberta may fare better overall than Saskatchewan and Manitoba, due to its robust economy and the high levels of human capital associated with that economy. However, social services and infrastructure are already under strain in this province, and climate change will only increase demands on these public resources. Saskatchewan and Manitoba both have a relatively higher number of small rural communities, higher proportions of some vulnerable populations, and lower median income levels.

In this context research in the areas of social vulnerability and adaptation to climate change in the Prairies is still incipient. There is an increasing urgency to develop a better understanding of the distribution of the different determinants of adaptive capacity among the three provinces and within each one of them. There is a special necessity to intensify research efforts and resources on the role of institutions in increasing adaptive capacity in the Prairies. In spite of the existence of a stable and robust institutional system at the level of government, there are indications that existing institutional arrangements are inadequate to face the challenge of climate change (Hurlbert, Corkal and Diaz, 2009). It is also necessary to increase our understanding about the role that social capital could play in reducing the vulnerability of prairie people. There are several studies that show that local forms of social capital in existence in urban and social settings, but there is limited information about the role played by more institutionalized forms of social capital, such as participation in formal organizations, in the fostering of a regional adaptive capacity.

REFERENCES

Adger, N. 2000. "Institutional Adaptation to Environmental Risk under the Transition in Vietnam." *Annals of the Association of American Geographers* 90, no. 4: 738–58.

Adger, N. 2001. *Social Capital and Climate Change* [online]. Tyndall Centre Working Paper. No. 8. Accessed at: *http://www.tyndall.ac.uk/content/social-capital-and-climate-change*, November 25, 2009.

——. 2003a. "Social Aspects of Adaptive Capacity." In B. Smit, R. Klein, and S. Huq (eds.). *Climate Change, Adaptive Capacity and Development*. Imperial College Press: London.

——. 2003b. "Social Capital, Collective Action, and Adaptation to Climate Change." *Economic Geography* 79, no. 4: 387–404.

Adger, W.N., N. Brooks, G. Bentham, M. Agnew and S. Eriksen. 2004. "New Indicators of Vulnerability and Adaptive Capacity." *Technical Report* 7. Tyndall Centre for Climate Change Research.

Apedaile, L.P. 1992. "The Economics of Rural-Urban Integration: A Synthesis of Policy Issues." In R. Bollman (ed.). *Rural and Small Town Canada*. Thompson Education Publishing Inc.

Arnfield, John A. 2003. "Two Decades of Urban Climate Research: A Review of Turbulence, Exchanges of Energy and Water, and the Urban Heat Island." *International Journal of Climatology* 23, no. 1: 1–26.

Bigio, Anthony G. 2002. "Cities and Climate Change." Draft Paper presented at the December 2002 conference, *The Future of Disaster Risk: Building Safer Cities*. World Bank, Disaster Management Facility.

Bradshaw, B., H. Doal and B. Smit. 2004. "Farm Level Adaptation to Climatic Variability and Change: Crop Diversification in the Canadian Prairies." *Climatic Change* 67: 119–47.

Canada Parliament. 2003. *Climate Change: We Are at Risk*. Standing Senate Committee on Agriculture and Forestry. Final Report. The Honourable Donald Oliver, Q.C., Chair; the Honourable John Wiebe, Deputy Chair.

City of Edmonton. 2005. Flood Prevention Program. Accessed at: *http://www.edmonton.ca/business/flood-prevention-program.aspx*, November 25, 2009.

Cross, John A. 2001. "Megacities and Small Towns: Different Perspectives on Hazard Vulnerability." *Environmental Hazards* 3, no. 2: 63–80.

Cutter, S., B.J. Boruff and W.L. Shirley. 2003. "Social Vulnerability to Environmental Hazards." *Social Science Quarterly* 84, no 2: 242–61.

Davidson, Debra J., Tim Williamson and John Parkins. 2003. "Understanding Climate Change Risk and Vulnerability in Northern Forest-based Communities." *Canadian Journal of Forest Research* 33, no. 11: 2252–61.

Department of Foreign Affairs and International Trade. 2003. *Fourth Annual Report on Canada's State of Trade*. Accessed at: *http://www.dfait-maeci.gc.ca/eet/trade/sot_2003/SOT_2003-en.asp*, July 20, 2005.

Desjardins, Ellen and Sanjay Govindaraj (eds.). 2005. *Proceedings of the Third National Food Security Assembly.* Region of Waterloo Public Health: Waterloo. September 30–October 2.

Diaz, H. and M. Nelson. 2006. "Rural Community, Social Capital and Adaptation to Climate Change." *Prairie Forum*, 30, no. 2: 289–312.

DuWors, E., M. Villeneuve, F.L. Filion, R. Reid, P. Bouchard, D. Legg, P. Boxall, T. Williamson, A. Bath and S. Meis. 1996. *The Importance of Nature to Canadians: Survey Highlights.* Environment Canada: Ottawa. Cat. No. En 47–311/1999E.

Epp, Roger and Dave Whitson. 2001. "Writing Off Rural Communities?" Pp. xiii–xxxv in R. Epp and D. Whitson (eds.). *Writing Off the Rural West: Globalization, Governments, and the Transformation of Rural Communities.* University of Alberta Press: Edmonton.

Ford, James D. and Barry Smit. 2004. "A Framework for Assessing the Vulnerability of Communities in the Canadian Arctic to Risks Associated with Climate Change." *Arctic* 57, no. 4: 289–400.

Field, John, Tom Schuller and Stephen Baron. 2000. "Human and Social Capital Revisited." In S. Baron, J. Field, and T. Schuller (eds.). *Social Capital. Critical Perspectives.* Oxford University Press: New York.

Glaeser, E. 2001. "The Formation of Social Capital." *Isuma* 2, no. 1: 34–40.

Haque, C.E. 2000. "Risk Assessment, Energy Preparedness and Response to Floods: The Case of the 1997 Red River Valley Flood, Canada." *Natural Hazards* 21, nos. 2–3: 225–45.

Harris, C., W. Mclaughlin, G. Brown, D.R. Becker. 2000. *Rural Communities in the Inland Northwest: An Assessment of Small Rural Communities in the Interior and Upper Columbia Basin.* USDA Forest Service, Pacific Northwest Research Station. General Technical Report: Portland, OR.

Hurlbert, M., D. Corkal and H. Diaz. 2009. "Governments and Adaptive Water Management in the South Saskatchewan River Basin." *Prairie Forum* 34, no. 1: 181–210.

IPCC (Intergovernmental Panel on Climate Change). 2001. *Climate Change 2001: Impacts, Adaptation, and Vulnerability Technical Summary.* A Report of Working Group II of the Intergovernmental Panel on Climate Change 2001, WMO and UNEP.

Last Link on the Left [The]. 2005. "The Money Trees." November. Accessed at *http://www.lastlinkontheleft.com/e0070509.html*, November 25, 2009.

Milne, Bruce. 2001. "The Politics of Development on the Sunshine Coast." Pp. 185–203 in R. Epp and D. Whitson (eds.). *Writing Off the Rural West: Globalization, Governments, and the Transformation of Rural Communities.* University of Alberta Press: Edmonton.

Mossler, M. 1996. "Environmental Hazard Analysis and Small Island States: Rethinking Academic Approaches." *Geographische Zeitschrift* 84, no. 2: 86–93.

Neudoerffer, R.C. 2005. "Lessons from the Past—Lessons for the Future: A Case Study of Community-based Adaptation on the Canadian Prairies." Conference presentation, *Adapting to Climate Change in Canada 2005*. Montreal, May 4. Accessed at: *http://adaptation2005.ca/abstracts/pdf/neudoerffercynthia.pdf*, June 5, 2007.

Newton, J. 1995. "An Assessment of Coping with Environmental Hazards in Northern Aboriginal Communities." *Canadian Geographer* 39, no. 2: 112–20.

Portes, A. 1998, "Social Capital: Its Origins and Applications in Modern Sociology." *Annual Review of Sociology* 24: 1–24.

Pittman, J. 2009. "The Vulnerability of the James Smith and Shoal Lake First Nations to Climate Change and Variability" [online]. M.Sc. Thesis, Program of Geography, Faculty of Graduate

Studies and Research, University of Regina. Accessed at: *http://www.parc.ca/mcri/pittman_thesis.php.* November 25, 2009.

Roach, Robert, Vien Huynh and Sarah Dobson. 2004. *Drop by Drop: Urban Water Conservation Practices in Western Canada.* Western Cities Project Report 29. Canada West Foundation: Calgary, AB.

Scott, D. and B. Jones. 2005. *Climate Change and Banff National Park: Implications for Tourism and Recreation.* Report prepared for the Town of Banff. University of Waterloo: Waterloo, ON.

Service Alberta. 2005. *Alberta Tourism Statistics and Trends.* Accessed at: *http://www.servicealberta.gov.ab.ca/*, November 25, 2009.

Soskolne, C.L., K.E. Smoyer-Tomic, D.W. Spady, K. McDonald, J.P. Rothe and J.D.A. Klaver. 2004. *Final Report: Climate Change, Extreme Weather Events And Health Effects in Alberta.* HPRP# 6795–15–2001/4400013. April 30, 2004. 485 pgs.

Statistics Canada. 2006. Summary Tables. Population urban and rural, by province and territory Accessed at *http://www40.statcan.gc.ca/101/cst01/demo62j-eng.htm*, November 25, 2009.

——. 2006a. Community Profiles. Accessed at *http://www12.statcan.ca/census-recensement/2006/dp-pd/prof/.* November 25,2009.

Stedman, Richard C., John R. Parkins and Thomas M. Beckley. 2004. "Resource Dependence and Community Well-being in Rural Canada." *Rural Sociology* 69, no. 2: 213–34.

Stroh Consulting. 2005. *Agriculture Adaptation to Climate Change in Alberta: Focus Group Results.* March 31. Camrose, Alberta.

Sygna, L. 2005. *Climate Vulnerability in Cuba. The Role of Social Networks.* CICERO Working Paper 2005:01. Accessed at: *http://www.cicero.uio.no/webnews/index.aspx?id=10532&lang=no*, November 25, 2009.

Torry, W.I. 1983. "Anthropological Perspectives on Climate Change." In R.S. Chen, E. Boulding and S.H. Schneider (eds.). *Social Science Research and Climate Change: An Interdisciplinary Appraisal.* D. Reidel Publ. Co.: Dordrecht, Holland.

Waller, D.H. and R.S. Scott. 1998. "Canadian Municipal Residential Water Conservation Initiatives." *Canadian Water Resources Journal* 23, no. 4: 369–406.

Weber, Marian, Debra J. Davidson and David Sauchyn. 2008. *Climate Change Vulnerability Assessment for Alberta.* Alberta Environment: Edmonton.

Wellstead, Adam. 2006. "Natural Resource Policy Complexity and Change." PhD dissertation, Department of Renewable Resources, University of Alberta.

Wheaton, E. with V. Wittrock, S. Kulshreshtha, G. Koshida, C. Grant, A. Chipanshi and B. Bonsal. 2005. *Lessons Learned from the Canadian Droughts Years of 2001 and 2002: Synthesis Report.* SRC Publication No. 11602–46E03. Saskatchewan Research Council: Saskatoon.

Wittrock, V, E.E. Wheaton and C.R. Beaulieu. 2001. *Adaptability of Prairie Cities: The Role of Climate. Current and Future Impacts and Adaptation Strategies.* Environment Branch, Saskatchewan Research Council SRC Publication No. 11296–1E01. June 2001.

CHAPTER 11

HEALTH

Justine Klaver-Kibria

INTRODUCTION

In the recently released Intergovernmental Panel on Climate Change (IPCC) Fourth Assessment Report (AR4), the "Health" chapter (Confalonieri et al., 2007) listed the climate change scenarios and/or major health concerns for global populations; however, each country and each region within a country will experience their own health challenges. Health, for the purpose of this chapter, is the state of physical, social and mental well-being of an individual. Population health is an approach to health that aims to improve the health of entire populations. Climate change can threaten the health of individuals as well as entire populations and sub-populations. Experiential health challenges of prairie populations to climate change will be determined by the significance of disturbance, the apparent vulnerability of prairie populations, the adaptive capacity of the population to overcome these challenges, and the ability to mitigate future negative health outcomes.

This chapter outlines the implications to health from climate change for the Canadian prairie population by addressing four significant issues. First, by determining how climate change will alter the external environment and, subsequently, human health. Second, by examining how certain segments of the prairie populations are more vulnerable to ill health from climate change. Third, by examining the adaptive capacity of the Prairie Provinces as it pertains to health and its ability to mitigate future health challenges. And fourth, the gaps in knowledge will be examined.

CLIMATE CHANGES AND HUMAN HEALTH

People on the Canadian Prairies have shaped their culture, livelihoods, economic structure and, to some extent, their health and well-being around prairie grassland and boreal forest ecosystems. We have come to expect certain long-term and short-term weather patterns, and we base future growth and opportunity on this. However, as explained in Chapter 4, these assumed weather and climate patterns are being altered under climate change and may have a profound impact on human health if we are not prepared. The purpose of this section is to link climate and weather to health outcomes, providing insight into the possible consequences to human health under climate change.

In the context of future climate conditions, extreme events such as increased drought conditions, increased average temperatures and heat waves, intense precipitation events, and changing ecosystems are important from the perspective of health. It must be noted that extreme cold will decrease under climate change, positively affecting health. However, the focus of this chapter is the challenges that lay ahead and the need of public health initiatives to shift focus under climate change. Therefore, this chapter will focus on the negative impacts of climate change on health, not the intricate balance between the positive and the negative effects of climate change on human health.

DROUGHT

The links between drought and human health can be made directly and indirectly. Discussed here are the mental health effects from drought where farming livelihoods are strained, water availability for urban centres and the decreasing quality of potable water sources giving rise to waterborne disease outbreaks, increasing dust production impacting respiratory health, and increasing wildfire occurrence.

Farming and Mental Health

Farming is one of the most stressful occupations in society (Fraser et al., 2005; Gregoire, 2002), and drought conditions only exacerbate the problem. Droughts have caused substantial decreases in crop yields and large export losses of agricultural products, leading to an increased need by farmers and cattle producers for assistance and crop insurance payouts (Chapter 7).

Several studies have documented the aspects of farming lifestyles that tend to increase personal stress and distress levels. Most of the literature concludes that financial concerns—stemming mainly from government policy/bureaucracy, market prices, hail, insect infestations, and farm expenses coupled with unpredictable incomes—are highly associated with the stress felt by farmers (Deary, Willcock and McGregor,1997; Simkin et al., 1998; May, 1998; Booth and Lloyd, 1999; Raine, 1999; Gregoire, 2002; Fraser et al., 2005; Soskolne et al., 2004). Unfortunately, most farmers do not experience only a single stressor, but rather multiple stressors at any given time.

Emotional stress among agricultural workers contributes to many health problems, such as suicide, depression and occupational injury (May, 1998; Raine, 1999; Carruth and Logan, 2002). Suicide among farm owners, managers and tenants is highly correlated to depression in the year before death, which in turn is correlated to financial pressures (Malmberg, Hawton and Simkin, 1997). A study of rural populations in southern Saskatchewan (Masley et al., 2000) found stress to be one of the most commonly reported health problems among men aged 18–55 years. Farmers in Manitoba self-reported significantly higher stress symptoms than non-farmers (Walker and Walker, 1988). Another study also found that farmers had more depression and anxiety than control groups (non-agricultural workers), but the results were inconclusive (Eisner, Neal and Scaife, 1998). Soskolne et al. (2004) also found that drought affected the mental health of farmers in Alberta.

Stress in agricultural occupations cascades into family life (Plunkett et al., 1999, Fraser et al., 2005). Farmwives experience a great deal of stress (Walker and Walker 1988; Deary et al., 1997; Carruth and Logan, 2002; Fraser et al., 2005). However, their stress encompasses not only the financial difficulties of farm life, but also interpersonal relations, conflicts and family concerns (Walker, Walker and MacLennan, 1986, Carruth and Logan, 2002, Fraser et al., 2005). Adolescents were also found to perceive greater family stress in times of farm economic crisis (Plunkett, Henry and Knaub, 1999; Fraser et al., 2005).

Water Availability and Water Quality

> "Water is essential for life. The amount of fresh water on earth is limited, and its quality is under constant pressure. Preserving the quality of fresh water is important for the drinking-water supply, food production and recreational water use. Water quality can be compromised by the presence of infectious agents, toxic chemicals, and radiological hazards."
> —WORLD HEALTH ORGANIZATION (2009)

All the major rivers in the western Prairie Provinces begin in the Rocky Mountains and wind their way through agricultural and urban areas as they head east and north (Schindler and Donahue, 2006). Surface waters can become increasingly polluted with wastes (including waterborne diseases) from agricultural fields, intensive livestock operations, processing plants, and municipalities (Coleman et al., 1974; Vanderpost and Bell, 1977; Menon, 1985; Lincoln, 2005). Drought decreases water levels, which leads to increased concentration of pathogens and toxins (e.g. fertilizers and other chemicals) in ground and surface waters used for domestic use (Charron et al., 2004; Lincoln, 2005; Kampbell et al., 2003; Zwolsman and van Bokhoven, 2007).

Compounding water quality issues in the Prairie Provinces is the probability of decreased water availability, especially in southern Alberta and Saskatchewan. As the population increases, drought decreases river flow, and glaciers continue to recede, there will be less water for all end users—communities, agriculture, and oil and gas exploration (Schindler and Donahue, 2006). Striking a balance between all users, while keeping in mind current water legislation, will be difficult. Abundant, clean water is essential to the health and well-being of populations all around the world.

Dust Production and Particulate Matter

The Prairie Provinces are often dry and windy; when combined with large-scale agricultural practices, atmospheric dusts result. Farming activities produce most of the aerosolized particulate matter in south-central Alberta (Green et al., 1990), whereas in Saskatchewan, dust from wind-blown soils and unpaved roads are significant sources (Environment Canada, 2006). Green et al. (1990) found that, within Alberta farming communities, there was bimodal peaking of total suspended particulates (TSP) that coincided with spring and fall maximal farming activities. Drought conditions will exacerbate dust production in the prairie region.

In general, air particles must be less than 10 microns in diameter to be inhalable and deposited in the respiratory tract (Health Canada, 1998). The major health effect from inhaling particulate matter is airway inflammation, which can manifest as acute or chronic forms of asthma, allergic rhinitis, bronchitis, hypersensitivity pneumonitis, and organic dust toxic syndrome (Rylander, 1986; do Pico, 1986, 1992; Lang, 1996; Simpson et al., 1998). These respiratory diseases often begin as acute attacks of respiratory distress, and with continued exposure to dust the disease may become chronic. Over time, these lung conditions can lead to irreversible lung damage (Rylander, 1986). Although windblown dust may be most noticeable in rural communities, urban populations may also experience added atmospheric dust of rural origin (Haller et al., 1999).

Wildfires

Drought exacerbates wildfire (Smoyer-Tomic et al., 2004). Electron microscope scanning of haze particles showed that 94% of the particles were below a diameter of 2.5 microns, at which they are inhalable and can aggravate the respiratory tract (Emmanuel, 2000). Wildfire smoke has been found to: increase hospital and physician visits for respiratory diseases and symptoms (e.g. asthma) and cardiovascular conditions (e.g. chest pain), especially for individuals with preexisting lung and heart conditions; and increase all-cause mortality (Bowman and Johnston, 2005; Johnston et al., 2002). Vulnerable populations include those with asthma or other respiratory diseases, airway hyper-responsiveness, cardiovascular disease, the elderly, children, pregnant women, and smokers (Ammann et al., 2001).

The economic costs associated with the health effects from wildfires can be high. The health costs associated with the Chisolm, Alberta, fire in 2001 had a large economic impact with cost estimates ranging between $4 and $22.9 million (Rittmaster et al., 2006). Not only are breathing difficulties signs of morbidity, but forest fires are especially apt to produce fright or other mental health problems. Sudden evacuation from forest fires can trigger anxiety and fear in the affected populations (Soskolne et al., 2004). The direct consequences of wildfires (e.g. evacuation and loss of property/lives) will inevitably cause more discomfort and need for public assistance, and raising insurance costs. Loss due to forest fires can increase Post-Traumatic Stress Disorder (PTSD) symptoms in children, adolescents and their parents (Jones et al., 2002; McDermott et al., 2005).

EXTREME RAINFALL EVENTS

Extreme rainfall events are expected to become more frequent and severe under climate change. Events expected to impact the health and well-being of prairie residents include rural and urban flash flooding and slow-rising river floods. These events have the potential to contaminate potable water sources with waterborne diseases and other toxins, as well as flood communities and neighbourhoods—taking a toll on mental health.

Waterborne Diseases

To cause a waterborne disease outbreak in a population, pathogens must enter the potable water system and be consumed by humans. Most prairie communities draw potable water from surface water sources such as rivers or lakes; however, some smaller communities and individual rural homes use ground water sources. It is relatively easy to contaminate surface and ground potable water sources, as they are open to the environment. Contaminated water must evade community water treatment processes and, subsequently, proceed to individual houses. The peak concentration of waterborne pathogens in surface waters occurs after an extreme rainfall event (Schijven and de Roda Husman, 2005).

The most common pathogens to cause waterborne disease outbreaks are those associated with enteric bacteria from fecal wastes of birds and cattle, as well as naturally occurring in water, enteric protozoa from wild and livestock animals, as well as human fecal wastes, or viruses mainly from human feces (Leclerc, Schwartzbrod and Dei-Cas, 2002). The Prairies are home to the vast majority of livestock in Canada: 71.8% of Canada's cattle and 41.4% of pigs (Statistics Canada 2008a; 2008b); and when waterborne disease outbreaks have occurred in human populations they have a high correlation with intense precipitation events and/or flooding, and run-off from areas that contain agricultural livestock (Millson et al., 1991; Bridgman et al.,1995; Charron et al., 2004; Schuster et al., 2005).

Curriero et al. (2001) found that 68% of the waterborne disease outbreaks were preceded by precipitation events above the 80th percentile. Similarly, Thomas et al., (2006) found that rainfall events greater than the 93rd percentile increased the odds of an outbreak of waterborne disease by a factor of 2.3 when compared to rainfall events less than the 93rd percentile. A case-crossover study in southern Alberta (Charron et al., 2005) found that each extra day of rain in the 42 days preceding the case or control outbreak increased the risk of hospitalization for gastrointestinal illness.

Mental Health

Health issues surrounding flooding can be short-term, such as the threat of infectious diseases from waterborne pathogens, injury, or mortality (drowning, heart failure). Fortunately, slow-rising riverine floods have a low potential for mortality, and the major health effects from floods may be longer-term, such as mental health issues (Post-Traumatic Stress Disorder [PTSD]), and moulds/mildew and the associated respiratory conditions from extremely wet conditions, during and after a flooding event (Square, 1997; Greenough et al., 2001).

In 1997, the Manitoba Red River flood resulted in approximately 28,000 Manitobans being evacuated from their homes (Square, 1997). Losing a home or witnessing it being destroyed, being evacuated on short notice, or being displaced for an extended period of time causes great anxiety (Soskolne et al., 2004). Evacuation orders have the greatest impact on individuals that have limited mobility, no transportation source, few economic resources, physical dependents (e.g. children), or have underlying medical conditions requiring regular medical attention; and often these are the elderly (Soskolne et al., 2004). Psychological problems can linger after a disaster, even once the community is on the way to recovery (Durkin et al., 1993; Phifer, 1990; Phifer, Kaniasty and Norris, 1988; Tyler and Hoyt, 2000; Ginexi et al., 2000).

Flooding causes economic loss, which in turn creates stress and hardship. Uncertainty about who is expected to pay for the loss is also a source of stress (Soskolne et al., 2004). The 1997 Manitoba Red River flood did cost society US$5 billion for the United States and Canada combined (IJC, 2000).

CHANGING ECOSYSTEMS

Climate change will inevitably change the flora and fauna patterns, densities and species, and these changes in the dominant prairie ecosystem can have implications to human health, namely rodent- and vector-borne diseases. For human infection of rodent- or vector-borne diseases to occur, several biological organisms rely on one another: reservoir (often mammals or birds), vector (insect or rodent); and pathogen (bacteria, virus, or protozoa). With respect to the various flora and fauna that play central roles in the spread and proliferation of vector- and rodent-borne diseases into human populations, climate change can help or hinder insect or rodent reproduction, behaviour, and survival, and consequently influence the reproductive capacity of their pathogens within. Below is a discussion of two diseases that are known to be influenced by climatic parameters—Hantavirus Pulmonary Syndrome (HPS) and West Nile Virus (WNV)—and are a threat to prairie population health.

Hantavirus

Hantaviruses are transmitted to humans via the inhalation of aerosolized hantavirus from rodent excreta and saliva (Stephen, Johnson and Bell, 1994; Gubler et al., 2001). The deer mouse is most often associated with HPS (Stephen, Johnson and Bell, 1994; Glass et al., 2000). The first Canadian case of HPS occurred in 1994 (Stephen, Johnson and Bell, 1994). However, HPS was not made a nationally notifiable disease until January 1, 2000. There were only 12 cases of HPS in Canada between 2001 and 2004, all of which were reported in Saskatchewan and Alberta (PHAC, 2006). In total, 40–50% of HPS cases die (PHAC, 2001).

Human cases of HPS may reflect the yearly and seasonal patterns of high rodent population densities (Mills et al., 1999). Large increases in rodent populations have been linked to mild wet winters, and to above-average rainfall followed by drought and higher-than-average temperatures (Engelthaler et al., 1998; Gubler et al., 2001)—weather conditions that are expected more often in the Prairie Provinces under climate change.

West Nile Virus

WNV is transmitted between its natural bird reservoirs by mosquitoes (primarily the *Culex* family). Humans are considered dead-end carriers of the virus, as they may become ill if bitten by an infected mosquito, but do not continue the transmission cycle (Craven and Roehrig, 2001; Huhn et al., 2003). WNV transmission only occurs in the months in which mosquitoes are active (Craven and Roehrig, 2001), and the climatologic conditions that favour the WNV are mild winters, coupled with prolonged drought and heat waves (Epstein, 2001; Huhn et al., 2003). Drought facilitates WNV propagation by forcing birds to congregate around ever fewer and shrinking water sites, allowing the virus to circulate more easily between them, and by reducing the numbers of natural mosquito predators, like frogs and dragonflies (Epstein 2001). Heat tends to speed up the viral development within the mosquito, increasing the probability that the virus will

Year	Cases Prairie Provinces	Cases Canada-wide	Percentage attributed to Prairie Provinces
2003	1,351	1,481	91.2%
2004	9	25	36.0%
2005	123	225	54.7%
2006	108	151	71.5%
2007	2,181	2,215	98.5%
2008	33	36	91.7%

Table 1. Cases of WNV reported in the Prairie Provinces compared to all of Canada. Source: Compiled from PHAC, 2008a; PHAC, 2008b.

mature and be transmitted (Epstein, 2001). Given that the climatological conditions that favour the WNV are the dominant scenarios expected for the Prairie Provinces under climate change, WNV could become a major public health threat in the future.

The majority of people (80%) infected with WNV are asymptomatic. Almost 20% of those infected show milder flu-like symptoms such as fever, headache, and body aches. Approximately one in 150 people infected with WNV show severe symptoms such as high fever, stupor, disorientation, coma, tremors, convulsions, vision loss, and paralysis, where some neurological effects may be permanent (CDC, 2006). Clinical syndromes include fever, meningitis, encephalitis, and flaccid paralysis (Huhn et al., 2003). West Nile infection tends to be equally distributed between the sexes and all age groups (Huhn et al., 2003). However, people over 50 years of age are at greater risk of developing severe illness (CDC, 2006). West Nile Virus Surveillance Data (Summary of Human Surveillance 2002–2008) have documented wide variations in the number of clinical cases of WNV from year to year. Table 1 indicates that the vast majority of cases of WNV are found in the Prairie Provinces, and especially profound are the outbreak years 2003 and 2007.

HIGHER AVERAGE TEMPERATURES AND HEAT WAVES

Climate change will give rise to higher average temperatures and heat waves, each having different consequences on the human health of prairie populations. Heat waves cause health-related morbidity and mortality, whereas increased average temperatures bring about more air pollution days, and can make the possibility of food-borne disease outbreaks more likely.

Heat Waves

As described in Chapter 4, the Canadian Prairies are expecting an increased occurrence of heat waves. It is the timing and magnitude of the heat event that will most likely determine the impact of a heat wave on human populations (Schneider et al., 2007). Thus, one ill-timed and extreme heat wave could have very grave consequences for the health and well-being of prairie populations. Extremely hot days and heat waves are especially dangerous to people living in cooler climatic regions like the Canadian Prairies, where extremely high temperatures occur infrequently or irregularly and populations are not accustomed to high heat.

Heat wave-related morbidity and mortality occurs directly and indirectly. Specific heat-related illnesses include heat edema, heat syncope, heat cramps, heat exhaustion and heat stroke. Factors that indirectly increase the risk of mortality and morbidity are co-morbid conditions, such as cardiovascular disease, diabetes, psychosis, drug usage (prescription or illicit), and demographic or social factors, such as age, isolation, disability, and living conditions (Hett and Brechtelsbauer, 1998). Heat waves increase the number of patients presented to hospitals and emergency departments, and are associated with "excess" deaths (Kalkstein and Smoyer, 1993; Kilborne 1999; Semenza et al., 1999). The most common heat-related illnesses giving rise to hospital admissions and excess deaths are heat exhaustion, heat stroke, and dehydration (Faunt et al., 1995; Kilborne, 1999; Semenza et al., 1999). Heat-related health outcomes can be exacerbated by the heat island effect in which temperatures are higher in urban centres than in surrounding rural areas (Koppe et al., 2004), and because of the substantial amount of thermal mass (e.g. pavement and concrete) cities retain heat at night, not allowing for night-time cooling and relief.

Foodborne Diseases

Higher average temperatures will play an important role in increasing the incidence of food-poisoning outbreaks in prairie populations. Several factors are responsible: 1) warmer temperatures, which are more conducive to accelerated growth of bacterial species in animal feces, raw animal products, and prepared foods; 2) longer summer periods, which will increase the time period during which bacterial species and their carriers (e.g. flies) can survive in the environment; and 3) methods of food preparation and patterns of food consumption in the summer months (e.g. BBQ) increase the probability of food-borne illnesses (Rose et al., 2001; D'Souza et al., 2004; Hall, D'Souza and Kirk, 2002; Kovats et al., 2004).

There is often a time-lag between ambient temperatures and the outbreak of food poisonings. Fleury et al. (2006) investigated the relationship between ambient temperature and confirmed cases of food-borne diseases in Alberta, and found that there was a positive association between ambient temperature and disease for every degree increase in weekly mean temperature prior (up to six weeks) to the outbreak. Furthermore, depending on the pathogen type, the log relative risk of weekly case counts increased by 1.2% for salmonella, 2.2% for *Campylobacter*, and 6.0% for *E.coli* for every degree increase in weekly mean temperature. Bentham and Langford (1995) estimated that, in England and

Wales, an additional 179,000 cases of food poisonings per year could be realized as a result of climate change.

Air Pollution

Although prairie cities have relatively low concentrations of air pollution, current pollution levels do affect prairie population morbidity and mortality (Burnett et al., 1997; Burnett, Cakmak and Brook, 1998; Duncan et al., 1998). Adverse health outcomes from air pollution include exacerbation of pre-existing respiratory conditions (e.g. asthma, hay fever, COPD), increased mortality from respiratory or cardiovascular distress, increased morbidity and hospital admissions for patients with cardiac or respiratory disease (Burnett et al., 1995; Burnett et al., 1999), initiation of respiratory diseases (especially in children) like asthma or hay fever, headaches, fatigue (Last, Trouton and Pengelly, 1998), and premature mortality (Burnett, Cakmak and Brook, 1998) of individuals with pre-existing medical conditions.

Air pollution episodes in the Prairies occur mainly during the winter and fall, where temperature inversions trap pollution near the ground; and in the summer, when hot, calm weather conditions are occurring and photochemical smog forms (CASA, 2007). Air pollution may be a more significant factor in determining ill health under climate change in the future as warmer temperatures enhance the production of secondary pollutants (e.g. ground-level ozone, acid aerosols, and complex organic particles) which are formed from primary pollutants (end products of fossil fuel combustion) via photochemical reactions (e.g. with water vapour and sunlight) (Bernard et al., 2001). In addition, population density is expected to increase, possibly encompassing previous industry, or introducing industry into highly populated areas, which will further exacerbate the effects of air pollution on population health.

VULNERABILITY

The previous section outlined how the changes expected under climate change can influence the health and well-being of prairie populations. However, the health impacts from climate change will not affect all prairie people equally, and those bearing a disproportionate amount of risk of ill health from climatic change are deemed more vulnerable. Individual vulnerabilities include disease status, socio-economic factors and demographic factors, whereas community vulnerabilities include infrastructure, emergency management, and population density (IPCC, 2001). Another vulnerability that is geographical in origin—which incorporates habitation on flood plains, rural-urban resources, and degraded landscapes—is noted because of its effects on community vulnerability, but will not be discussed here (IPCC, 2001).

VULNERABLE POPULATIONS

There are segments of the prairie population that will incur greater suffering, loss and ill health than others. Below is a discussion of the various population groups that will incur a disproportionate burden of the ill-health outcomes based on individual and personal characteristics. These include those with underlying health concerns, children, the elderly, those with lower socio-economic status, and First Nations peoples.

Elderly

Chronological age alone is not a good predictor of vulnerability, as many older people continue to have good health and mobility, and remain very socially active into their older years (Powell 2006). However, the vulnerability of the 65+ demographic age group is important because the general vulnerability of the elderly is higher than the younger age groups. The aging are more likely to have the underlying medical conditions (i.e. respiratory ailments and cardio-vascular disease) that are associated with increased risk of negative health impacts from extreme heat and air pollution (Delfino et al., 1997; Last, Trouton and Pengelly, 1998; Kilborne et al., 1982; Kenney and Hodgson, 1987; McGee-hin and Mirabelli, 2001; Kilborne 1999; Basu and Samet, 2002; Diaz et al., 2002; Worfolk, 2000).

The elderly may be most at risk for heat-related illnesses and death owing to the association of aging with decreased heat tolerance, alterations in thermoregulatory capacity in general, and taking drug therapy which can further reduce thermoregulatory capacity (Kenney and Hodgson 1987; Flynn, McGreevy and Mulkerrin, 2005; Worfolk, 2000). Also, the elderly are more likely to have health problems requiring regular medical attention, or be associated with limited mobility and inability to care for oneself (Turcotte and Schellenberg, 2007) and, therefore, rely more heavily on health services (Powell, 2006). This is problematic because an extreme event (e.g. evacuation as a result of a forest fire) can exacerbate existing conditions and disrupt the management of their illness.

An aging population is more susceptible to food-borne illnesses (Bentham and Langford, 2001; Rose et al., 2001). People over 50 years of age are at greater risk of developing severe outcomes from WNV infection (CDC, 2006). Vulnerable populations (especially the elderly) could extend the reservoir for opportunistic water-borne pathogens (Levin et al., 2002; Theron and Cloete, 2004). Social isolation and loneliness tend to increase as people age, as networks of family and friends tend to decrease in size. During an extreme event, the elderly may not have a close relative or neighbour from whom to ask help, and some elderly people could remain hidden from the formal social services network (Powell, 2006).

Children

Children are vulnerable to climate change in a variety of ways, such as undeveloped physiology and metabolism, behaviours that are more risky, immature perception of danger, rapid growth and development, exposures per unit of body mass are higher (water and air), and a tendency to have a high intake of single foods (Tamburlini, 2002; Health Canada, 2005a). Children may be more vulnerable to heat waves and air pollution effects of climate change because their thermoregulatory capacity is underdeveloped until one year of age, and their higher ratio of surface area to body mass (up to age five) makes them more vulnerable to heat stress. As well, they take in more air pollutants because their lung volume to weight ratio is higher than adults in addition to spending more time outdoors (Longstreth, 1999; Mathieu-Nolf, 2002; Health Canada, 2005a).

Because of children's immature immune systems, poor perception of risk, and food eating habits, an increased prevalence of food-borne and water-borne diseases associated with increased average temperature and extreme hydrological events could disproportionally affect children (Pond, 2002; Jermini, 2002). In his review of hazard literature, Shrubsole (1999) found that children's reaction to a natural hazard depended, in part, on the ability of their parents to deal with it, and should the parent handle the disaster period adversely, so will the child. However, it was also found that that children are more at risk from natural hazards because they have lower coping capabilities, and perceive the world differently than adults according to their level of cognitive and emotional development (Shrubsole, 1999).

Therefore, extreme events can increase the psychological trauma experienced by children during and after the event (Shea, 2003; McDermott, Lee and Gibbon, 2005; Heinz Center, 2002). Children can experience any number of psychological-related illnesses due to the trauma they have experienced during a disaster. These can be as extreme as losing recently developed skills, eating and sleeping disorders, and behavioural issues (Heinz Center, 2002).

Children are dependent on others for their well-being and safety, and thus any amount of vulnerability that affects their caregiver, such as evacuation, directly affects children as well. Thus, children not only have increased vulnerability to the health risks from climate change, they carry the vulnerability of their caregivers (e.g. lower socio-economic status). For example, children living in low-income families are more vulnerable to heat waves due to poor buildings and lack of air conditioning, as well as being unable to physiologically cope as well as adults. Moreover, children generally do not have the capacity or resources to change their situation, thus they cannot go to cooling centres on their own, may not understand how to relieve the stresses from heat (e.g. more fluid intake), or for that matter, change their living conditions to mitigate the effects of the next heat wave.

Underlying Health Conditions

People with underlying health problems, such as chronic or acute illness, mental illness (including addictions) and limited mobility, as well as the immune-compromised, are more likely to be adversely affected by extreme events such as heat waves, forest fires, floods, dust, and air pollution. These types of illnesses often require regular medical attention and would be exacerbated by extreme heat or the disruption in life caused by extreme rainfall events (Mokdad et al., 2005). This heightened stress includes not being able to access required medications or life sustaining treatments, or the impairment or damage of necessary health equipment, such as ventilators or oxygen generators (Powell, 2004). Those with limited mobility will require aid from others to evacuate or reach a safer location during an extreme weather event.

Severe mental illness, including addictions, has a serious impact on a person's ability to function effectively and cope with the simplest activities of everyday life. Being impaired will inhibit good decision-making regarding health and safety before, during or after an extreme event. In addition, individuals requiring medication for health conditions or are addicted to drugs, are less capable of thermoregulation during extreme heat (McGeehin and Mirabelli, 2001).

Lower Socio-economic Status and Education

Two key determinants of health are socio-economic status and education. Those with higher socio-economic status and education tend to have better overall health (PHAC, 2003). Education equips people with skills and knowledge to better their outcomes during extreme weather and to access the information and resources to do so. Canadians with low literacy levels are more likely to be poor, unemployed and generally less healthy (PHAC, 2003). People with low education levels and who live in poverty have a harder time staying healthy under "normal" conditions, thus when an extreme event occurs they may be rendered helpless.

Low income and less educated people are more likely to be located in less desirable locations with poor or damaged infrastructure, which can increase the risk of negative outcomes from extreme events. They will experience higher losses, and have fewer resources available to recover (Hutton, 2001; Shrubsole, 1999). They are also less likely to be able to afford adaptation measures, such as air conditioners (Koppe et al., 2004). Income and status allows for greater control over life's circumstances, especially stressful situations, and the discretion to act to better their outcome (PHAC, 2003).

Homelessness is often associated with drug and alcohol addiction, and mental illness (Fischer and Breakey, 1991; Hwang, 2001). Homeless people are more vulnerable to extremes in temperature and rainfall as they are more exposed (Hwang, 2001).

First Nations Peoples

The overall health of Aboriginal people is poorer when compared to other Canadians. Compared to the general population, First Nations and Inuit people have high rates of chronic and contagious diseases, plus shorter life expectancy. For example, the prevalence of heart disease is 1.5 times higher and diabetes is 3 to 5 times higher, and 15% of new HIV and AIDS infections occur in Aboriginal people (Health Canada, 2007; Health Canada, 2005a). Alcohol consumption has been identified as a serious problem in Aboriginal communities, and there is an increasing use of prescription or illicit drugs by Aboriginal people (Health Canada, 2005b). Aboriginal people have lesser formal education than other Canadians, and almost half are at or below the poverty line (Health Canada, 2005c; Indian and Northern Affairs, 2008).

First Nations people have many individual characteristics of vulnerability: poorer health; less education; lower income; many underlying health problems; and a higher percentage of the population are children, as children younger than 14 represented one third of the Aboriginal population compared to 19% in the non-Aboriginal population (Statistics Canada, 2003). Compounded by the infrastructural deficits, water quality issues, and the relative isolation of First Nations people, they will likely bear a disproportionate burden of ill health under climate change.

COMMUNITY VULNERABILITY

Lessening community vulnerability will ultimately reduce individual vulnerability and negative human health outcomes. Infrastructure and public health services, and their ability to withstand an extreme weather event, are crucial for decreasing community vulnerability.

Infrastructure

Infrastructure is the complex urban structures—physical facilities and systems that support human activity and provide services—that improve quality of life. Impacts from an extreme event may be exacerbated by weaknesses in infrastructure (Henstra et al., 2004), where facilities can be damaged or destroyed, affecting health and well-being. Infrastructure can also intensify the effects of an extreme event, e.g. the urban heat island effect, combined storm-sewer systems which can become overwhelmed during intense precipitation, and the age and deterioration of infrastructure which may no longer be appropriate for today's climate (Henstra et al., 2004), let alone the future intensification of extreme weather under climate change.

During extreme heat events, temperatures are higher in urban centres (heat island effect) than in surrounding rural areas (Koppe et al., 2004). Living on higher floors of a multi-storey building can amplify risk or exacerbate pre-existing health conditions in extreme heat events (Smoyer, 1998, Kilborne et al., 1982). During the July 1999 heat wave in the midwestern US, a record-setting usage of electric power was observed; in addition, a burned-out transformer caused 72,000 residents to lose power in the peak of heat stress; buckling highways lead to road closures; and small communities with well-water systems faced problems in meeting water demands (Palecki, Changnon and Kunkel, 2001).

Soskolne et al. (2004) found that service interruption and infrastructure damage were two of the most commonly reported indirect health outcomes associated with extreme events in Alberta. This included washed out roads, disruption of power, telephone and water services (e.g. flooded water treatment plants and rationing of water in drought-prone communities), closure of medical facilities (e.g. hospitals), and water contamination. One particularly dangerous situation occurred in Fort McMurray, where the only road in and out of the community was closed because of a forest fire, making evacuation impossible, medical emergencies more dangerous, and the delivery of basic supplies was prevented (Soskolne et al., 2004).

Public Health Services

Primary health care services, including basic emergency services, are the first points of contact with the health care system. Secondary services are more specialized care, such as hospitals (Health Canada, 2006). Hospitals and emergency centres will not only have to cope with additional acute injuries and illnesses from an extreme event (e.g. heat wave), but also with individuals that can no longer manage their pre-existing health condition (e.g. cardiovascular disease) that has been amplified by an extreme event (Powell, 2006).

An extreme event can quickly become a disaster if health services infrastructure is not prepared. This may require the relocation or retrofitting public health infrastructure, such as hospitals, clinics, and nursing homes that could be severely damaged from extreme events, a continuity of services for current or emergent individuals during times of internal system disruptions and external community impacts, and an alternate system of communication (F/P/T Network on Emergency Preparedness and Response, 2004).

ADAPTATION AND MITIGATION OF FUTURE ILL HEALTH FROM CLIMATE CHANGE

As explained in previous sections, the health impacts from climate change are complex and many issues work in synergy to further impact the health and well-being of prairie people. Because of this multifaceted nature of climate change and human health, adaptation must be met by various sectors working together. Table 2 is a summary of the various climate-sensitive health outcomes expected to impact prairie populations. The table identifies the different sectors outside the health system that can help alleviate the negative health outcomes from

Climate Sensitive Health Outcomes	Adaptations Outside the Health Care System	Current Health System Capacity	Additional Capacity Needed or Research Gaps
Drought-related Stress/Anxiety in agricultural workers	Adoption of drought-resistant crops. Improved irrigation infrastructure.	Health care facilities and programs, education, public outreach	Linking mental health to agri-economic and farm employment statistics and drought years, health-weather warning system
Dust-related health outcomes	Community's soil conservation programs	Health care facilities	Education and awareness for populations at risk, link dust levels with weather variables and health outcomes, health-weather warning system
Wildfire-related health outcomes	Community evacuation plans, wildfire buffer zones, and suppression strategies	Smoke advisories and evacuation measures for affected communities and vulnerable populations	Link known health outcomes to particulate matter levels, address the current capacity to manage expected increase in patients
Drought/flood water quality issues and WBD	Improved technology for water treatment.	Health care facilities, education, public outreach, water quality monitoring, NNDD*	Link water quality, outbreak data, boil water orders to weather variables locally and distally (e.g., watershed), precipitation-health advisory for water treatment plants, public outreach
Drought-induced water quantity	Improvement of community infrastructure, Local conservation programs	not applicable	Future water needs for an expanding population. Better public outreach about the need to conserve water and health
Increasing average temperatures	Heat-resistant crops	Health care facilities, education, public outreach, NNDD*	Link foodborne pathogens (along the food processing chain) to weather variables, temperature-health advisory for points along the food processing chain, public outreach
Air pollution and respiratory illnesses	Mitigation of fossil fuel combustion by-products departments of environment, conservation, infrastructure, energy	Health care facilities, education, public outreach, warning systems	Link respiratory illness variables to weather variables and air pollution levels, the use of air mass analysis, address the current capacity to manage expected increase in patients
Flooding and post-traumatic stressdisorder/ stress/anxiety	Communities flood mitigation Emergency Preparedness Canada, Red Cross	Health care facilities, education, public outreach, warning systems	Additional community support for flood prevention, address the current capacity to manage expected increase in patients
West Nile Virus (WNV) and Hantavirus Pulmonary Syndrome (HPS)	Departments of environment responsible for reservoir monitoring and surveillance	Health care facilities, education, public outreach, NNDD*	Further research into weather conditions and ecosystem change connected to disease prevalence, weather-health advisories

Table 2. Adaptation measures associated with other sectors outside the health care system that have the potential to alleviate the negative health impacts from climate change

*National Notifiable Disease Database

climate change, as well as the current capacity the health system has to manage the known weather-related health outcomes. For example, drought has a negative mental health effect on farmers. The current health care system can offer a variety of mental health services, education and public outreach initiatives to support farmers during tough times; however, to mitigate future ill-health outcomes from drought, agriculture itself must adapt to a changing climate.

The final column in Table 2 looks at how the health sector could be improved to reach a higher level of adaptation to climate change in the future. This includes gaps in knowledge and research initiatives such that the health care system can play a larger role in understanding the health consequences from climate change, and how to mitigate negative health outcomes in the future. Health care is a defining characteristic of Canadian culture. Substantial capacity (e.g. monitoring or surveillance measures) is available and may need only modification to make current systems more applicable to climate change. For example, building the capacity to link current climate-sensitive health outcomes (e.g. respiratory illnesses) to weather and climate variables will allow researchers to better determine how changes in climate might affect respiratory illness in the future, and attempt to provide health services that will mitigate outcomes in the future.

As evident in Table 2, the health sector cannot act alone in alleviating negative health outcomes from climate change. The health care system does a good job at treating illnesses that are climate sensitive, but is separated from the adaptation and mitigation processes. This is evident as climate change or global warming is not a topic addressed by provincial departments of Health. Subsequently, climate change is an important topic for provincial governmental departments of Environment or Conservation, Infrastructure, Agriculture, Energy, Science and Technology, Mining, and Sustainable and Natural Resources. In other words, public health systems are good at treating the ill-health outcomes expected with climate change, but are not involved in their prevention, or the mitigation of climate change itself (e.g. reduction of greenhouse gases).

Health can be a powerful tool and motivator for people to take climate change more seriously. Therefore, the health sector needs to further collaborate and partner with other sectors and departments to play a greater role in adaptation to climate change itself, as well as mitigating as much ill health as possible from climate change. However, it is important to note that the Public Health Agency of Canada (*http://www.phac-aspc.gc.ca*) is actively collecting and analyzing various health data types, as well as attempting to link other data sources (e.g. agricultural and weather data) to health data, which could play an important role for health and climate change research in the Prairie Provinces. The main roles of the health sector should be: 1) have a health liaison in each affected sector that climate change can have a direct or indirect impact on health; 2) work with the various sectors to integrate relevant data, such that future negative health outcomes can be better predicted and prepared for; 3) continue to research the linkages between the various sectors and health as it pertains to a changing climate; and 4) create a sector and/or weather-specific health warning system, in addition to increasing public outreach on the health effects from climate change. For example, health research can determine the relationship between temperature and risk of food-borne diseases. This information could be translated into a set of guidelines for a temperature warning system for various points along the food processing chain, along with a public outreach campaign that is temperature-dependent. However, these ideals can only be met with close collaboration with agricultural departments and industries.

GAPS IN KNOWLEDGE

Gaps in knowledge are those areas within the health sector that require strengthening, outside the collaboration between the various sectors involved in adaptation and mitigation. The primary area of concern is research, followed by public outreach, and identification of vulnerability.

Meaningful climate change and human health research needs to be based on sound climate scenarios and plausible human health outcomes. First, there is a need to determine baseline prevalence or incidence rates of climate-sensitive health outcomes at specific spatial (e.g. ecozones or urban centres) and temporal (e.g. daily, monthly, or seasonally) variables, specific to the prairie region. Second, a better understanding of the relationship between weather variables and mediating variables that impact health and well-being indirectly (i.e. air pollution and dust levels, intake water quality from water treatment plants) is required. This will need to be eventually linked to health outcomes. Third, the relationship between weather variables and climate-sensitive health outcomes, such as temperature-mortality relationships (i.e. the temperature of minimum mortality) or food-borne disease outbreaks, needs to be better understood in the context of the prairie region. Fourth, it is necessary to assess the capacity of the health care system to handle an increased demand for various health services during severe weather events. Fifth, the use of ecological level variables (e.g. air masses, vegetation cover, vector densities) to aid in the understanding and possible surveillance of climate-sensitive diseases must be investigated. Finally, a climate-sensitive surveillance system, incorporating current data, as well as new data to be collected, needs to be implemented in the prairie region.

Public outreach is needed to follow up research outcomes. As mentioned previously, human health can be a motivator for change, and change must come from all levels—from individuals changing their lifestyles to industry changing best practices based on weather conditions. Climate change and its health effects is a huge concept, and it is at times confusing for the public. Intelligent and directed communication and education campaigns aimed at a specific population (e.g. farmers during drought only) can have maximal effects with little confusion.

Those populations and places that are vulnerable today will likely become more vulnerable in the future under climate change, if appropriate adaptation is not attained. Thus, identification of vulnerable populations and places within communities and regions is paramount. Each community will have to identify its vulnerability with respect to climate change outcomes (e.g. drought) and, subsequently, its vulnerable places, infrastructure and economies. The health sector will then need to work closely with municipalities to identify vulnerable populations within each community, and devise a health support structure appropriate for each community. By addressing the needs of vulnerable populations within each region or community specifically, the prairie region as a whole can be better equipped to lessen climate change's impacts on human health.

SUMMARY

The Canadian prairie region will likely experience climate changes that lead to drought, increased average temperatures, extreme precipitation, and changing ecosystems. Negative health consequences include, food-borne diseases, vector-borne diseases, ill mental health, exacerbation of acute (e.g. asthma) and chronic (e.g. cardiovascular diseases) conditions, and even mortality. Vulnerable populations will likely carry most of the ill-health burden from climate change.

The health sector plays a large role in relieving the negative health outcomes expected by climate change, but plays only a small part in the mitigation of the ill-health effects associated with climate change. Thus, the health sector needs to work more closely with all the sectors that are negatively affected by climate change, as these sectors can alleviate much of the human health burden associated with climate change via appropriate adaptations. Various gaps in knowledge (additional research, followed by public outreach, and identification of vulnerability within communities and regions) need to be addressed to further the health sector's ability to mitigate the ill-health effects from climate change.

REFERENCES

Ammann, H., R. Blaisdell, M. Lipsett, S.L. Stone and S. Therriault. 2001.*Wildfire Smoke: A Guide for Public Health Officials.* University of Washington, 27pp.

Basu, R. and J.M. Samet. 2002. "Relationship Between Elevated Ambient Temperature and Mortality: A Review of the Epidemiological Evidence." *Epidemiologic Reviews* 24, no. 2: 190–202.

Bentham, G. and I.H. Langford. 1995. "Climate Change and the Incidence of Food Poisoning in England and Wales." *International Journal of Biometeorology* 39: 81–86.

——. 2001. "Environmental Temperatures and the Incidence of Food Poisoning in England and Wales." *International Journal of Biometeorology* 45: 22–26.

Bernard, S.M., J.M. Samet, A. Grambsch, K.L. Ebi and I. Romieu. 2001. "The Potential Impacts of Climate Variability and Change on Air Pollution-related Health Effects in the United States." *Environmental Health Perspectives* 109 (supplement 2): 199–209.

Booth, N.J. and K. Lloyd. 1999. "Stress in Farmers." *International Journal of Social Psychiatry* 46, no. 1: 67–73.

Bowman, D.M.J.S. and F.H. Jonhston. 2005. "Wildfire Smoke, Fire Management, and Human Health." *EcoHealth* 2: 76–80.

Bridgman, S.A., R.M. Robertson, Q. Syed, N. Speed, N. Andrews and P.R. Hunter. 1995. "Outbreak of Cryptosporidiosis Associated With a Disinfected Groundwater Supply." *Epidemiology & Infection* 115, no. 3: 555–66.

Burnett, R.T., J.R. Brook, M. Smith-Doiron, D. Stieb and S. Cakmak. 1999. "Effects of Particulate and Gaseous Air Pollution on Cardiorespiratory Hospitalizations." *Archives of Environmental Health* 54, no. 2: 130–39.

Burnett, R.T., J.R. Brook, W.T. Yung, R.E. Dales and D. Krewski. 1997. "Association Between Ozone and Hospitalizations for Respiratory Diseases in 16 Canadian Cities." *Environmental Research* 72: 24–31.

Burnett, R.T., S. Cakmak and J.R. Brook. 1998. "The Effect of the Urban Ambient Air Pollution Mix on Daily Mortality Rates in 11 Canadian Cities. *Canadian Journal of Public Health* 89, no. 3: 152–56.

Burnett, R.T., R. Dales, D. Krewski, R. Vincent, T. Dann, J.R. Brook. 1995. "Associations Between Ambient Particulate Sulfate and Admissions to Ontario Hospitals for Cardiac and Respiratory Diseases." *American Journal of Epidemiology* 142, no. 1: 15–22.

Carruth, A.K. and C.A. Logan. 2002. "Depressive Symptoms in Farm Women: Effects of Health Status and Farming Lifestyle Characteristics, Behaviors, and Beliefs." *Journal of Community Health* 27, no. 3: 213–28.

CASA (Clean Air Strategic Alliance). 2007. *Air Quality Index* [Online]. [accessed February 2009]. Available from World Wide Web: *http://www.casadata.org/airqualityindex/index.asp*

CDC (Centres for Disease Control and Prevention). 2006. *West Nile Virus: What You Need to Know* [Online]. CDC fact sheet. Department of Health and Human Services. [accessed February 2009]. Available from World Wide Web: *http://www.cdc.gov/ncidod/dvbid/westnile/wnv_factsheet.htm*

Charron, D.F., T. Edge, M.D. Fleury, W. Galatianos, D. Gillis, R. Kent, A.R. Maarouf, C. Neudoerffer, C.J. Schuster, M.K. Thomas, J. Valcour and D. Waltner-Toews. 2005. *Final Report: Links between Climate, Water, and Waterborne Illness, and Projected Impacts of Climate Change.* HPRP File No. 6795-15-2001/4400016c. 164 pgs.

Charron, D.F., M.K. Thomas, D. Waltner-Toews, J.J. Aramini, T. Edge, R.A. Kent, A.R. Maarouf and J. Wilson. 2004. "Vulnerability of Waterborne Diseases to Climate Change in Canada: A Review." *Journal of Toxicology and Environmental Health, Part A* 67: 1666–77.

Coleman, R.N., J.N. Campbell, F.D. Cook and D.W.S. Westlake. 1974. "Urbanization and the Microbial Content of the North Saskatchewan River." *Applied Microbiology* 27, no. 1: 93–101.

Confalonieri, U., B. Menne, R. Akhtar, K.L. Ebi, M. Hauengue, R.S. Kovats, B. Revich and A. Woodward. 2007. "Human Health." pp. 391–431 in M.L. Parry, O.F. Canziani, J.P. Palutikof, P.J. van der Linden and c.e. Hanson (eds.). *Climate Change 2007: Impacts, Adaptation and Vulnerability. Contribution of Working Group II to the Fourth Assessment Report of the Intergovernmental Panel on Climate Change.* Cambridge University Press: Cambridge, UK.

Craven, R.B. and J.T. Roehrig. 2001. "West Nile Virus." *JAMA* 286, no. 6: 651–53.

Curriero, F.C., J.A. Patz, J.B. Rose and S. Lele. 2001. "The Association between Extreme Precipitation and Waterborne Disease Outbreaks in the United States, 1948–1994." *American Journal of Public Health* 91, no. 8: 1194–99.

Deary, I., J. Willcock and J. McGregor. 1997. "Stress in Farming." *Stress Medicine* 13: 131–36.

Delfino, R.J., a.m. Murphy-Moulton, R.T. Burnett, J.R. Brook and R.M. Becklake. 1997. "Effects of Air Pollution on Emergency Room Visits for Respiratory Illnesses in Montreal, Quebec." *American Journal of Respiratory and Critical Care Medicine* 155: 568–76.

Diaz, J., A. Jordan. R. Garcia, C. Lopez, J.C. Alberdi, E. Hernandez and A. Otero. 2002. "Heat Waves in Madrid 1986–1997: Effects on the Health of the Elderly." *International Archives of Occupational and Environmental Health* 75: 163–70.

do Pico, G.A. 1986. "Report on Disease." *American Journal of Industrial Medicine* 10: 261–65.

——. 1992. "Hazardous Exposures and Lung Diseases among Farm Workers." *Clinics in Chest Medicine* 13, no. 2: 311–28.

D'Souza, R.M., N.G. Becker, G. Hall and K.B.A. Moodie. 2004. "Does Ambient Temperature Affect Foodborne Disease?" *Epidemiology* 15, no. 1: 86–92.

Duncan, K., T. Guidotti, W. Cheng, K. Naidoo, G. Gibson, L. Kalkstein, S. Sheridan, D. Waltner-Toews, S. MacEachern and J. Last. 1998. "Canada Country Study: Impacts and Adaptation. Health Sector." In G. Koshida and W. Avis (eds.). *Volume VII: National Sectoral Volume*. Environment Canada.

Durkin, M.S., N. Khan, L.L. Davidson, S.S. Zaman and Z.A. Stein. 1993. "The Effects of a Natural Disaster on Child Behaviour: Evidence for Posttraumatic Stress." *American Journal of Public Health* 83, no. 11: 1549–53.

Eisner, C.S., R.D. Neal and B. Scaife. 1998. "Depression and Anxiety in Farmers." *Primary Care Psychiatry* 4, no. 2: 101–05.

Emmanuel, S.C. 2000. "Impact to Lung Health from Forest Fires: The Singapore Experience." *Respirology* 5: 175–82.

Engelthaler, D.M., c.e. Levy, T.M. Fink, D. Tanda and T. Davis. 1998. "Short Report: Decrease in Seroprevalence of Antibodies to Hantavirus in Rodents from 1993–1994 Hantavirus Pulmonary Syndrome Cases." *American Journal of Tropical Medicine and Hygiene* 58: 737–38.

Environment Canada. 2006, *Air Pollution Sources in the Prairies and Northern Region* [online]. [accessed January 2009]. Available from World Wide Web: *http://www.ec.gc.ca/cleanair-airpur/Regional_Clean_Air_Online/Prairie_and_Northern_Region/Air_Pollution_Sources_in_the_Prairies_and_Northern_Regions-WS44A2C6B9-1_En.htm*

Epstein, P.R. 2001. "West Nile Virus and the Climate." *Journal of Urban Health: Bulletin of the New York Academy of Medicine* 78, no. 2: 367–71.

Faunt, J.D., P. Henschke, T.J. Wilkinson, M. Webb, P. Aplin and R.K. Penhall. 1995. "The Effete in the Heat: Heat-related Hospital Presentations During a Ten Day Heat Wave." *Australian and New Zealand Journal of Medicine* 25: 117–21.

Fischer, P.J. and W.R. Breakey. 1991. "The Epidemiology of Alcohol, Drug, and Mental Disorders among Homeless Persons." *American Psychologist* 46, no. 11: 1115–28.

Fleury, M., D.F. Charron, J.D. Holt, O.B. Allen and A.R. Maarouf. 2006. "A Time Series Analysis of the Relationship of Ambient Temperature and Common Bacterial Enteric Infections in Two Canadian Provinces." *International Journal of Biometerology* 50: 385–91.

Flynn, A., C. McGreevy and E.C. Mulkerrin. 2005. "Why Do Older Patients Die in a Heatwave?" *Quarterly Journal of Medicine* 98: 227–29.

Fraser, c.e., K. B. Smith, F. Judd, J.S. Humphreys, L.J. Fragar and A. Henderson. 2005. "Farming and Mental Health Problems and Mental Illness." *International Journal of Social Psychiatry* 51, no. 4: 340–49.

F/P/T Network on Emergency Preparedness and Response. 2004. *National Framework for Health Emergency Management: Guideline for Program Development*. Prepared for the Conference of F/P/T Ministers of Health. F/P/T Network for Emergency Preparedness and Response.

Ginexi, E.M., K. Weihs, S.J. Simmens and D.R. Hoyt. 2000. "Natural Disaster and Depression: A Prospective Investigation of Reactions to the 1993 Midwest Floods." *American Journal of Community Psychology* 28, no. 4: 495–518.

Glass, G.E., J.E. Cheek, J.A. Patz, T.M. Shields, T.J. Doyle, D.A. Thoroughman, D.K. Hunt, R.E. Enscore, K.L. Gage, C. Irland, C.J. Peters and R. Bryan. 2000. "Using Remotely Sensed Data to Identify Areas of Risk for Hantavirus Pulmonary Syndrome." *Emerging Infectious Disease* [serial online] 6, no. 3: 238–47.

Green, F.H.Y., K. Yoshida, G. Fick, J. Paul, A. Hugh and W.F. Green.1990. "Characterization of Airborne Mineral Dusts Associated with Farming Activities in Rural Alberta, Canada." *International Archives of Occupational and Environmental Health* 62: 423–30.

Greenough, G., M. McGeehin, S.M. Bernard, J. Trtanj, J. Riad and D. Engelberg. 2001. "The Potential Impacts of Climate Variability and Change on Health Impacts of Extreme Weather Events in the United States." *Environmental Health Perspectives* 109 (supplement 2): 191–98.

Gregoire, A. 2002. "The Mental Health of Farmers." *Occupational Medicine* 52, no. 8: 471–76.

Gubler, D.J., P. Reiter, K.L. Ebi, W. Yap, R. Nasci and J.A. Patz. 2001. "Climate Variability and Change in the United States: Potential Impacts on Vector- and Rodent-borne Diseases." *Environmental Health Perspectives* 109 (suppl 2): 223–33.

Hall, G.V., R.M. D'Souza and M.D. Kirk. 2002. "Foodborne Disease in the New Millennium: Out of the Frying Pan and into the Fire." *MJA* 177: 614–18.

Haller, L., C. Claiborn, T. Larson, J. Koenig, G. Norris and R. Edgar. 1999. "Airborne Particulate Matter Size and Distributions in an Arid Urban Area." *Journal of Air and Waste Management Association* 49: 161–68.

Health Canada. 1998. *National Ambient Air Quality Objectives for Particulate Matter: Executive Summary. Part 1: Science Assessment Document.* Minister, Public Works and Government Services: Ontario, Canada.

——. 2005a. "Diabetes among Aboriginal (First Nations, Inuit, and Metis) People in Canada: The Evidence." Health Canada, Ottawa, Ontario, March 10, 2000a. In Health Canada. *Your Health and a Changing Climate: Information for Health Professionals.* Minister of Health: Ottawa, Ontario.

——. 2005b. *Literature Review- Evaluation Strategies in Aboriginal Substance Abuse Programs: A Discussion* [online]. [accessed January 2009]. Available from World Wide Web: *http://www.-hc-sc.gc.ca/fniah-spnia/pubs/substan/_ads/literary_examen_review/index-eng.php*

——. 2005c. *First Nations, Inuit, and Aboriginal Health: Statistical Profile on the Health of First Nations in Canada* [online]. [accessed February 2009]. Available from World Wide Web: *http://www.hc-sc.gc.ca/fniah-spnia/pubs/aborig-autoch/stats_profil-eng.php*

——. 2006. *Canada's Health Care System: How Health Care Services are Delivered* [online]. [accessed February 2009]. Available from World Wide Web: *http://www.hc-sc.gc.ca/hcs-sss/pubs/system-regime/2005-hcs-sss/del-pres-eng.php*

——. 2007. First Nations and Inuit Health: Diseases and Health Conditions [online]. [accessed January 2009]. Available from World Wide Web: *http://www.hc-sc.gc.ca/fnih-spni/diseases-maladies/index_e.html*

Heinz Center. 2002. *Human Links to Costal Disasters.* The H. John Heinz III Center for Science, Economics and the Environment: Washington DC.

Henstra, D., P. Kovacs, G. McBean and R. Sweeting. 2004. *Backgrounder Paper on Disaster Resilient Cities.* Institute for Catastrophic Loss Reduction for Infrastructure Canada: Ottawa, Ontario.

Hett, H.A. and D.A. Brechtelsbauer. 1998. "Heat-related Illnesses: Plan Ahead to Protect Your Patients." *Postgraduate Medicine* 103, no. 6: 107–08.

Huhn, G.D., J.J. Sejvar, S.P. Montgomery and M.S. Dworkin. 2003. "West Nile Virus in the United States: An Update on an Emerging Infectious Disease." *American Family Physician* 68, no. 4: 653–60.

Hutton, D. 2001. *Psychosocial Aspects of Disaster Recovery: Integrating Communities into Disaster Planning and Policy Making*. Institute for Catastrophic Loss Reduction, Paper Number 2.

Hwang, SW. 2001. "Homelessness and Health." CMAJ 164, no. 1: 229–33.

Indian and Northern Affairs. 2008. *Social Development Program* [online]. [accessed February 2009]. Available from World Wide Web: *http://www.ainc-inac.gc.ca/ai/mr/is/sdpr-eng.asp*

IPCC (Intergovernmental Panel on Climate Change). 2001. Working Group I. *Third Assessment Report: Climate Change 2001: The Scientific Basis, Summary for Policy Makers*. Cambridge University Press: Cambridge, UK.

IJC (International Joint Commission). 2000. *Living with the Red: A Report to the Governments of Canada and the United States on Reducing Flood Impacts in the Red River Basin*. International Joint Commission: Washington D.C. and Ottawa.

Jermini, M.F.G. 2002. "Foodborne diseases." In G. Tamburlini, O.V. Ehrenstein and R. Bertollini (eds.). *Children's Health and Environment: A Review of Evidence*. Environmental Issues Report No. 29. European Environmental Agency (EAA): Copenhagen, Denmark.

Johnston, F.H., a.m. Kavanagh, D.M.J.S. Bowman and R. Scott. 2002. "Exposure to Bushfire Smoke and Asthma: An Ecological Study." *Medical Journal of Australia* 176: 535–38.

Jones, R.T., D.P. Ribbe, P.B. Cunningham, J.D. Weddle and A.K. Langley. 2002. "Psychological Impact of Fire Disaster on Children and their Parents." *Behavior Modification* 26, no. 2: 163–86.

Kalkstein, L.S. and K.E. Smoyer. 1993. "The Impact of Climate Change on Human Health: Some International Implications." *Experentia* 49: 969–79.

Kampbell, D.H., Y.J. An, K.P. Jewell and J.R. Masoner. 2003. "Groundwater Quality Surrounding Lake Texoma During Short-term Drought Conditions." *Environmental Pollution* 125, no. 2: 183–91.

Kenney, W.L. and J.L. Hodgson. 1987. "Heat Tolerance, Thermoregulation and Ageing." *Sports Medicine* 4: 446–56.

Kilborne, E.M., K. Choi, T.S. Jones, S.B. Thacker, the Field Investigation Team. 1982. "Risk Factors for Heatstroke: A Case-control Study." *JAMA* 247, no. 24: 3333–36.

Kilborne, E.M. 1999. "The Spectrum of Illness During Heat Waves." *American Journal of Preventative Medicine* 16, no. 4: 359–60.

Koppe, C., S. Kovats, G. Jendritzky and B. Menne. 2004. "Heat-Waves: Risks and Responses." *Health and Global Environmental Change*. Series No. 2. World Health Organization: Copenhagen, Denmark.

Kovats, R.S., S.J. Edwards, S. Hajat, B.G. Armstrong, K.L. Ebi and B. Menne. 2004. "The Effect of Temperature on Food Poisoning: A Time-series Analysis of Salmonellosis in Ten European Countries." *Epidemiology and Infection* 132: 443–53.

Lang, L. 1996. "Danger in the Dust." *Environmental Health Perspectives*. 104, no. 1: 26–30.

Last, J., K. Trouton and D. Pengelly. 1998. *Taking Our Breath Away: the Health Effects of Air Pollution and Climate Change*. David Suzuki Foundation: Vancouver, British Columbia.

Leclerc, H., L. Schwartzbrod and E. Dei-Cas. 2002. "Microbial Agents Associated with Waterborne Diseases." *Critical Reviews in Microbiology* 28, no. 4: 371–409.

Levin, R.B., P.R. Epstein, T.E. Ford, W. Harrington, E. Olson and E.G. Reichard. 2002. "U.S. Drinking Water Challenges in the Twenty-first Century." *Environmental Health Perspectives* 110 (suppl. 1): 43–52.

Lincoln, R. 2005. "Water: The Effects of Climate Change on the Availability and Quality of Drinking Water." In Paul R. Epstein and Evan Mills (eds.). *Climate Change Futures: Health, Ecological and Economic Dimensions*. The Center for Health and the Global Environment, Harvard Medical School: Massachusetts.

Longstreth, J. 1999. "Public Health Consequences of Global Climate Change in the United States: Some Regions May Suffer Disproportionately." *Environmental Health Perspectives* 107 (suppl. 1): 169–79.

Malmberg, A., K. Hawton and S. Simkin. 1997. "A Study of Suicide in Farmers in England and Wales." *Journal of Psychosomatic Research* 43, no. 1: 107–11.

Masley, M.L., K.M. Semchuck, A. Senthilselvan, H.H. McDuffie, P. Hanke, J.A. Dosman, A.J. Cessna, M.F.O. Crossley, a.m. Irvine, a.m. Rosenberg and L.M. Hagel. 2000. "Health and Environment of Rural Families: Results of a Community Canvass Survey in the Prairie Ecosystem Study (PECOS)." *Journal of Agricultural Safety and Health* 6, no. 2: 103–15.

Mathieu-Nolf, M. 2002. "Poisons in the Air: A Cause of Chronic Disease in Children." *Clinical Toxicology* 40, no. 4: 483–91.

May, J.J. 1998. "Clinically Significant Occupational Stressors in New York Farmers and Farm Families." *Journal of Agricultural Safety and Health* 4, no. 1: 9–14.

McDermott, B.M., E.M. Lee, M. Judd and P. Gibbon. 2005. "Posttraumatic Stress Disorder and General Psychopathy in Children and Adolescents Following a Wildfire Disaster." *Canadian Journal of Psychiatry* 50, no. 3: 137–43.

McGeehin, M.A. and M. Mirabelli. 2001. "The Potential Impacts of Climate Variability and Change on Temperature-related Morbidity and Mortality in the United States." *Environmental Health Perspectives* 109 (suppl. 2): 185–89.

Menon, A.S. 1985. "Salmonellae and Pollution Indicator Bacteria in Municipal and Food Processing Effluents and the Cornwallis River." *Canadian Journal of Microbiology* 31: 598–603.

Mills, J.N., T.G. Ksiazek, C.J. Peters and J.E. Childs. 1999. "Long-term Studies of Hantavirus Reservoir Populations in the Southwestern United States: A Synthesis." *Emerging Infectious Diseases* 5, no. 1: 135–42.

Millson, M., M. Bokhout, J. Carlson, L. Spielberg, R. Aldis, A. Borczyk and H. Lior. 1991. "An Outbreak of *Campylobacter jejuni* Gastroenteritis Linked to Meltwater Contamination of a Municipal Well." *Canadian Journal of Public Health* 82: 27–31.

Mokdad, A.H., G.A. Mensah, S.F. Posner, E. Reed, E.J. Simones, M.M. Engelgau and the Chronic Diseases and Vulnerable Populations in Natural Disasters Working Group. 2005. "When Chronic Conditions Become Acute: Preventions and Control of Chronic Diseases and Adverse Health Outcomes During Natural Disasters." *Prevention of Chronic Diseases* [serial online]. [accessed January 2009]. Available from World Wide Web: *http://www.cdc.gov/pcd/issues/2005/nov/05_0201.htm*.

Palecki, M.A., S.A. Changnon and K.E. Kunkel. 2001. "The Nature and Impacts of the July 1999 Heat Wave in the Midwestern United States: Learning from the Lessons of 1995." *Bulletin of the American Meteorological Society* 82, no. 7: 1353–67.

Phifer, J.F. 1990. "Psychological Distress and Somatic Symptoms after Natural Disaster: Differential Vulnerability Among Older Adults." *Psychology and Aging* 5, no. 3: 412–20.

Phifer, J.F., K.Z. Kaniasty and F.H. Norris. 1988. "The Impact of Natural Disaster on the Health of Older Adults: A multiwave Prospective Study." *Journal of Health & Social Behavior* 29, no. 1: 65–78.

Plunkett, S.W., C.S. Henry and P.K. Knaub. 1999. "Family Stressor Events, Family Coping, and Adolescent Adaptation in Farm and Ranch Families." *Adolescence* 34, no. 133: 149–71.

Pond, K. 2002. "Waterborne Gastrointestinal Diseases." In G. Tamburlini, O.v. Ehrenstein and R. Bertollini (eds.). *Children's Health and Environment: A Review of Evidence*. Environmental Issues Report No. 29. European Environmental Agency (EAA): Copenhagen, Denmark.

Powell, S. 2006. Draft Background Paper. Prepared for the Public Health Agency of Canada Invitational Meeting on Emergency Preparedness and Seniors; February 26–28, Division of Aging and Seniors, Public Health Agency of Canada, Toronto, Canada.

PHAC (Public Health Agency of Canada). 2001. *Material Safety Data Sheet—Infectious Substances: Hanatvirus* [Online]. [accessed February 2009]. Available from World Wide Web: *http://www.phac-aspc.gc.ca/msds-ftss/msds74e-eng.php*

——. 2003. *What Makes Canadians Healthy or Unhealthy?* [Online]. [accessed February 2009]. Available from World Wide Web: *http://www.phac-aspc.gc.ca/ph-sp/determinants/ determinants-eng.php#income*

——. 2006. *Hantavirus Pulmonary Syndrome: Notifiable Diseases On-line* [Online]. [accessed March 2009]. Available from World Wide Web: *http://dsol-smed.phac-aspc.gc.ca/dsol-smed/ndis/disease2/hantavirus_e.html*

——. 2008a. *West Nile Virus MONITOR. Human Surveillance (2002 to 2007)* [Online]. [accessed Jan 2009]. Available from World Wide Web: *http://www.phac-aspc.gc.ca/wnv-vwn/mon-hmnsurv-archive_e.html*

——. 2008b. *West Nile Virus MONITOR. 2008 Human Surveillance* [Online]. [Accessed Jan 2009]. Available from World Wide Web: *http://www.phac-aspc.gc.ca/wnv-vwn/mon-hmnsurv_e.html*

Raine, G. 1999. "Causes and Effects of Stress on Farmers: A Qualitative Study." *Health Education Journal* 58, no. 3: 259–70.

Rittmaster, R., W.L. Adamowicz, B. Amiro and R.T. Pelletier. 2006. "Economic Analysis of Health Effects from Forest Fires." *Canadian Journal of Forest Research* 36: 868–77.

Rose, J.B., P.R. Epstein, E.K. Lipp, B.H. Sherman, S.M. Bernard and J.A. Patz. 2001. "Climate Variability and Change in the United States: Potential Impacts onWwater- and Foodborne Diseases Caused by Microbiologic Agents." *Environmental Health Perspectives* 109 (suppl. 2): 211–20.

Rylander, R. 1986. "Lung Diseases Caused by Organic Dusts in the Farm Environment." *American Journal of Industrial Medicine* 10: 221–27.

Schijven, J.F. and a.m. de Roda Husman. 2005. "Effect of Climate Change on Waterborne Diseases in The Netherlands." *Water Science and Technology* 51, no. 5: 79–87.

Schindler, D.W. and W.F. Donahue. 2006. "An Impending Water Crisis in Canada's Western Prairie Provinces." *Proceedings of the National Academy of Sciences of the United States of America* 103, no. 19: 7210–16.

Schneider, S.H., S. Semenov, A. Patwardhan, I. Burton, C.H.D. Magadza, M. Oppenheimer, A.B. Pittock, A. Rahman, J.B. Smith, A. Suarez and F. Yamin. 2007. "Assessing Key Vulnerabilities and the Risk from Climate Change." Pp. 779–810 in M.L. Parry, O.F. Canziani, J.P. Palutikof, P.J. van der Linden and c.e. Hanson (eds.). *Climate Change 2007: Impacts, Adaptation and Vulnerability. Contribution of Working Group II to the Fourth Assessment Report of the Intergovernmental Panel on Climate Change.* Cambridge University Press: Cambridge, UK.

Schuster, C.J., A.G. Ellis, W.J. Robertson, D.F. Charron, J.J. Aramini, B.J. Marshall and D.T. Medeiros. 2005. "Infectious Disease Outbreaks Related to Drinking Water in Canada, 1974-2001." *Canadian Journal of Public Health* 96, no. 4: 254–58.

Semenza, J.C., J.E. McCullough, D. Flanders, M.A. McGeehin and J.R. Lumpkin. 1999. "Excess Hospital Admissions During the July 1995 Heat Wave in Chicago." *American Journal of Preventative Medicine* 16, no. 4: 269–77.

Shea, K.M. 2003. "Global Environmental Change and Children's Health: Understanding the Challenges and Finding Solutions." *Journal of Pediatrics* 143: 149–54.

Shrubsole, D. 1999. *Natural Disasters and Public Health Issues: A Review of the Literature with a Focus on the Recovery Period.* ICLR Research, Paper Series, No. 4. Institute for Catastrophic Loss Reduction.

Simkin, S., K. Hawton, J. Fagg and A. Malmberg. 1998. "Stress in Farmers: A Survey of Farmers in England and Wales." *Occupational and Environmental Medicine* 55: 729–34.

Simpson, J.C.G., R.M. Niven, C.A.C. Pickering, a.m. Fletcher, L.A. Oldham and H.M. Francis. 1998. "Prevalence and Predictors of Work Related Respiratory Symptoms in Workers Exposed to Organic Dusts." *Occupational and Environmental Medicine* 55: 668–72.

Smoyer, K.E. 1998. "Putting Risk in its Place: Methodological Considerations for Investigating Extreme Event Health Risk." *Social Science and Medicine* 47, no. 11: 1809–24.

Smoyer-Tomic, K.E., J.D.A. Klaver, C.L. Soskolne and D.W. Spady. 2004. "Health Consequences of Drought on the Canadian Prairies." *EcoHealth* 1 (suppl. 2): 144–54.

Soskolne, C.L., K.E. Smoyer-Tomic, D.W. Spady, K. McDonald, J.P. Rothe and J.D.A. Klaver. 2004. *Final Report: Climate Change, Extreme Weather Events And Health Effects in Alberta.* HPRP# 6795–15–2001/4400013. 485 pgs.

Square, D. 1997. "Hospital Evacuated, Mental Health Issues Dominated as Manitoba Coped with Flood of Century." *Canadian Medical Association Journal* 156: 1742–45.

Statistics Canada. 2003. *2001 Census: Analysis Series. Aboriginal Peoples of Canada: A Demographic Profile.* Minister of Industry 2003: Ottawa, Ontario.

——. 2008a. *Cattle Inventories, by Province*, CANSIM, table 003–0032 [online]. Catalogue no. 23–012-X. Statistics Canada [accessed January 2009]. Available from World Wide Web: *http://www40.statcan.ca/10, 1/cst01/prim50a.htm*

——. 2008b. *Pig Inventories, by Province* (quarterly), CANSIM, table 003–0004 [online]. Catalogue no. 23–010-X. Statistics Canada [accessed January 2009]. Available from World Wide Web: *http://www40.statcan.ca/101/cst01/prim51a.htm*

Stephen, C., M. Johnson and A. Bell. 1994. "First Reported Cases of Hantavirus Pulmonary Syndrome in Canada." *Canada Communicable Diseases Report* 20, no. 15: 121–25.

Tamburlini, G. 2002. "Children's Special Vulnerability to Environmental Health Hazards: An Overview." In G. Tamburlini, O.v. Ehrenstein and R. Bertollini (eds.). *Children's Health and Environment: A Review of Evidence.* Environmental Issues Report No. 29. European Environmental Agency (EAA): Copenhagen, Denmark.

Theron, J. and T.E. Cloete. 2004, "Emerging Waterborne Infections: Contributing Factors, Agents and Detection Tools." *Critical Reviews in Microbiology* 28, no. 1: 1–26.

Thomas, M.K., D.F. Charron, D. Walnter-Toews, C. Schuster, A.R. Maarouf and J.D. Holt. 2006. "A Role of High Impact Weather Events in Waterborne Disease Outbreaks in Canada, 1975–2001." *International Journal of Environmental Health Research* 16, no. 3: 167–80.

Turcotte, M. and G. Schellenberg. 2007. *A Portrait of Seniors in Canada: 2006.* Statistics Canada, Minister of Industry: Ottawa, Canada.

Tyler, K.A. and D.R. Hoyt. 2000. "The Effects of an Acute Stressor on Depressive Symptoms Among Older Adults: The Moderating Effects of Social Support and Age." *Research on Aging* 22, no. 2: 143–64.

Vanderpost, J.M. and J.B. Bell. 1977. "Bacteriological Investigation of Alberta Meat-packing Plant Wastes with Emphasis on *Salmonella* Isolation." *Applied and Environmental Microbiology* 33: 538–45.

Walker, J.L. and L.J.S. Walker. 1988. "Self Reported Stress Symptoms in Farmers." *Journal of Clinical Psychology* 44, no. 1: 10–16.

Walker, J.L., L.S. Walker and p.m. MacLennan. 1986. "An Informal Look at Farm Stress." *Psychological Reports* 59: 427–30.

Worfolk, J.B. 2000. "Heat Waves: Their Impact on the Health of Elders." *Geriatric Nursing* 21, no. 2: 70–77.

World Health Organization (WHO). 2009. *Water* [Online]. [Accessed Jan 2009]. Available from World Wide Web: *http://www.who.int/topics/water/en/*

Zwolsman, J.J.G. and A.J. van Bokhoven. 2007. "Impact of Summer Droughts on Water Quality of the Rhine River—A Preview of Climate Change?" *Water Science & Technology* 56, no. 4: 45–55.

CHAPTER 12

GOVERNMENT INSTITUTIONS AND CLIMATE CHANGE POLICY

Margot Hurlbert and Darrell R. Corkal

INTRODUCTION

A changing climate impacts many human social and economic activities. Impacts will affect industry, agriculture, health, rural and urban water supplies, power generation, recreation, and other important areas of economic activity. Increased water scarcity, more frequent extreme events (e.g. droughts, floods, storms), and changes or disturbances in climate regimes will impact ecosystems with such events as forest fires; insect outbreaks (e.g. mountain pine beetle); vector-borne disease, such as West Nile virus; and the introduction of non-native plants and animals. All of these impacts will place significant pressure on the environment as well as on human activities, such as farming, agriculture, forestry, development, and industry in the Prairie Provinces (Sauchyn and Kulshreshtha, 2008). Such climate change impacts will undoubtedly affect government institutions with a mandate relating to this broad range of activities. Arguably, the disparate effects of climate change leave all government institutions to contemplate impacts on mandates, programs and policies.

This chapter provides an overview of climate change planning at mid-year 2009 for the three Prairie Provinces and the federal government. The chapter summarizes these governments' overarching climate change strategies, plans and policies relating to climate change mitigation and adaptation. A brief discussion of government institutions that have a central role in relation to climate change, and their role in helping society adapt to climate change, is provided along with a review of the literature on effective climate change policy. These findings are synthesized, and observations are provided to help strengthen adaptive capacity and effectiveness of government policy responses to climate change. Increased citizen engagement, long-term comprehensive planning, and designing flexibility in climate change plans are suggested options that will build resilience and help society cope with climate change and climate variability.

GOVERNMENT INSTITUTIONS, ENVIRONMENT, AND CLIMATE CHANGE ROLES

A review will be made of the government institutions with mandates relating directly to the issue of climate change and their climate change roles in the three Prairie Provinces and the federal government. The institutions discussed below either have a direct or related mandate concerning climate or its effects. The linkage to climate is most often related to the natural or indoor environment—interactions between climate and air, water, land, biodiversity, or associated effects on human health through air, food and water intake. It is important to consider the multitude of provincial and federal institutions in the Prairie Provinces with linkages in their program and policy mandate to climate change requiring due consideration of emerging climate change information.

THE FEDERAL LEVEL—THE GOVERNMENT OF CANADA

At the federal level of government, two departments have been tasked with key lead climate change responsibilities: Environment Canada has a specific role in relation to mitigation of greenhouse gases, while Natural Resources Canada has specific responsibilities in relation to adaptation to a changing climate. Because of the breadth and depth of potential climate change impacts, many other federal government agencies and departments also have inescapable roles in relation to climate change. For example, the departments of agriculture, fisheries and health can each expect to be impacted by the effects of climate on their various activities. Several of these key institutions will be mentioned in this overview to help illustrate the complexity and diversity of the government actors involved and potentially affected by climate change impacts.

Environment Canada has a mandate to preserve and enhance the quality of the natural environment, conserve and protect Canada's water resources, forecast weather and environmental change, enforce regulations relating to transboundary waters (international and inter-provincial), and coordinate environmental policies and programs for the federal government. Environment Canada employs about 6,800 people with an annual budget of just over $1 billion. Approximately 60% of its workforce and 80% of its budget relates to science and technology activities (Environment Canada, 2008a, 2008b). Because of Environment Canada's mandate, changing weather and, hence, climate change itself are central to aspects of its mandate. In addition to its longstanding environmental protection mandate, Environment Canada is responsible for overseeing the federal government's plan on the reduction of greenhouse gas emissions and the regulatory framework supporting this plan. The central regulatory tool in the reduction of greenhouse gas emissions proposed by the federal government is the Clean Air Act (Bill C-30 tabled in 2006 as "An Act to amend the Canadian Environmental Protection Act, 1999" in the First Session of the 39th Parliament), which will be the responsibility of Environment Canada. Environment Canada also oversees other important legislative tools which could be significant in relation to climate change; examples include the conducting of environmental assessments which are required prior to the building of a new development, conducting water research related to water quantity and quality, monitoring trans-boundary flows, and interacting with the provincial ministers of the environment to establish the *Canadian Environmental Quality Guidelines*. These environmental guidelines relate to such things as air, soil and water quality, and associated ecosystems. Environment Canada has also assumed responsibility for interacting with key industry sectors respecting greenhouse gas emissions through the Large Final Emitters Group.

The Large Final Emitters Group was established in 2002 as part of Natural Resources Canada, but with the release of *Project Green: Moving Forward on Climate Change* in 2005 (Environment Canada, 2009b), the group was transferred to Environment Canada given the Government's intent to regulate Large Final Emitters under the Canadian Environmental Protection Act, 1999, S.C. 1999, c. 33. This group, made up of staff of Environment Canada, is responsible for working with key industry sectors to reduce annual greenhouse gas emissions. Through its discussions with industry, provinces and territories, and other stakeholders, the Large Final Emitters Group is tasked with designing policies and measures that are effective in encouraging reductions and are administratively efficient and clear, while still maintaining the competitiveness of Canadian industry. The group was originally created under the purview of the Minister of Natural Resources.

Natural Resources Canada promotes the responsible use of natural resources (minerals, metals, energy, forests, earth sciences) for the protection of human health, the environment and the landmass, and conducts water research (including groundwater research and mapping activities) and other research related to forestry, mining and energy sectors. It is organized into sectors which include Canadian Forest Service, Energy Policy Sector, Energy Technology and Programs Sector, Earth Sciences Sector, Minerals and Metals Sector, and Corporate Management Sector. There is also an Office of the Chief Scientist tasked with developing a departmental profile, vision and direction for NRCan science and technology, and providing advice to the government on national and global scientific trends and developments.

Many other federal departments have a mandate relating directly or indirectly to climate change. The principal departments are the remaining three of the

5NR (five natural resource departments): Environment Canada, Natural Resources Canada (both previously discussed), Health Canada, Fisheries and Oceans, and Agriculture and Agri-Food Canada (Morrison and Gee, 2001). Health Canada has many responsibilities relating to public safety, including safe work environments, safe drinking water, and establishing the Guidelines for Canadian Drinking Water Quality in partnership with federal-provincial-territorial governments. A specific climate change and health office dedicated to the impacts of climate change has been created. Health Canada also co-leads the Canadian Environmental Protection Act with Environment Canada. Fisheries and Oceans Canada is responsible for fisheries and the protection and restoration of fish habitat. Agriculture and Agri-Food Canada (AAFC) promotes a competitive agricultural industry and agricultural practices that protect the environment. Because of agriculture's use and dependence on water, climate and land resources, this is a key sector and department that will most significantly be affected by any climate variability and climate change. Agriculture and Agri-Food Canada undertakes extensive agricultural and agri-environmental research, and promotes adoption of best (beneficial) agricultural management practices to protect the environment. The Prairie Farm Rehabilitation Administration (PFRA), as a part of AAFC, was created as a special agency in 1935 in response to severe climate impacts on the Prairies caused by the multi-year droughts of the 1920s and 1930s. Its mandate related to improving the security of prairie land and water resources to achieve greater economic security of agriculture in the semi-arid prairie region; its role recently evolved to include national agri-environmental activities related to promoting farm practices that afford a competitive profitable agricultural sector with improved protection of the environment from potential agricultural contamination risks. PFRA continues to be active some 74 years later, albeit under the new name of Agri-Environment Services Branch (AESB).[1] AESB's mandate is national in scope and focused on agri-environmental sustainability and innovation to promote a competitive profitable agricultural sector.

Many other federal departments have mandates indirectly affected by climate change. A few examples are presented. Industry Canada has a mandate related to industry and development. Activities of industry and development have both implications in respect of emission mitigation or reduction and adaptation to future climate in respect of industrial activities. Parks Canada has a mandate for water resource management and protection within national parks. Indian and Northern Affairs Canada has shared responsibilities with First Nations Band Councils for the environment, land and water on First Nations reserve communities and many water acts relating to northern Canada. Both these agencies will have implications on both their land and water portfolio in respect of climate change implications with significant effects on Canada's water resources.

THE PROVINCIAL LEVEL IN THE PRAIRIE PROVINCES

The provincial governments of the three Prairie Provinces have similar challenges within the bureaucratic structure in responding to climate change because of the breadth and depth of the challenge and the pre-existing government structure of institutions created prior to today's focus on climate change and water. Each province's response to climate change has been somewhat different in both form and substance. Generally, one particular provincial department is tasked with climate change and a few others have supporting mandates or programs.

In Alberta, Alberta Environment is tasked with addressing climate change. After establishing its 2002 climate change plan, and after a series of public meetings, Alberta renewed its climate change plan early in 2008 (Alberta Environment, 2008b). Alberta Environment is also tasked with both the protection and wise use of the environment and natural resources. Most activities affecting the environment must be reviewed through environmental assessment by Alberta Environment. Alberta Environment also carries the bulk of the responsibilities relating to both water and the environment, which includes surface and groundwater allocation, flow regulation, water supply and flood forecasting, pollution control, regulation of municipal potable water systems, and developing watershed management plans in conjunction with local groups. Alberta Environment also establishes and enforces drinking water and waste water objectives and legislation.

Other Alberta government departments also have a stake in climate change because of the impacts on the subject matter under their jurisdiction. For instance, the Alberta Ministry of Sustainable Resource Development is responsible for management of public lands, fish and wildlife, and forestry, and the Ministry of Agriculture and Rural Development promotes beneficial agricultural management practices and environmental farm planning to protect the environment, and supports and conducts agricultural research as well as water supply projects to meet farm water quality needs. A changing climate impacts all of these activities and associated programs. Infrastructure and Transportation administers grant programs which support the development of municipal water and wastewater systems. Regional Health Authorities (funded and controlled by Alberta Health and Wellness) advise clients on their private water system, and monitor municipal drinking water supplies and recreational water sources. Alberta Research Council supports partners and customers in innovative applied research and technology development in sectors such as energy, agriculture, environment, and forestry. Climate change as a topic, as well as the implications on all of these research areas, will require focus by Alberta Research Council.

In Saskatchewan, the Ministry of the Environment has been central in developing a climate change strategy. This ministry protects and manages Saskatchewan's natural resources; leads environmental water quality monitoring and enforces environmental protection guidelines; sets water quality regulations and objectives (including those for drinking water and waste water); and retains legislative responsibility for enforcing municipal drinking water regulations. Unlike Alberta, which manages both climate and water resources under the environment department, Saskatchewan manages its water resources by a separate Crown corporation, the Saskatchewan Watershed Authority. The Saskatchewan Watershed Authority was created to manage water in a holistic fashion aimed at protecting water source; the agency was established in 2002 as a direct response to the North Battleford drinking water disease outbreak in 2001. Saskatchewan Watershed Authority is responsible for protecting source water by promoting stewardship, water management, water allocations and diversions, and developing watershed and source water protection plans in partnership with watershed groups. Other departments also play a key role in responding to climate change, including the Ministry of Agriculture in respect of farming, improving farming practice, enforcing the Agricultural Operations Act, promoting environmental farm planning, and encouraging the adoption of agricultural beneficial management practices (BMPs) to protect water resources and the environment. The Ministry of Health is involved with drinking water quality, water quality testing, and public and environmental health protection. The cost-recovery Crown corporation, SaskWater, is responsible for water development, treatment and water infrastructure, and assisting Saskatchewan communities in meeting their water needs. The Saskatchewan Research Council is responsible, in part, for climate and water research. Climate change impacts on water will affect all of these government institutions.

Manitoba differs in one major respect from both Alberta and Saskatchewan. Climate and Green Initiatives, a branch of Manitoba Science, Technology, Energy and Mines, has the responsibility for coordinating the Manitoba government's climate change plan. Significant change did occur in Manitoba when government agencies were reorganized in 1999. Manitoba Water Stewardship was created and is now responsible for: the management, control, and licensing of water; and water infrastructure; it also includes fisheries and the Office of Drinking Water. These branches or activities were previously the responsibility of Manitoba Environment, which no longer exists. Manitoba Conservation was created by amalgamating a number of departments and agencies who shared some similar responsibilities. Their common link was the environment. The amalgamation of departments with more traditional names is referred to in Manitoba Conservation's first annual report (1999–2000), which now includes sections from the former ministries of Natural Resources (branches of Water Resources, Parks and Natural Areas, Forestry, Fisheries, Wildlife, Policy), Energy and Mines, and Environment (Manitoba Conservation, 2000). Manitoba Conservation is responsible for environmental permitting, pollution prevention, environmental assessments and licensing, and sustainable resource management. The Department of Health also has an environmental health unit, and has responsibilities for public health protection, safe drinking water, food safety, and reviewing the health components of environmental risk assessments.

FEDERAL AND PROVINCIAL SHARED JURISDICTIONS

Provincial and federal ministries appear to overlap or duplicate climate and water mandates and services. In reality, the jurisdictions of natural resource management are shared by mutual agreement, and often by joint programming and funding. The potential duplication is resolved in practice in two ways. Firstly, matters which are within federal jurisdiction such as First Nation lands, international and inter-provincial waters, and inter-provincial undertakings are governed by the federal government departments and legislation; matters within provincial control, such as provincial lands and natural resources (including water management), are governed by the provincial legislation and policy of the provincial government departments. Secondly, in the event that it is not clear whether a particular issue or activity falls under provincial or federal government jurisdiction, there are many federal and provincial agreements resolving these issues which generally provide that the strictest legislation shall apply. For example, under administrative or equivalency agreements with many provinces, federal environment legislation is held in abeyance providing provincial legislation is equivalent to, or more stringent than, that expressed in federal Acts.

Agriculture, water and environmental issues are commonly shared jurisdictional matters between the federal and provincial governments. An example of one of the most significant government policy responses to a climate stress occurred in 1935 when the federal government created the Prairie Farm Rehabilitation Administration (PFRA) (now called the Agri-Environment Services Branch). The goal was to rehabilitate agricultural land devastated by years of drought—creating a regional and national crisis. In 1935, the federal decision to create a new agency caused some tension between federal and provincial orders of government (Marchildon, 2009). Its agricultural land and water stewardship programs continue to be shared between the federal and provincial governments, and local proponents. These programs have been instrumental policy responses promoting adaptation to climate impacts in a water-scarce region. Successful adaptations have been developed and promoted, including conservation tillage to maintain soil structure and soil moisture; carefully managing community pastures for rotational cattle grazing to safeguard ecologically-sensitive grasslands; planting of trees as shelterbelts to reduce wind erosion and loss of soil organic matter; promoting permanent grass cover on sensitive lands not suitable for field cropping; field irrigation and agricultural diversification to enhance sustainable agriculture in the dry prairies; and developing secure agricultural and rural water supplies for the agricultural sector and rural populations in the region.

CLIMATE CHANGES POLICIES AND STRATEGIC PLANNING

The Prairie Provinces and the federal government have responded to climate change in a variety of manners. The initial focus was on greenhouse gas emissions (because of their relationship to human-induced climate change) and, as a result, policies have been aimed at reducing greenhouse gas emissions through new technology, or changed practices in either a regulatory or voluntary manner. While mitigation is important, it is now recognized that a focus must also be placed on adaptation to cope with a changing climate. Some governments have responded with overarching climate change policies, and some respond with a particular program or policy within a department or ministry whose mandate or activities are affected by climate change. Often there is a combination of these two approaches because of the potentially significant effects of climate change on the environment, economic activities and practices, and the well-being of society.

A synopsis of the climate policies and government leadership is as follows:

- CANADA: In March, 2008, Environment Canada released "Turning the Corner—Taking Action to Fight Climate Change" (Environment Canada, 2008c), which provides an overview of federal initiatives, published under Canada's ecoACTION website. The plan seeks to force industry to reduce greenhouse gas emissions, will set up a carbon emissions trading market, and establish a market price for carbon. Environment Canada also publishes annually the National Inventory Report Greenhouse Gas Sources and Sinks in Canada, complete with provincial and sector emission summary tables. Natural Resources Canada plays a key role in the assessment reports on climate change impacts and adaptations, with key reports published in 2004 and 2008 (Natural Resources Canada, 2004; Natural Resources Canada, 2008a).

- ALBERTA: Leadership is provided by Alberta Environment. Alberta's 2008 Climate Change Strategy (Alberta Environment, 2008b) follows the 2002 Climate Change Action Plan. The plan is focused on conservation, energy efficiency, carbon capture and storage, and the greening of energy production. Alberta's Climate Change Central is a public-private partnership that was established in 2002 to help Alberta use energy more efficiently (Alberta Environment, 2002).

- SASKATCHEWAN: Leadership is provided by Saskatchewan Environment. Following the November 2007 election, Saskatchewan initially maintained previous targets, but as of spring 2009 it is developing a new climate change plan under its Go Green initiatives (released in the form of Bill 95, The Management and Reduction of Greenhouse Gases Act, in May of 2009). It is expected to be similar to Alberta's approach (Government of Saskatchewan, 2009a, J. Wood in *The Star Phoenix,* 2009).

- MANITOBA: Leadership on climate change is provided by Science, Technology, Energy and Mines, under its Climate and Green Initiatives programming. The Manitoba Climate Change Action Plan was adopted in 2002 and renewed in 2008 (Government of Manitoba, 2002; Government of Manitoba, 2008b). The plan details Manitoba's programs and activities in relation to climate change mitigation and adaptation.

CLIMATE CHANGE MITIGATION POLICIES

The federal government and each of the three Prairie Provinces have committed to the reduction of greenhouse gases with targets not always in accordance with those assumed by Canada in the Kyoto agreement. Table 1 outlines the federal and provincial governments' mitigation targets as well as those set by the Kyoto protocol. The nature of the climate change mitigation problem is especially significant for Alberta and Saskatchewan. Both provinces produce large volumes of oil and gas, and rely on thermal electrical generation from natural gas or coal. These factors result in a large quantity of greenhouse gas emissions per province as well as per capita. In contrast, Manitoba does not have a large oil and gas sector and relies mostly on hydro-electricity for power generation.

Until January 2008, Alberta had linked its emission reduction target to economic activity. Its target had been stated as 50% of 1990 levels of specified gas emissions relative to provincial gross domestic product (GDP) by 2020. Moving to an absolute mitigation target is a promising development. Alberta has created Climate Change Central, a public-private, not for profit organization made up of public and private leaders to study technology needed to meet emission targets (Climate Change Central, 2008).

In 2007, Saskatchewan's emissions targets had been established as: stabilizing emissions at 71 Mt by 2010, achieving a 32% reduction from year 2004 emissions of 71 Mt by 2020, and achieving an 80% reduction from year 2004 by 2050 (by calculation, the net emissions would be 14 Mt). Saskatchewan's population has been relatively stable at around 1 million for the last 50 years. The province experienced rapid economic growth in 2008, driven by a resource sector that includes the development of oil and gas. Such aggressive mitigation targets would be challenging to achieve with an expanding economy and population. As of spring 2009, Saskatchewan's targets have now been revised to achieve a 20% reduction from 2006 levels (Government of Saskatchewan, 2009b).

Manitoba adopted Kyoto targets, and as a member of the Western Climate

Government	Type of Target	Stated Target[a]	Net Target Value[b]	Timeline
Alberta (ALBERTA ENVIRONMENT, 2008B)	Intensity-based/ Absolute	20 Mt reduction relative to business as usual	~220 Mt	2010
		50 Mt reduction relative to business as usual	~ 250 Mt	2020
		200 Mt reduction (50% below business as usual & 14% below 2005 level of 205 Mt)	176 Mt	2050
Saskatchewan (GOVERNMENT OF SASKATCHEWAN, 2009A)	Absolute	20% lower than 2006 level of 72 Mt)	58 Mt	2020
Manitoba (GOVERNMENT OF MANITOBA, 2002, 2008C)	Absolute	6% below 1990 level of 18 Mt	17 Mt	2012
		15% below 2005 level of 20 Mt (regional target from Western Climate Initiative)	17 Mt	2020
Canada (ENVIRONMENT CANADA: 2008A-D, 2007A,B; AND 2006A,B)	Absolute/ emphasizing intensity reductions	20% below 2006 level of 721 Mt; a 330 Mt reduction from projected levels	577 Mt	2020
Kyoto target for Canada (ENVIRONMENT CANADA, 2008C AND 2008D; ENVIRONMENT CANADA, 2007B, ENVIRONMENT CANADA, 2006A)	Absolute	6% below 1990 level of 592 Mt	558 Mt	2012

Table 1: Greenhouse Gas Target Levels, as reported in July, 2009

a. Numeric value for reference year was extracted from Canada's National Inventory Report 1990–2005—Greenhouse Gas Sources and Sinks in Canada (except where actually reported for Alberta and Canada).

b. Net Target Value was calculated by determining the "% reduction" multiplied by the "numeric value for reference year" (except where actually reported for Alberta and Canada), and then subtracting the numeric reduction from the emissions for the Reference Year to obtain the net emissions target value.

Initiative[2] with seven US states and four Canadian provinces, has agreed to an aggregate emissions reduction goal of 15% below 2005 levels by 2020 (Western Climate Initiative, 2008).

The Government of Canada's targets under the Kyoto Protocol (6% below 1990 levels or 558 Mt by 2012) are not going to be attained. The current national target is 20% below 2006 by year 2020 (577 Mt). The Government of Canada tracks actual greenhouse gas emissions and trends across the country, by province and sector, in the Environment Canada's National Inventory Report (Environment Canada, 2007a). As of 2006, the country's emissions were at 721 Mt of carbon dioxide, or about 29% above Canada's Kyoto target of 558 Mt.

Canada has tracked and reported on greenhouse gas emissions since 1990. In Canada's Fourth National Report on Climate Change (Environment Canada, 2006b), the federal government adopted an integrated approach to the reduction of greenhouse gas emissions and air pollutant emissions. It was noted that billions of dollars in funding had been provided since 1992 in addressing the commitment to reduce greenhouse gases. However, as of 2004, these emissions were 26.6% above 1990 levels. The report did outline a change in voluntary reductions to a regulatory approach pursuant to the Clean Air Act, Bill C-30. Initiatives, such as transportation, new standards of energy efficiency, emissions trading, a technology investment fund, and investment in CO_2 capture and storage, were outlined. Regulatory requirements to reduce GHG emissions on vehicles are expected to become effective by 2011 (Environment Canada, 2009a).

In March 2008, Canada released the document, "Turning the Corner—Taking Action to Fight Climate Change" (Environment Canada, 2008c; Environment Canada 2008d). The Government of Canada's climate change focus is three-pronged: forcing industry to reduce greenhouse gas emissions, setting up a carbon emissions trading market, and establishing a market price for carbon. The plan confirms that the federal government has provided $1.5 billion in new funding to the provinces and territories to support their climate change initiatives. Canada's plan is further referenced in "A Climate Change Plan for the Purposes of the Kyoto Protocol Implementation Act, 2007" (Environment Canada, 2007b).

The federal government plan places regulatory requirements through legislation for the reduction of greenhouse gases on big industry, including oil and gas producers, iron and steel producers, forestry production, and electricity generation. These sectors comprise the Large Final Emitters Group referenced above. All sectors must reduce emissions from 2006 levels by 18% by 2010, with 2% annual reductions thereafter until 2020.

Firms may contribute to the Technology Fund for credits against each firm's emissions intensity targets. The fund will promote the development and deployment of low-emission technologies. Alternatively, over the next 10 years, firms can invest in pre-certified qualifying projects, and there is also the ability to purchase or use domestic offsets (voluntary projects to reduce or sequester greenhouse gas emissions independent of regulated mandates to do so). This provision has received criticism for allowing the continued release of greenhouse gases (De Souza, 2008). In June of 2009, the federal government released a draft Offset System guideline for a cap and trade of GHG emissions. This system would allow the registration of offset projects which, once implemented and verified as reducing GHG emissions, would be certified, and offset credits would be issued (Environment Canada, 2009c). So far, the three Prairie Provinces have not pursued this mechanism.

The federal climate change plan requires improvements in emissions performance every year beginning with those plants that began operation in 2004. New oil sands plants and coal-fired power plants coming into operation in 2012 or later face the toughest requirements of any sector. All coal-fired plants constructed in 2012 will be required to meet targets based on the use of clean technologies, such as carbon capture and storage, a requirement that will drive the development of new clean technologies to reduce greenhouse gas emissions. Existing coal-fired plants will have to meet tough new emissions standards. Mandatory renewable fuel content will be legislated as well as fuel consumption standards for vehicles and energy efficiency requirements for commercial and consumer products, including a ban on inefficient incandescent light bulbs.

The Prairie Provinces and the federal government have invested in initiatives over the last decade in an effort to reduce greenhouse gas emissions. The main areas of focus have been clean electricity, carbon capture, and biofuels. Some of the ongoing projects and initiatives by province are described below.

Alberta has invested $11 million in a research project on generating clean electricity from coal and using CO_2 for enhanced oil recovery (Bollinger and Kari, 2008, 16). Alberta invests research money through its Alberta Energy Research Institute overseeing the Energy Environment Technology Fund. These research projects focus on technology relating to such things as alternative energy sources and clean energy, and "scientifically defensible" environmental policy (Alberta Environment, 2008a). Alberta implemented the "Climate Change and Emissions Management Act" in 2003, with revisions in 2007, and recognizes this as "the first jurisdiction in North America to impose comprehensive regulations requiring large facilities in various sectors to reduce their greenhouse gas emissions" (Alberta Environment, 2009). The Act includes penalty regulations, gas reporting and emitting regulations, and the province has established an Alberta-based offset credit system, and technology fund.

Saskatchewan, with cost-sharing from the federal government, invested in a Petroleum Technology Research Centre and the International Test Centre for Carbon Dioxide Capture using solvent absorption and post-combustion capture starting in 1999. This project updated and refurbished an existing demonstration plant at Boundary Dam power station. The Saskatchewan government is also partnering with industry in financing the Weyburn–Midale CO_2 monitoring and storage project, which assesses the viability of enhanced oil recovery with CO_2 (Bollinger and Kari, 2008: 16). Saskatchewan and the federal government are also funding a clean coal initiative, carbon capture carbon sequestration project, four ethanol plants (grain-based), and one biodiesel (canola) plant. Saskatchewan is working with the oil and gas industry to improve efficiency and competitiveness, and reduce emissions from flaring, venting, and fugitive emissions.

Saskatchewan also now has a draft Bill with similar features to Alberta's legislation before its Legislature. The draft bill (Bill 95, The Management and Reduction of Greenhouse Gases Act) provides for a technology fund to manage carbon compliance payments from regulated emitters as well as a climate research and development corporation, a charitable foundation, and an Environment Corporation, to provide financial assistance for carbon capture storage, energy conservation and other initiatives (Government of Saskatchewan, 2009c).

In Manitoba, the government is minimizing natural and fugitive gas flaring and venting by integrating amendments to provincial drilling and production regulations (Government of Manitoba, 2002). As well, Manitoba is establishing a Hydrogen Centre of Expertise and innovating hybrid buses, and at the same time expanding its ethanol plant at Minnedosa (Government of Manitoba, 2009). It should be noted, however, that the oil and gas sector in Manitoba is relatively small when compared to Alberta and to Saskatchewan; Manitoba has developed extensive hydroelectricity capability, and is complementing this now with wind energy.

All three provinces have invested in alternative energy, and have a combination of wind power projects, methane gas projects from landfill, "standard offer" programs encouraging small-scale environmentally friendly power, and biomass and hydrogen energy. Saskatchewan is considering in its "vision for the future" large-scale hydroelectric power on the Saskatchewan and Churchill rivers and nuclear power generation close to the Alberta border. All three provinces have also implemented targets for renewable fuel integration, and provide financial incentives for biofuel promotion. All three provinces are also part of an effort by all of Canada's premiers to "Take Action on Climate Change" and work together to collectively produce an additional 25,000 MW of renewable energy by 2020, and create a Climate Registry to measure GHG emissions (The Council of the Federation, 2009).

In October 2007, the federal government also announced $5 million funding for ecoAgriculture Biofuels Capital Initiative ("ecoABC") for the construction of an additional ethanol production plant in Saskatchewan (as well as one in Alberta). This is one of many biofuel initiatives of the federal government, and is a response to laws increasing the minimum average renewable fuel content.

The federal Office of Energy Efficiency (part of Natural Resources Canada) offers grants to improve household energy efficiency (Natural Resources Canada, 2008b). Similar programs are offered by the governments of Alberta (Climate Change Central, 2008), Saskatchewan (Government of Saskatchewan, 2008) and Manitoba (Manitoba Hydro, n.d.).

These initiatives, and many others, support the three Prairie Provinces and the federal government's efforts to reduce greenhouse gas emissions. Sometimes they are part of a formal plan to address climate change, documented in a central overarching plan, and sometimes they are not. A discussion of the overarching general climate change plans which exist in the three Prairie Provinces and within the federal government allows an assessment of the degree and nature of climate change planning.

CLIMATE CHANGE PLANS

In addition to the mitigation efforts described above, climate change plans have been evolving in Canada. As early as 1998, an office titled the "Climate Change Secretariat" was created which reported to both Natural Resources Canada and Environment Canada in response to Canada's commitment to the 1997 Kyoto Protocol. Shortly thereafter, the Canada Country Study (CCS): Climate Impacts and Adaptation (Maxwell, Mayer and Street, 1997) was completed. This was the first Canadian report to assess the potential impacts of climate change and variability together with an impact analysis based on scenarios of future climate change.

In 2002, the Liberal Government (led by Prime Minister Jean Chrétien) issued a climate change strategy through the Climate Change Secretariat. The Climate Change Secretariat, however, wound down operations in 2004 when the Climate Change Action Fund ended.

In 2002, the federal government also produced a climate change assessment report on Canadian perspectives related to climate change impacts and adaptations titled, "Climate Change Impacts and Adaptation: A Canadian Perspective" (Natural Resources Canada, 2004). This document identified several knowledge gaps and research needs in the area of climate change. Some identified needs were: improving understanding of factors that influence adaptation decision-making and how to designate responsibility for action; better understanding the linkages between science and policy and how to strengthen them; building the understanding of current capacity of water management structures and institutions to deal with projected climate change impacts and the associated social, economic and environmental costs and benefits of future adaptations; and improving our understanding of environmental justice and equity consequences.

Canada's Fourth National Report on Climate Change, released by Prime Minister Harper's Conservative Government in 2006 outlined Canada's planned actions to meet its commitments under the United Nations Framework Convention on Climate Change (Environment Canada, 2006b). In 2006, the report committed $150 million over six years to vulnerability assessment in respect of climate change impacts and the development of adaptation assessments. Focus was also on providing knowledge and understanding, and tools to integrate climate change into planning and policy processes. The most recent climate change mitigation plan issued by the federal government, "Turning the Corner—Taking Action to Fight Climate Change" (Environment Canada, 2008c; Environment Canada, 2008d), was described above under mitigation plans.

One of Natural Resources Canada's current programs is the Climate Change Impacts and Adaptation Program which encompasses two main activities. The first is funding for research and activities to improve the knowledge of risks and opportunities presented by climate change; the second is assessing the knowledge of Canada's vulnerability to climate change through coordination and publication of scientific assessments.

With respect to the first activity of the Climate Change Impacts and Adaptation Program, to improve knowledge of risks and opportunities presented by climate change, a project database is available online (Natural Resources Canada, 2008c). In 2005, an intergovernmental working group, which included the three Prairie Provinces and the federal government, developed a National Climate Adaptation Framework which had eight recommendations aimed at enhancing planning for adaptation to climate change. Recommendations included ensuring policy responded to climate change, enhancing capacity to respond to climate change, and promoting and coordinating research on impacts and planning for climate change (ICCIAWG, 2005).

Current climate change strategies tend to focus on mitigation. Effects from climate change are expected to be unavoidable even in the unlikely event of achieving global mitigation targets. This looming challenge emphasizes the need for new adaptation efforts. Significant adaptation successes have occurred in the past. In fact, modern prairie agriculture is profitable largely because of sustainable land and water management practices developed after the 1920s–30s droughts. With anticipated future climate scenarios, new adaptations will be required to cope with increasing vulnerabilities caused by climate variability. This will undoubtedly require significant research and technology transfer efforts to help local decision-makers and rural communities reduce climate risk and seize new opportunities from a changing climate.

With respect to the second main activity of the Climate Change Impacts and Adaptation Program, Natural Resources Canada has produced assessment reports. The first assessment was *Climate Change Impacts and Adaptations: A Canadian Perspective,* published in 2004 (as discussed at the beginning of this section). It focused on potential impacts of climate change and adaptation options, based on research conducted from 1997 to 2002. In March 2008, this program released a second assessment report, *From Impacts to Adaptation: Canada in a Changing Climate 2007*, which reflects the advances made in understanding Canada's vulnerability to climate change during the past decade (Natural Resources Canada, 2008a). The report, written by experts across the country, takes a regional approach based on published scientific and technical literature identifying knowledge gaps, advances in understanding adaptation, and highlights recent and ongoing adaptation initiatives. A chapter dedicated to the Canadian Prairies (Natural Resources Canada, 2008a) emphasizes key climate change and climate variability risks, and the need for improved adaptive capacity.

In summer 2008, Natural Resources Canada announced a call for proposals for Regional Adaptation Collaboratives (Natural Resources Canada, 2008d). This $35 million program is designed to encourage regional adaptation initiatives; it is part of the $85 million of federal programming, announced in December 2007, to help Canadians increase their capacity to adapt to a changing climate. Projects are slated to begin in summer 2009, and terminate in 2011.

Both Alberta and Saskatchewan have developed overarching policies which have implications for adaptation to climate change. For instance, in 2007, Saskatchewan's then Premier Lorne Calvert's New Democrat Party Government issued an "Energy and Climate Change Plan" which was a cross-governmental vision in response to climate change (Government of Saskatchewan, 2007). One of the highlights of this plan was the development of a province-wide climate change adaptation strategy, which included working with research organizations and supporting critical local research on climate change and adaptation.

In the fall of 2007, the Saskatchewan Party (led by Brad Wall) won the election after over a decade of government by the provincial New Democratic Party (led by Lorne Calvert and, previously, by Roy Romanow). The new government is reviewing, developing and implementing new measures relating to climate change and its effects. As a result, the New Democrat's "Energy and Climate Change Plan" was withdrawn, although Calvert's emission targets were initially retained, and new projects (as outlined above in mitigation section) were announced to reduce greenhouse gases. In May 2009, Saskatchewan's targets were revised to be more consistent with Government of Canada approaches to achieve a 20% emissions reduction below year 2006 emissions levels (Government of Saskatchewan, 2009b). Saskatchewan Environment continues to support green initiatives with its "Go Green" campaign. The new policy, released in spring 2009, is similar to Alberta's levy on polluters, and has a centrepiece technology fund (J. Wood in *The StarPhoenix*, 2009; Government of Saskatchewan, 2009c). It focuses on regulating GHG emitters, research for improved technologies, carbon capture and storage, and energy conservation. The current premier has also expressed support for a national strategy on forestry issues relating to climate change, and also plans to take a leading role in crop science relating to climate change (Wood, 2008).

Alberta initially developed an overarching plan to respond to climate change in 2002. Since this time, the Alberta government held consultations with the public, stakeholders and experts, and released an updated plan in January 2008 titled, "Alberta's 2008 Climate Change Strategy, Responsibility, Leadership, Action" (Alberta Environment, 2008b). This plan, overseen by Alberta Environment, takes a three-pronged approach to greenhouse gas mitigation actions. It focuses on carbon capture and storage (contributing to 70% of Alberta's emissions reductions), conservation and using energy efficiently (12%) and greening energy production (18%). The plan acknowledges that climate change will affect Alberta in the foreseeable future despite mitigation efforts, and that the province needs to be ready to anticipate the risks, and adapt accordingly. Details of adaptation measures associated with the January 2008 plan have not yet been released, although Alberta's Environment Business Plan 2008–11 refers to the implementation of a Climate Change Adaptation Strategy (Alberta Finance and Enterprise, 2008). Alberta's Progressive Conservative Government obtained an eleventh consecutive majority government on March 4, 2008 (Canadian Broadcasting Corporation, 2008). The longevity of this government has meant planning for climate change has been influenced consistently by the Progressive Conservative Party in Alberta—most recently led by Ralph Klein (from 1992–2006) and Ed Stelmach (2006–present). Even though the Conservative Government has been in power since 1971, climate change policies have varied throughout the years.

Manitoba has a climate change plan called "Kyoto and Beyond" (Government of Manitoba, 2002). The plan is overseen by the Department of Science, Technology, Energy and Mines. Manitoba's "Green and Growing Strategy" is a strategic framework for a green, prosperous, growing province (Government of Manitoba, 2008a). Green initiatives through government are promoted, developed, facilitated and coordinated. Some of these initiatives are:

a. promoting awareness, programs and funding opportunities for climate change action;
b. facilitating partnerships between government and the Manitoba community to develop green initiatives that result in economic, environmental and social benefits;
c. advancing climate change mitigation and adaptation research; and
d. developing regional partnerships for climate change action.

Manitoba has had a New Democrat Party Government, led by Gary Doer from 1999 to 2009. The 2007 election left the party with a majority government and another four-year term. In contrast to Alberta and Saskatchewan, the longevity of this government has resulted in a comparatively consistent and focused climate change plan, the central theme of which is sustainable development.

The development of climate change plans is clearly a challenge facing all orders of government. Perhaps by necessity, climate change and adaptation strategies are evolving. However, programs and policy tools that vary too quickly present a problem for local stakeholders, industry, and all orders of government. Establishing baseline long-term plans would be beneficial for all decision-makers, including industry and government. Within their respective climate change plans (as of year 2009), Alberta, Saskatchewan, Manitoba, and the federal government have acknowledged the importance of adaptation to climate change and the need for funding adaptation research leading to adaptations to cope with climate vulnerabilities. Adaptation strategies to cope with climate change impacts can exist under the auspices of an overarching climate change plan, or as smaller-scale ad hoc policies and programs, as they have occurred in the Prairie Provinces and within the federal government for decades (e.g. in the form of a variety of policies such as drought plans or strategies, or emergency responses to climatic events such as flooding). The breadth of planning and developing policies relating to the changing environment is vast, and considerable mention is made in climate change plans of each respective government. However, much work is yet to be done. Planning for adaptation to climate change by the provinces and the federal government appears to be in its infancy. Mitigation efforts have been ongoing for some time, but Canada is a long way from attaining its Kyoto targets. This renders adaptation plans all the more pressing.

PLANNING FOR THE UNCERTAIN: THE NEED FOR ADAPTIVE POLICIES

The government plays a key role in responding to climate change and, indeed, to climate variability (IISD, 2006). Government, through its policy, and facilitated through laws, regulations and programs, is an arbitrator or facilitator of societal decisions relating to economic development, requirements for protecting the environment, natural resources management and governance, and helping society respond to extreme weather events (such as providing drought or flood relief or recovering from severe storms). Climate change and climate variability certainly impact society and the environment. While uncertainty surrounds the timing, duration and magnitude of the impacts of climate change, there is an increasing consensus that the long-term trend is heading towards the warming of the earth's climate. Thus, climate change brings a new era of challenge for all orders of government (local, provincial, federal).

Conventional government policy has promoted scientific consensus and precise predictors of events, minimized conflict, and presumed certainty in seeking best action from a set of obvious alternatives (IISD, 2006: 34). The goal of conventional policy has traditionally been the consistent application of programs by government bureaucrats in order to achieve equilibrium of outcomes across constituents. Very little was left to the discretion of the bureaucrat, adaptation to the situation, or input by the constituent. The uncertainty of climate change renders the construct of such conventional government policies and programs both too restrictive and potentially inadequate to rise to the challenges posed by climate. The variability of weather, the changes in ecosystems, and the associated environmental effects on industry, agriculture, development and society in general, preclude the all-encompassing, precise predictions and actions of conventional policy approaches.

Developing policies in relation to climate change and its effects (which include water, drought, flooding, environment, and economic policies) requires a paradigm shift. For effective climate, environment and water policies, conventional top-down, bureaucratically-determined and applied policies need to be replaced with more flexible and adaptive policies, which are scientifically-based, developed in consultation with an informed civil society, and consequently driven through a bottom-up approach. Such policies need to have some baseline focus and stability, but also need to be adaptive and responsive to surprises and changes at all levels—i.e., locally, provincially, and nationally. Timely and effective feedback to capture the environmental impacts and societal inputs should be gathered continually as a means to review, modify and improve the desired policy outcomes. The World Bank has concluded that in enhancing sustainable development and human well-being, the institutional environment must be able to pick up signals about needs and problems from its constituents. This involves generating information, giving citizens a voice, responding with feedback, and fostering learning (World Bank, 2003). This finding is also supported by adaptive policy literature, institutional learning literature, and organizational learning models from the private sector (IISD, 2006; Smith and Stirling, 2008). This literature also concludes that adaptive policy capable of responding to the challenge of climate change must be able to respond to anticipated conditions as well as unanticipated conditions. This is achieved not through specific solutions necessarily, and not through tightly focused institutions established on the basis of an efficient model or plan, but rather by ensuring the existence of processes and frameworks that enable solutions to be identified and implemented in a changing and flexible manner. Such uncertainty will no doubt lead to less predictable approaches, and perhaps more clumsy, but more resilient and adaptive institutions.

In the context of adaptive capacity, effective governance requires flexibility to deal with the unanticipated conditions that may result from the impacts of climate change. The role of governance organizations and instruments includes implementing an enabling environment that strengthens civil society to deal successfully with the challenges of climate change, and by applying specific policies (resource mobilization and allocation, and incentives and disincentives). Transitioning to a system of governance for sustainability with increased challenges of socio-technical change and sustainable goal prioritization is not new (Smith and Stirling, 2008). Adaptive capacity, to be successful, must allow for the identification and resolution of communities' problems, and the ability to satisfy their needs in a fair, efficient and sustainable manner. Thus, the fundamental contribution of governance is to reduce the vulnerabilities of society, including urban and rural communities. While urban communities may be more vulnerable to catastrophic climate events (e.g. floods, tornados, hurricanes), rural society is likely more vulnerable to chronic events (e.g. short- and long-term drought, water stress, pests, seasonal shifts in temperature and rain). Major long-term environmental impacts from climate change may pose an even greater challenge to economic sustainability (e.g. catastrophic impacts on forestry from the mountain pine beetle). These dynamics are truly a challenge for all governments and policy-makers, who must balance the interests of all constituents and the environment. The capacity of governments to reduce vulnerabilities rests on their ability to anticipate problems in a holistic manner, and to manage risks and challenges using processes and means that balance social, economic, and environmental interests.

SYNTHESES

By themselves, present-day government organizations and policies are not able to achieve long-term climate strategies (Hurlbert, Corkal and Diaz, 2009). Governments develop and implement policies typically with a four-year mandate cycle; even when a government is re-elected, the present-day issues often mean that policies do not necessarily remain consistent. A new government may not endorse a previous government's climate change policy. Shorter-term visions and policy responses pose significant challenges in building capacity to reduce vulnerabilities from climate change and climate variability. Such environmental issues require a long-term outlook, one that likely extends a minimum of 10 to 50 years. Yet, simply setting a long-term goal will not hold a current or future government accountable to meet the goal. Furthermore, the ubiquitous nature of climate change makes it impossible to address the issue in a simple or centralized manner. A plethora of government institutions have mandates which are significantly affected by climate change. The effective coordination of these institutions poses a significant challenge. This challenge exists within each level of government and is multiplied across the provincial and federal governments (not to mention the municipal and First Nations governments not directly discussed in this chapter).

Adaptation to climate change will also be exacerbated by increasing population and the increased demands for development, especially in areas characterized by a scarcity of natural resources, resulting in increasing social tensions and conflicts. Lastly, climate change is cloaked in the uncertainty of precise future impacts and social responses under different governments. Government policy has historically responded to certainty and issues that are (or are at least "perceived to be") more immediate in nature. For example, variable precipitation may result in droughts or floods, and while this is a natural feature of the prairie climate, these events are viewed as "extreme." When a climate crisis develops and is severe enough, ad hoc programming may be developed to assist the affected parties, as was the case when the prairie region was devastated by the droughts of the 1920s and 1930s. A proactive response with a longer-term vision might involve the development of drought or flood preparedness plans, and developing adaptive practices such as strengthening agricultural resilience and seeking new economic opportunities in the face of even wider future climate variability than what is presently experienced.

These challenges, the Canadian electoral system, the challenges in setting and acting on long-term goals, the ubiquitous nature of climate change and environmental impacts, and the uncertainty of precise future climate change impacts and social responses, cannot be changed. These challenges do represent the complex nature of climate change for federal and provincial governments, which are struggling globally to decide what actions to take. Complicating this even further is the fact that regional approaches are only a small part of a global problem requiring global action. However, new approaches to government policy-making can be implemented and improved, and in doing so, more effective responses to these challenges are likely to evolve. The evolution of improved long-term environmental policies will require long-term vision, with base strategies that balance economic, social and environmental concerns. A joint commitment by all orders of government to a long-term vision will hopefully reduce the risk of unfulfilled commitments or recurring issues. Designing and implementing effective long-term climate and environment policies may avert potential crises, particularly when science and good governance arrangements can foresee or anticipate potential impacts and possible solutions. Some improvements are listed below.

ATTENTION TO CLIMATE IN LONG-TERM GOVERNMENT POLICY

Building society's resilience to climate impacts requires the establishment of baseline plans that stand the test of time for all orders of present and future governments, yet are flexible enough to be improved as necessary. Governments need to have a comprehensive integrative consideration of climate in the governance agenda. Because of the ubiquitous nature of climate change and the diversity of expected impacts of natural climate variability, overarching long-term strategies and responses are required. Overarching policies represent an increasing level of awareness and concern about climate change. The absence of a comprehensive governance approach to climate change translates into a lack of homogeneity within the multitude of government departments and ministries regarding climate change relevance and its proper integration into action agendas. For many departments in the Prairie Provinces and federal government, climate change is an important challenge, and attention has been directed towards understanding the phenomenon, its impacts upon water and other natural resources, the economy, and identifying viable adaptation strategies. To date, all orders of government have been primarily focused on mitigation targets. While mitigation is essential, new adaptation strategies and actions must also be developed.

There is a need to more concretely integrate the variable climate of the region into the governance agenda, especially for decisions that will have long-term impact. In an assessment of the adaptive capacity of a prairie city to climate change, it was concluded that although institutions were aware that the weather pattern could affect surface water supplies and needs of a water conservation program, this awareness had not translated into a significant integration of climate change issues into the institutional agenda (SDCCWG, 2005). In another study of several prairie cities, similar findings were made, and it was concluded that current awareness and adaptation strategies are inadequate, leaving cities vulnerable. Climate change will amplify current stresses, including transportation difficulties, compromised air quality, infrastructure problems, and even reduced social cohesion within communities—resulting in increased crime. The study found that the risks of damage or destruction to people and infrastructure (such as water, sewage and storm water systems, roads, bridges, and buildings) from extreme weather events were not being properly planned for (Wittrock, Wheaton and Beaulieu, 2001).

Information about climate is essential in order to effectively respond. Most research into climate change has focused on general trends and average conditions, but climate's biggest impacts on humanity have been a result of extreme events. The present notion of variability in climate change scenarios is based on past events, and this assumption may be flawed under future conditions where climate is influenced by global warming. A better understanding of climate variability—now and under various climate change scenarios—is needed. This is a significant contributor to not being able to translate comprehensive climate change policy into a comprehensive action plan to address climate change issues, and to manage the risks and opportunities of new climate conditions.

There have been significant increases in the institutional research capacity in the area of climate change and water resources within the Prairie Provinces and the federal government. For example, climate change is now an environmental priority among the Agriculture and Agri-Food Canada initiative Growing Forward established in 2008 (Agriculture and Agri-Food Canada, 2008). Climate change research programs have been developed in regional universities such as the Prairie Adaptation Research Collaborative at the University of Regina and the Water Institute for Semiarid Ecosystems based at the University of Lethbridge, the Alberta and Saskatchewan Research Councils, and in various government agencies. This increasing production of knowledge, fundamental to assessing impacts and vulnerabilities, and to developing adaptive capacities, still requires ongoing funding to ensure it remains leading edge, and proper channels of integration into governance policies, management, programs, and practices need to be found.

FOCUS ON ADAPTATION TO CLIMATE CHANGE AND CLIMATE VARIABILITY

Increased emphasis is needed on developing appropriate adaptation responses to climate impacts. A limited consideration to the development of stronger adaptive capacity is evident on the climate change agenda of all three Prairie Provinces and the federal government. As of 2009, most of the attention has been given to mitigation issues, while attention to climate change adaptation is still in its infancy. The Government of Alberta states adaptation is a component of its climate change plan, but focuses funding on mitigation; the previous Saskatchewan government had a climate change plan which focused mainly on mitigation of carbon dioxide emissions, and contained only a statement of intent to develop an adaptation strategy in the future. The current Saskatchewan government has released an impressive framework, but the legislation requires passing, and implementation of the legislative provisions. Manitoba's climate change plan does state that adaptation is a component in response to climate change, but the specifics to achieve adaptation are unclear.

Currently, there is limited support for adaptation research or implementation of adaptation techniques or technologies. Canada's Regional Adaptation Collaboratives program is expected to commence in summer 2009, and terminate in winter 2011. Alberta's adaptation strategy is pending release in 2009. What is clear is that more emphasis on adaptation and on the interrelationship with development decisions, in light of both economics and climate change predictions, is required. Developments in this area for provinces and regions will certainly contribute to a common understanding and purpose of what is required to create a suitable and comprehensive policy approach to climate change, avoiding political cleavages within the policy community and tensions within governance (Wellstead, Davidson and Stedman, 2002).

CIVIC ENGAGEMENT FOR CLIMATE CHANGE POLICY DEVELOPMENT

Citizen engagement and feedback are essential requirements to achieve successful adaptations to cope with climate stress. Increasingly, it is being recognized in the literature that climate change policy in relation to topics such as water governance and economic development decisions, in order to foster sustainability, will require significant engagement of civil society (World Bank, 2003; RIFWP, 2007). This engagement should occur at the overarching climate change policy development stage. With respect to climate change, the federal government consulted widely on its regulatory framework for the reduction of greenhouse gas emissions pursuant to the Clean Air Act. The results of these consultations still await regulatory enactment.

Meaningful citizen engagement would facilitate the formulation of policies implementing publicly compatible solutions. In this manner, the values and priorities of the public will be incorporated into policy. The values and priorities of the public may relate to economic savings, environmental protection, and adaptations. The benefits must be perceived, understood, and not believed to impact lifestyle. Public engagement requires more than initial consultation in policy formulation—it requires continuous feedback and monitoring to ensure objectives are met in a flexible manner. Adopting a process of engaging civil society in decision-making has the potential to achieve environmental protection, sustainable development, and healthy economic prosperity, cornerstones for a truly prosperous and balanced society.

RESPONSIVE POLICY DESIGN AND IMPLEMENTATION

Long-term policies require appropriate flexibility and a clear vision for implementation. The requirement for certainty in climate change predictions is problematic in strengthening adaptive capacity to a range of climate conditions, where changes, surprises, and uncertainty are part of the nature of the problem (IISD, 2006). Some government agencies express a concern with climate change, but they are reluctant to integrate the issue into their agendas, claiming that scenarios of climate change are plagued with uncertainty and, consequently, that they cannot yet be considered. This is a traditional orientation to policy response. A new adaptive view of policy making and response must be embraced, as outlined above. Important components are the consistent and regular engagement of civil society in making decisions, and ensuring the existence of processes and frameworks that enable solutions to be identified, implemented, and revised, as necessary.

Legal instruments, such as environmental legislation in both Alberta and Saskatchewan, have statements endorsing principles of precaution, or regulation and rule making prior to scientific certainty. The endorsement of the precautionary principle is a positive sign, but other than the endorsement in principle found in non-binding "guiding principles," it would appear that actual rules, regulations, policies, and strategies are not being developed in advance of scientific proof and certainty. There is a need to pay more attention to the principle of flexibility in policy and management approaches, to be prepared to experiment, and to find ways to prevent governance processes and approaches from becoming too rigid. Ultimately, policies and programs need a long-term vision, and yet be flexible and efficient, able to adapt to new knowledge, and structured to build the capacity of affected communities and minimize negative environmental impacts. Without such flexibility, insufficient responses and adaptations may occur and may not address the natural wide range of climate variability, let alone the potentially more severe future impacts from climate change. Once again, introducing the discourse of adaptation is an important milestone, but much work is needed to realize the significant benefits available for full adaptation research and policy responses to implement new adaptation processes.

CONCLUSION

This chapter has reviewed Canadian government policy in respect of both mitigation and adaptation to climate change and variability. Too often a focus on Canada's Kyoto commitment and impending inability to achieve Kyoto's greenhouse gas emissions targets provides a "red herring" which overshadows constructive discussion of policy development. Since their inception, the federal government and the provincial governments of the Prairie Provinces have responded to, and facilitated, adaptation to climate variability. An overview of government institutions and current policies playing a role in adaptation to climate shows that an important foundational base exists. However, this by no means warrants complacency.

Mitigation measures aimed at reducing levels of greenhouse gas emissions are receiving considerable government focus. Continued efforts are required in light of Canada's looming inability to meet Kyoto commitments. Canada will not achieve the Kyoto target without imposing significant costs to the Canadian economy (Environment Canada, 2007b). This is the tension between balancing greenhouse gas mitigation targets and economic activities of present-day society, while planning for, and respecting the needs of, future generations. It is complicated by the fact that one region cannot solve a national or global problem by itself.

Adaptation to climate change impacts is necessary today, and will be even more essential in the future. Historic adaptation successes have led to modern and sustainable agricultural practices in the Prairies. However, with the effect of new climate scenarios, new adaptation efforts can be expected. To achieve resilience and greater economic security, it will be necessary to undertake a greater number of new adaptation initiatives, research projects, and activities. Adaptation initiatives must consider the most vulnerable geographical locations and communities; this would indeed include rural communities, and climate change impacts to the environment itself, which cannot defend its own interests.

In drafting policies responding to climate change, all orders of Canadian government can be expected to focus more on climate and climate variability in an environment of greater uncertainty than historical practice. The development of baseline climate change and adaptation plans would be helpful if they can provide guidance for long-term durations. Greater continuous interaction with civil society in this exercise will influence policy formulations, and possibly create a new era of transition in governance, perhaps with more clumsy and administratively complicated or messy institutions. Institutions that actively engage citizen participation may be less predictable, but with appropriately-designed governance mechanisms that allow for flexibility, these institutions will have the potential to be more responsive, more resilient, and more capable of helping society adapt.

As Canada and the vulnerable prairie region confronts climate impacts, strengthened adaptive capacity for government institutions and society is essential. Increasing citizen engagement, and adopting long-term planning with built-in flexibility and feedback loops, will prove to be essential responses to climate change, and will help build capacity for a prosperous, healthy Canadian society and environment.

ENDNOTES

1. After roughly 70 years, PFRA evolved into a national agency, and was known briefly as the Prairie Farm Rehabilitation Administration and Environment Branch. In April, 2009, the agency's name became Agri-Environment Services Branch, a branch of Agriculture and Agri-Food Canada.
2. The Western Climate Initiative includes British Columbia, Manitoba, Ontario, Quebec, Arizona, California, Montana, New Mexico, Oregon, Utah, and Washington as partners. Observers include Saskatchewan, Alaska, Colorado, Idaho, Kansas, Nevada, Wyoming, and the Mexican states of Baja California, Chihuahua, Coahuila, Nueva Leon, Sonora, and Tamaulipas.

REFERENCES

Agriculture and Agri-Food Canada. 2008. *Governments Announce Completion of the Growing forward Multilateral Framework* [online]. July 11 News Release. Government of Canada: Ottawa. [accessed September 2, 2009]. Available at: *http://www.agr.gc.ca/cb/index_e.php?s1=n&s2=2008&page=n80711*

Alberta Environment. 2002. *Albertans and Climate Change: A Strategy for Managing Environmental and Economic Risks* [online]. Alberta Government: Edmonton [accessed May 26, 2008]. Available at: *http://environment.gov.ab.ca/info/library/5895.pdf*

——. 2008a. *Government of Alberta, Energy Environment Technology Fund* [online]. Government of Alberta: Edmonton [accessed May 26, 2008]. Available at: *http://www.environment.alberta.ca/2264.html*

——. 2008b. *Alberta's 2008 Climate Change Strategy, Responsibility, Leadership, Action* [online]. Government of Alberta: Edmonton [accessed February 25, 2009]. Available at: *http://environment.gov.ab.ca/info/library/7894.pdf*

——. 2009. *Legislation/ Guidelines* [online]. Government of Alberta: Edmonton [accessed February 23, 2009]. Available at: *http://www.environment.alberta.ca/3.html*

Alberta Finance and Enterprise. 2008. *Government and Ministry Plans 2008–11: Environment Busi-*

ness Plan 2008–11 [online]. Government of Alberta: Edmonton [accesseed July 25, 2008]. Available at: *http://www.finance.alberta.ca/publications/budget/budget2008/envir.pdf*

Bollinger, Jillian and Robert Kari. 2008. *Building on Our Strengths, An Inventory of Current Federal, Provincial and Territorial Climate Change Policies*. Canada West Foundation: Calgary.

Canadian Broadcasting Corporation. 2008. *Progressive Conservatives Elect Historic 11th Consecutive Majority Government* [online]. March 4, 2008 [accessed May 28, 2008]. Available at: *www.cbc.ca/canada/albertavotes2008/story/2008/03/03/election-call.html#sky/200x250*

Climate Change Central. 2008. *Furnace Rebates are Heating up Energy Efficiency in Alberta* [online] [accessed May 26, 2008]. Available at: *http://www.climatechangecentral.com/media/news/2004/furnace-rebates-are-heating-energy-efficiency-alberta*

Council of the Federation (The). 2009. *Canada's Premiers: Taking Action on Climate Change, A Report on Progress Achieved Since August 2007* [online] [accessed June 15, 2009]. Available at: *http://www.councilofthefederation.ca/pdfs/CCCommPiece_0718-FINAL.pdf*

De Souza, Mike. 2008. "Oilsands, Emissions Could Triple in Alberta" *The Leader-Post*, March 10, D2.

Environment Canada. 2006a. *Information on Greenhouse Gas Sources and Sinks, Canada's 2006 Greenhouse Gas Inventory—A Summary of Trends* [online]. Government of Canada: Ottawa [accessed January 14, 2009]. Available at: *http://www.ec.gc.ca/pdb/GHG/inventory_report/2006/som-sum_eng.pdf*

——. 2006b. *Canada's Fourth National Report on Climate Change. Actions to Meet Commitments Under the United Nations Framework Convention on Climate Change* [online]. Government of Canada: Ottawa [accessed July 24, 2008]. Available at: *http://www.ec.gc.ca/climate/home-e.html*

——. 2007a. *National Inventory Report—1990–2005 Greenhouse Gas Sources and Sinks in Canada* [online]. Government of Canada: Ottawa [accessed July 24, 2008]. Available at: *http://www.ec.gc.ca/pdb/ghg/inventory_report/2005_report/tdm-toc_eng.cfm*

——. 2007b. *A Climate Change Plan for the Purposes of the Kyoto Protocol Implementation Act—2007* [online]. Government of Canada: Ottawa [accessed July 24, 2008]. Available at: *http://www.ec.gc.ca/doc/ed-es/p_123/CC_Plan_2007_e.pdf*

——. 2008a. *Information on Greenhouse Gases, Sources and Sinks* [online]. Government of Canada: Ottawa [accessed May 26, 2008]. Available at: *http://www.ec.gc.ca/pdb/ghg/ghg_home_e.cfm*

——. 2008b. *Environment Canada 2008–2009 Report on Plans and Priorities* [online]. [accessed July 24, 2008]. Available at: *http://www.tbs-sct.gc.ca/rpp/2008–2009/inst/doe/doe00-eng.asp*

——. 2008c. *Turning the Corner—Taking Action to Fight Climate Change* [online] [accessed July 24, 2008]. Available at: *http://www.ec.gc.ca/doc/virage-corner/2008–03/brochure_eng.html*

——. 2008d. *Turning the Corner: Regulatory Framework for Industrial Greenhouse Gas Emissions, 2008* [online]. Government of Canada: Ottawa [accessed February 9, 2009]. Available at: *http://www.ec.gc.ca/doc/virage-corner/2008–03/pdf/COM-541_Framework.pdf*

——. 2009a. *Notice of Intent to Develop Regulations Limiting Carbon Dioxide* [online]. Government of Canada: Ottawa [accessed June 15, 2009]. Available at: *http://www.ec.gc.ca/Ceparegistry/documents/notices/g1–14314_n1.pdf*.

——. 2009b. *Project Green—Moving Forward on Climate Change* [online]. Government of Canada: Ottawa [accessed January 14, 2009]. Available from World Wide Web: *http://www.team.gc.ca/english/publications/team-200305/project-green.asp*

——. 2009c. *Canada's Offset System for Greenhouse Gases* [online]. Government of Canada: Ottawa [accessed 15 June, 2009]. Available from World Wide Web: *http://www.ec.gc.ca-/creditscompensatoires-offsets/default.asp?lang=En&n=0DCC4917–1*

Government of Manitoba. 2002. *Kyoto and Beyond, A Plan of Action to Meet and Exceed Manitoba's Kyoto Targets, Province of Manitoba Climate Change Action Plan 2002* [online] [accessed July

24, 2008]. Available at: *http://www.gov.mb.ca/stem/climate/key_documents.html*

——. 2008a. *Green and Growing Strategy* [online]. Government of Manitoba: Winnipeg [accessed May 26, 2008]. Available at: *www.gov.mb.ca/stem/clmate/about_us/index.html*

——. 2008b. *Next Steps: 2008 Action on Climate Change* [online]. Government of Manitoba: Winnipeg [accessed May 31, 2008]. Available at: *www.gov.mb.ca/beyond_kyoto/index.html*

——. 2008c. *The Climate Change and Emissions Reduction Act, 2008* [online]. Government of Manitoba: Winnipeg [accessed February 12, 2009]. Available at: *http://web2.gov.mb.ca/laws/statutes/ccsm/c135e.php*

——. 2009. *Acting on Energy and Climate Change* [online]. [accessed June 15, 2009]. Available at: *http://www.gov.mb.ca/greenandgrowing/acting.html*

Government of Saskatchewan. 2007. *Energy and Climate Change Plan 2007*. Premier's Office: Regina.

——. 2008. *Go Green Saskatchewan Energy Conservation* [online]. Government of Saskatchewan: Regina [accessed May 26, 2008]. Available at: *http://www.environment.gov.sk.ca/Default.aspx?DN=260faea7-f3b6-4085-a085-0a9b8861c542*)

——. 2009a. *Go Green Saskatchewan* [online]. Government of Saskatchewan: Regina accessed 14 January 14, 2009]. Available at: *http://www.environment.gov.sk.ca/Default.aspx?DN=9192fbe8-23fe-4077-ac7d-30b7b269bdbf*

——. 2009b. *Saskatchewan Takes Real Action to Reduce Greenhouse Gases Emissions* [online]. Government of Saskatchewan: Regina [accessed July 8, 2009]. Available at: *http://www.gov.sk.ca/news?newsId=387f7573-1e28-4155-b0ca-06fd17b0d38e*

——. 2009c. "Summary of Bill #95 introduced to the Saskatchewan Legislature on May 11, 2009 Respecting the Management and Reduction of Greenhouse Gases and Adaptation to Climate Change" [online]. Government of Saskatchewan: Regina [accessed June 15, 2009]. Available at: *http://www.gov.sk.ca/adx/aspx/adxGetmedia*

Hurlbert, Margot, Darrell R. Corkal and Harry Diaz. 2009. "Government and Civil Society: Adaptive Water Management in the South Saskatchewan River Basin." *Prairie Forum* 34, no. 1: 181–210.

ICCIAWG (Intergovernmental Climate Change Impacts and Adaptation Working Group). 2005. *National Adaptation Climate Change Framework* [online]. Co-Chairs Randy Angle, Ministry of Environment, Government of Alberta, and Paul Egginton, Climate Change Impacts and Adaptation Directorate, Natural Resources Canada Government of Canada: Ottawa [accessed May 26, 2008]. Available at: *http://www.conferenzacambiamenticlimatici2007.it/site/_Files/documentazione/CanadaFrameAdapt.pdf*

IISD (International Institute for Sustainable Development). 2006. *Designing Policies in a World of Uncertainty, Change and Surprise, Adaptive Policy-Making for Agriculture and Water Resources in the face of Climate Change* [online]. Phase 1 Research Report [accessed July 24, 2008]. Available at: *http://www.iisd.org/PUBLICATIONS/pub.aspx?id=840*

Manitoba Conservation. 2000. *Manitoba, Annual Report 1999–2000, Conservation* [online(. Government of Manitoba: Winnipeg [accessed October 5, 2009] Available at: *http://www.manitoba.ca/conservation/annual-reports/conservation/2000-report.pdf*

Manitoba Hydro. N.d. *Power Smart Savings, Rebates & Loans, Manitoba Hydro* [online]. Government of Manitoba: Winnipeg [accessed September 2, 2009]. Available at: *http://www.hydro.mb.ca/savings_rebates_loans.shtml*

Marchildon, Gregory P. 2009. "The Prairie Farm Rehabilitation Administration: Climate Crisis and Federal-Provincial Relations during the Great Depression." *The Canadian Historical Review* 90, no. 2: 275–301.

Maxwell, Barrie, Nicola Mayer and Roger Street (eds.). 1997. *Canada Country Study: Climate Impacts and Adaptation. National Summary for Policy Makers.* Environment Canada: Ottawa.

Morrison, Heather A. and Jon Gee. 2001. "DIAGNOSTIQUE: Federal Water Policy and Ontario

Region." As reported with updated information in *Federal Government Departments' Roles and Mandates for Water in the Prairie Provinces, A Brief Summary*. Federal Prairie Water Quality Workshop, Nov. 6–9.

Natural Resources Canada. 2004. *Climate Change Impacts and Adaptation: A Canadian Perspective* [online]. Climate Change Impacts and Adaptation Directorate Natural Resources Canada, Her Majesty the Queen in Right of Canada: Ottawa [accessed May 26, 2008]. Available at: *http://adaptation.nrcan.gc.ca/perspective/summary_1_e.php*

——. 2008a. *From Impacts to Adaptation: Canada in a Changing Climate 2007* [online]. Government of Canada: Ottawa [accessed May 26, 2008]. Available at: *http://www.adaptation.nrcan.gc.ca/assess/index_e.php*

——. 2008b. *Retrofit Your Home and Qualify for a Grant!* [online]. Government of Canada: Ottawa [accessed May 26, 2008]. Available at: *http://www.oee.nrcan.gc.ca/residential/personal/retrofit-homes/retrofit-qualify-grant.cfm*

——. 2008c. *Climate Change Impacts and Adaptation Project Database* [online]. Government of Canada: Ottawa [accessed May 26, 2008]. Available at: *http://adaptation.nrcan.gc.ca/projdb/index_e.php*

——. 2008d. *Regional Adaptation Collaboratives* [online]. Government of Canada: Ottawa [accessed July 24, 2008]. Available at: *http://www.adaptation.nrcan.gc.ca/collab/index_e.php*

RIFWP (Rosenburg International Forum on Water Policy). 2007. *Report of the Rosenburg International Forum on Water Policy to the Ministry of Environment,* Province of Alberta, February 2007[online]. University of California, Division of Agriculture and Natural Resource, University of California: Berkeley [accessed May 26, 2008]. Available at: *http://www.waterforlife.gov.ab.ca/docs/Rosenberg_Report.pdf*

Sauchyn, D. and S. Kulshreshtha, 2008, "The Prairies" [online]. in D.S. Lemmen, F.J. Warren, J. Lacroix and E. Bush (eds.). *From Impacts to Adaptation Canada in a Changing Climate 2007*. Government of Canada: Ottawa [accessed May 26, 2008]. Available at: *http://www.adaptation.nrcan.gc.ca/assess/2007/index_e.php*

SDCCWG (Social Dimension of Climate Change Working Group). 2005. *Social Dimensions of the Impact of Climate Change on Water Supply and Use in the City of Regina*. Research Report, University of Regina and Canadian Plains Research Center: Regina.

Smith, A. and A. Stirling. 2008. *Transitions—Social-Ecological Resilience and Socio-Technical Transitions: Critical Issues for Sustainability Governance.* Steps Centre Working Paper 8: Brighton.

StarPhoenix. 2009. "Gov't Moves Ahead with Climate Change Plan." By James Wood in *The Saskatoon StarPhoenix,* February 20, 2009.

Wellstead, A., D. Davidson and R. Stedman. 2002. *Assessing the Potential for Policy Responses to Climate Change* [online]. PARC Project 011 [accessed May 26, 2008]. Available at: *http://www.parc.ca/research_pub_communities.htm*

Western Climate Initiative. 2008. *U.S. States, Canadian Provinces Announce Regional Cap-and-Trade Program to Reduce Greenhouse Gases* [online]. September 23, 2008 Press Release [accessed February 12, 2009]. Available at: *http://www.westernclimateinitiative.org/*

Wittrock, V., E.E. Wheaton, and C.R. Beaulieu. 2001. *Adaptability of Prairie Cities: The Role of Climate Current and Future Impacts and Adaptation Strategies.* Limited Report, Environment Brach, Saskatchewan Research Council, Publication No. 11296–1E01.

Wood, James. 2008. "Climate Change, Provinces to Combine on Water Conservation." Saskatchewan News Network, *The Leader-Post,* January 30, A4.

World Bank. 2003. *World Development Report. Sustainable Development in a Dynamic World, Transforming Institutions, Growth and Quality of Life*. World Bank and Oxford University Press: Washington.

3. STUDIES

in Impacts and Adaptations

CHAPTER 13

RURAL VULNERABILITY TO CLIMATE CHANGE IN THE SOUTH SASKATCHEWAN RIVER BASIN

Johanna Wandel, Jeremy Pittman and Susana Prado

INTRODUCTION

The IPCC defines vulnerability as the degree to which a geophysical, biological or socio-economic system is susceptible to and unable to cope with the adverse impacts of climate change (Schneider et al., 2007). Implicit in this definition is an emphasis on a stress (an aspect of climate change relevant to the system) in the context of a system's ability to manage stress (adaptive capacity). Adaptive capacity is unequal across and within societies (Adger et al., 2007), and it can be expected that both the relevant stresses and adaptive strategies employed to manage these are specific to particular communities, sectors or livelihoods. Consequently, an understanding of rural vulnerability to climate change in the context of the Canadian Prairies requires consideration of the ways in which various stakeholder groups in the region are exposed and sensitive to climate change, and the specific strategies they have developed to manage these stresses.

This chapter presents a summary of rural agricultural stakeholders with empirical illustrations selected from a series of case studies conducted by the Institutional Adaptation to Climate Change (IACC) Project.[1] This project included several in-depth community vulnerability assessments in the South Saskatchewan River Basin, Canada, from 2005 to 2007. This paper brings together insights from three of the Canadian communities.

CONCEPTUAL MODEL OF VULNERABILITY

Vulnerability is highly context and scale-specific (Smit and Wandel, 2006). Recent years have seen the evolution of the concept from a "residual vulnerability" to a "process-based" approach. Residual vulnerability is derived by projecting changes in climate and anticipating adaptations in light of these changes—vulnerability is the difference between impact and adaptation. A residual vulnerability approach consequently treats vulnerability as the "end point" of an analysis, and is useful for estimates of particular changes under an everything-else-remains-equal scenario (Füssel, 2007; O'Brien et al., 2004). However, resource-use decisions are rarely made independently of other stresses and opportunities. Process-based vulnerability assessments are more suited to identifying which attributes of climate change are of importance to a system of interest, and how these are managed in the context of other stresses (Wandel, Wall and Smit, 2007). In process-based approaches, the analysis begins with the identification of how climate-related factors influence individuals, communities or economic sectors, and what ability exists to manage changes in these. In this type of assessment, potential climate change effects are treated as hazards; i.e., they are relevant if they exert some negative influence over a valued attribute of the system (Füssel, 2007). The research presented here adopts a process-based vulnerability perspective in order to identify what aspects of climate change rural stakeholders in the South Saskatchewan River Basin are sensitive to and what adaptive strategies exist to manage these.

Conceptually, the discussion is organized around the two components of vulnerability exposure-sensitivities and adaptive capacities. This categorization (based on Smit and Wandel, 2006) defines exposure as the likelihood of a system experiencing a particular condition and sensitivity as the particular conditions of the system which make the particular change relevant. *Exposure-sensitivities* are thus those attributes of climate change that are likely to have an effect (positive or negative) on stakeholders within the system. The actions taken by stakeholders to deal with negative effects or capitalize on opportunities are adaptations. The suite of adaptations available to a group of stakeholders and their use, alone or in combination with other management decisions, collectively, is considered an *adaptive strategy*. Stakeholders' abilities (managerial, financial, technical, etc.) to implement adaptive strategies determine their *adaptive capacities*. Vulnerability is viewed as a function of exposure-sensitivities and adaptive strategies and is, in other words, the stresses a particular system is under and its ability to manage these.

Both exposure-sensitivities and adaptive strategies/capacities, while frequently felt and managed at the individual level, are not independent of larger social, political and economic forces. For instance, while crop producers operate within an environment largely determined by growing conditions related to weather and climate, farm management decisions are also based on a host of factors, including commodity prices, labour availability, and institutional agricultural support. Increasingly, vulnerability scholarship recognizes that successful adaptation is not to climate alone and needs to be "mainstreamed" into other policies (Klein et al., 2007; Kok and de Coninck, 2007).

THE STUDY AREA

A DESCRIPTION OF THE SSRB

The SSRB, at 172,920 square kilometres, is Canada's largest dryland watershed, with more than 30% of the watershed classified as arid (Hammond et al., 1998; Toth et al., 2009). Although it contains areas of the Montane Cordillera and Boreal Plains Ecozones, 80% of the basin consists of the Prairie Ecozone (Toth et al., 2009). Much of the SSRB falls within the area known as Palliser's Triangle, which is characterized by semi-arid climate within the Prairie Ecozone. Precipitation (P) is less than Potential Evapotranspiration (PET) of the Prairie Ecozone, with values well below 0.5 P/PET in the eastern portion of the basin south of Saskatoon. While precipitation values are as low as 282 mm per annum in the central part of the SSRB, the Montane Cordillera receives well over 600 mm (Toth et al., 2009). Consequently, most of the flow in the tributaries of the South Saskatchewan River originates in the Rocky Mountains.

Natural vegetation in the Prairie Ecozone includes aspen forest, shrub cover and grasslands, with agriculture as the dominant land use. Although rural residents account for less than 20% of the total population of the basin, farming accounts for 87% of the land use (Bruneau et al., 2009). An analysis of rural vulnerability in the SSRB thus primarily focuses on agricultural producers as stakeholders.

More than half (56%) of the farm operations in the basin primarily rely on grain production, although this is not uniform throughout the region: in Alberta, 53% of farms are livestock operations, compared to only 24% in Saskatchewan (Bruneau et al., 2009). By land area, cropping is the most frequent agricultural land use at 46%, followed by pasture (41%) and fallow/other uses (13%) (Bruneau et al., 2009). As can be expected given the concentration of livestock in the western part of the basin, cropping accounts for a greater proportion of Saskatchewan's land use than Alberta's (53% and 42% respectively) (Bruneau et al., 2009).

Given high evapotranspiration and low natural precipitation in the Prairie Ecozone, lack of water is a limiting factor in productive agriculture. In the context of the South Saskatchewan River Basin, it is not possible to consider climate-related exposure sensitivities without considering surface water. A substantial part of the SSRB is considered "non-contributing"; i.e., it produces no net runoff. Consequently, there is a high reliance on water that originates outside the Prairie Ecozone in the Montane Cordillera of the Canadian Rockies. The South

Saskatchewan's main tributaries, the Bow, Oldman and Red Deer rivers deliver the bulk of both irrigation and potable water to many communities in the basin.

The Rocky Mountain rivers are fed primarily by snowmelt, and thus have a high degree of seasonal variability related to spring and summer temperatures in the upper part of the basin. Furthermore, snow accumulation in winter is important for total volume, and too little snow or too warm a winter (preventing accumulation) can translate into too little water or water at the wrong time of the year. Consequently, communities in the SSRB are exposed and sensitive to the temperature and precipitation regime in the Montane Cordillera in addition to local weather patterns.

While it is possible to generalize enough to argue that a large portion—and, in particular, the agriculturally productive Prairie Ecozone—have numerous exposure-sensitivities related to moisture availability, it can also be expected that SSRB communities are exposed to a varying array of stresses and have developed diverse management strategies. The prairie portion of the basin is split between two provinces, Alberta and Saskatchewan, and includes a range of livelihood types. Consequently, this chapter outlines some of the relevant stresses for ranchers, dryland farmers, irrigated farmers, and non-farm rural households in the region based on three case studies conducted under the IACC project.

THE CASE STUDY COMMUNITIES

Research summarized in this chapter is based on in-depth vulnerability case studies conducted in three localities: Taber (Alberta), Hanna (Alberta) and Outlook (Saskatchewan). These three communities represent very distinct portions of the SSRB, both in location and climate and water-related exposure-sensitivities.

The Town of Taber (population 7,591) is located 50 km east of Lethbridge, just south of the Oldman River. The Taber Municipal District (population 6,280) straddles the Oldman River and borders the Bow River on the northeast. Taber Municipal District is representative of the 5.1% of the SSRB which is irrigated land, and the Municipal District includes three of Alberta's large Irrigation Districts (IDs): the Taber ID, St. Mary's River ID and Bow River ID. Collectively, these three IDs account for over 250,000 hectares of irrigated land.

Irrigation development in the Taber Municipal District dates back to 1915, with the establishment of the first Irrigation District, the Taber ID, under the Alberta Irrigation Districts Act. The Oldman and Bow rivers have experienced insufficient water supply to meet Alberta's needs in recent history and, in particular, during the drought of 2001–2002, prompting innovative water-sharing arrangements among license holders in parts of the Oldman River Basin (Rush et al., 2004; McGee, 2007).

The regional economy of Taber relies heavily on agriculture, which accounts for the largest share of employment in the Town and Municipal District. The combination of a favourable climate and availability of irrigation has meant that Taber produces a range of diversified specialty crops, notably corn, sugar beets, vegetables, and potatoes. Local processing of these crops accounts for further employment in local potato, sugar beets and vegetable processing plants.

Water allocation in Alberta is administered through water licenses, and the date associated with a particular license determines its priority in times of water shortage. While the established irrigation districts have very early dates (i.e. high priority numbers), it is not possible to "hoard" water, as allocated volumes must be used in order to be maintained. Consequently, even established IDs may have portions of their licenses with very junior priorities. In dry years (arguably the times when irrigation is most crucial to producers), junior license holders have no entitlement for any water. This situation can be expected frequently for junior license holders, as both the Bow and Oldman rivers are fully (if not over) allocated and the Province of Alberta placed a moratorium on new water licenses on these basins in 2006. Furthermore, in the case of Taber, some of the processing industries, which agricultural producers rely on, have much more junior license priority than the Irrigation Districts. The combination of junior and senior licenses within IDs and the reliance on more junior processing license holders led to some innovative adaptation strategies in this area during the recent 2001–2002 droughts.

Research in Taber consisted of 31 semi-structured interviews which represented agricultural producers, local authorities, business and service industries, and retired residents in both the Town and the Municipal District during the summer of 2006 (Prado, 2008).

The Town of Hanna (population 2,847) is surrounded by Alberta's Special Area No. 2 (population 2,074). Alberta's Special Areas comprise a tract of just over two million hectares north of the Red Deer River. This region of Alberta is part of the dry mixed-grass natural grasslands, characterized by treeless prairie with shrubs and trees only in relatively moist areas such as coulees and river valleys. Most of the area receives less than 250 mm precipitation per annum. Evaporation generally exceeds precipitation throughout the study area and, consequently, solonetzic soil groupings are common in parts of the study area, including 30% or more of all land in the western portion.

Hanna and the Special Areas represent a unique institutional situation in the SSRB. The Special Areas were established as a direct result of drought and poor harvests in the 1920s and 1930s (Marchildon, 2007). The area was settled at the beginning of the 20th century and reached an all-time population high of 29,689 in 1921. In accordance with the Dominion Lands Act, settlers broke the native shortgrass prairie sod and established a wheat farming culture based on relatively small (65 ha) parcels.

Poor harvests and wind erosion during the 1920s and 1930s meant that many settlers were unable to meet tax obligations. The title of land in tax arrears reverts to the Crown. Tax recovery through loss of title and out-migration was so widespread in non-irrigated southeastern Alberta at this time that the provincial government replaced regional government in this region with a Crown agency which manages the remaining Crown (never claimed) and tax recovery lands under the Special Areas Act of 1938. Unlike the remainder of Alberta's counties and municipal districts, regional government falls directly under the authority of the province, and is administered by a provincially appointed Special Areas Board (Marchildon, 2007).

Despite depopulation and out-migration in the last century, the Special Areas remain a predominantly agricultural landscape. The Special Areas administration administers agricultural land with a view to drought via grazing and farming leases and community pastures. The governance framework established by the Special Areas Act and the Agricultural Service Board Act (1945) allows the Special Areas Board to devote resources to prevent and mitigate some of the impacts of drought. For example, the Special Areas Agricultural Service Board has undertaken extensive research into, and promotion of, zero tillage initiatives to decrease wind erosion during dry periods, and recommended particular mixes of native grasses for reseeding grazing leases. To generate research and extension with particular relevance to the drought-prone Special Areas, the Agricultural Service Board established the Dryland Applied Research Association (now known as the Chinook Applied Research Association, CARA).

Primary industries remain as the dominant source of employment within the Special Areas, with agriculture as the largest source of employment in the rural areas of the region. Over the past several decades, ranching has increased in importance at the expense of dryland (non-irrigated) farming. Furthermore, there is one very small irrigation project managed by the Special Areas (approx. 30 irrigators, most of whom only irrigate a small proportion of their operations) on Berry and Deadfish creeks in Special Area 2. The entire area was hard hit by the recent droughts, as severe moisture stress meant poor crop growth, and grasshopper outbreaks decimated much of what did grow. In addition, the area's increasing orientation toward ranching meant that climatic exposure-sensitivities were compounded by the difficulty beef producers faced in shipping to the United States as a result of BSE outbreaks beginning in 2003.

Although dryland farmers and most ranchers rely primarily on spring runoff and precipitation, surface water from the Red Deer River plays a key role in the area. The irrigation project's water is supplied via a raw water pipeline and surface water canal operated by Alberta Environment and Sheerness Generating Station. In addition, potable water for the Town of Hanna, surrounding villages, some rural water co-operatives and bulk tank filling stations is provided by the Henry Kroeger Water Commission in Hanna. The Henry Kroeger Commission receives raw water via the Alberta Environment/ Sheerness pipeline. Most rural residents have wells, but overall water quality in many of these is poor due to sodium and sulfate loads exceeding Canadian drinking water guidelines.

Primary research in Hanna and Special Area 2 consisted of 47 semi-structured interviews with producers, industry, community residents, and local institutional representatives in the fall of 2006 (Young and Wandel, 2007; Wandel, Young and Smit, 2009).

The Town of Outlook (population 1938, in 2006) is located southwest of Saskatoon on the South Saskatchewan River, just downstream of the Gardiner Dam and Lake Diefenbaker. Similar to Taber and the Special Areas, livelihoods in Outlook and the surrounding Rural Municipality (RM) of Rudy (population 434) are centred on agriculture and its related industries. Like Taber, there is some agricultural processing via a hay dehydration plant and an extraction plant for essential herb oils.

Like Hanna and Taber, Outlook derives its potable water by treating surface water obtained directly from Lake Diefenbaker. Similar to Hanna, treated water is pumped to rural areas through pipelines. Agricultural producers rely on a mixture of surface water and precipitation for their operations, with some supplementing supplies from wells (which are rarely of sufficient quality for potable water). There is access to irrigation water from Lake Diefenbaker, but the majority of farmers continue to "dryland" farm (i.e. crop farming without supplemental irrigation). Saskatchewan currently has no statutory scheme for water rights, and water licenses are granted at the discretion of the Saskatchewan Watershed Authority (Hurlbert, 2009). Currently, evaporation from Lake Diefenbaker accounts for the largest consumptive water use, and irrigation is underdeveloped relative to supply (Weiterman and Thauberger, 2006).

Fieldwork in Outlook consisted of 34 semi-structured interviews with respondents representing agriculture, business and the service industry. All the interviews were completed in 2006 (Pittman, 2008).

CLIMATE-RELATED EXPOSURE-SENSITIVITIES AND ADAPTIVE STRATEGIES IN THE SSRB

The three communities studied for this research, like much of the South Saskatchewan River Basin, are heavily based on agriculture. However, Taber, Hanna and Outlook represent very diverse situations: Taber is heavily reliant on production in the Irrigation Districts, Hanna has shifted to a predominant ranching mode, and Outlook includes a large number of dryland farmers. Despite these differing situations, the communities have similar exposure-sensitivities to climate and other stresses.

CLIMATE-RELATED EXPOSURE-SENSITIVITIES

Currently, agricultural producers are exposed and sensitive to temperature in a number of ways, including heat waves, temperature fluctuations, and early and late frosts. However, temperature and moisture requirements go hand in hand, as an increase in temperature during the growing season can generally be offset if moisture availability is also increased.

The western portion of the SSRB, including Taber and Hanna, are frequently exposed to chinook winds. The "chinooks" are foehn winds, originating in the Rocky Mountains, that can raise temperature by as much as 30 °C over the course of a few hours. Chinooks can be problematic for cattle producers as freeze-thaw cycles can melt snow which subsequently re-freezes as ice. This ice crust can be difficult on cattle hooves. Vegetable producers in Taber report that crops in long-term storage are sensitive to chinook-related temperature fluctuations, as these reportedly decrease sugar content in crops that are not stored in a climate-controlled environment.

The chinooks present particular problems for moisture management in both crop and livestock farming. Closed snow cover throughout the winter months means that the soil is protected from the drying effect of wind. However, chinooks can melt enough snow that the ground is now "open," and thus stored soil moisture is lost. Furthermore, if the chinook melts accumulated snow in moisture-stressed areas without widespread irrigation, the snow's contribution to soil moisture is generally inferior to what it could achieve during a slow, gradual spring melt. During rapid warming such as during a chinook, the soil does not warm up enough to thaw the ground, and thus runoff is not absorbed and the accumulated precipitation is lost, save for what can be captured in surface storage facilities such as dugouts.

In addition to freeze-thaw, absolute temperature in winter represents an important exposure-sensitivity. The southern Prairies have, historically, frequently experienced extreme cold snaps during the winter, and crop farmers count on these to control insect and crop pests from year to year. Lack of periods of extreme colds (-25 °C) results in year-to-year survival of such pests, which ultimately results in lower yields. Cattle producers, however, face an increase of 30–50% in feed requirements if temperatures are below -20 °C.

In the late winter and early spring, ranchers are particularly sensitive to extreme cold during the calving season (March to May), as such temperatures can affect the health of the cow and her calf and contribute to an increase in pneumonia among newborn calves. Temperatures during snow melt also present exposure-sensitivities for farmers and ranchers: if the snow melt is slow and gradual, soil moisture increases and contributes to better crop and grass growth for crop and dryland farmers throughout the growing season. However, a slow and gradual snow melt coupled with low snow accumulation—either through low absolute precipitation or as a result of repeated "chinooking off"—may mean that ranchers' dugouts will not fill and, thus, there may be insufficient stock water throughout the year.

Later in the spring, all producers need adequate heat for crop germination and growth. However, the need for sufficient warmth is closely tied to adequate moisture, as both are needed. Ideally, farmers require rain and mild temperatures after seeding. If it is both cold and dry, the seed lies dormant until such conditions are reached. However, if conditions are both wet and cold, seeds may become unviable. Similarly, if the plants germinate and do not receive moisture after this, they will fail. This is an issue for both field crops and tame pastures, though native pastures are more resilient to these conditions.

Crop farmers are particularly sensitive to late frosts if they occur post-emergence of the plant. Too much heat come summer can be a problem. Dryland farmers are sensitive to extreme heat (30°C and above) as it dramatically increases crops' moisture requirements, thus, extreme heat coupled with dry conditions poses substantial problems. However, even those who rely on irrigation to meet crop moisture requirements have difficulty with extreme heat as this hastens crop ripening. If crops ripen before they are fully developed, yield and quality decrease.

Crops and pastures require moisture throughout the growing season, though the crucial times vary by enterprise type. Ranchers need moisture early in the season for hay germination, establishment and growth. After haying in mid-summer, they are far less sensitive to low moisture than field crop producers whose crop develops throughout the growing season. Even among ranchers, those relying on native grasses are less sensitive to low moisture than those who seed tame grasses. Both native and tame grasses, as well as field crops, are subject to drying winds during hot summers with inadequate moisture, but native grasses will retain their protein value even if they dry on the stalk.

Hot summer days are linked to the formation of thunderstorms. When these are accompanied by hail—a form of precipitation welcomed by no one—all agricultural producers can be affected. Although hail is particularly damaging to pre-harvest field crops, even pasture areas can be decimated by localized hail events. Similarly, wind, whether accompanying thunderstorms or on its own, can be problematic as it contributes to the drying effect of hot temperatures and, in cases of high winds, can damage crops and buildings. Irrigators are particularly sensitive to high winds, as irrigation equipment such as pivots is particularly sensitive to wind damage.

Later in the growing season, early frosts (August and September) can be problematic for both dry land and irrigated farmers. A frost event before frost-sensitive crops such as vegetables have been harvested will affect the quality of the crop. Irrigated farmers are particularly exposed and sensitive to early frosts, as their crops tend to ripen later than dryland crops, and thus the pre-harvest window is longer.

Although the SSRB is particularly exposed and sensitive to lack of moisture, too much precipitation can—and does—pose significant problems. Producers in Taber and, to a lesser extent, in Outlook have reported flooding during extreme rain events in the growing season. One notable rain event in southern Alberta in June 2006 dumped more than an average season's worth of precipitation in a short time. For both dryland and irrigated farmers, this means crop losses. Ranchers, however, are far more resilient to excess moisture at any stage of the growing season, as grass is generally more resilient to these events.

CLIMATE-RELATED ADAPTATION STRATEGIES AND ADAPTIVE CAPACITY

Agricultural producers call on a range of management strategies to deal with climate-related exposure-sensitivities. Generally, agricultural producers rely on changing the timing of their operations, diversifying their crop and livestock mix to spread or change climate-related exposure-sensitivities, constructing additional infrastructure to cope with altered moisture regimes, or using institutionally supported risk reduction strategies.

Farmers have always adjusted the timing of their operations based on microclimatic conditions. For example, seedbed preparation requires that fields are free of snow and adequately dry to support the weight of tillage equipment without leading to excessive soil compaction. Similarly, seeding is often adjusted to weather forecasts. Farmers will often re-seed in situations where seed does not germinate or germinates poorly, or if the plants are damaged or destroyed post-emergence. Re-seeding places additional stress on farm profitability due to the increased costs of seed, fuel and operator time.

A common adaptation strategy among producers is to diversify the operation to include fewer moisture-sensitive crops as well as livestock or, in the case of several producers in the Hanna area, abandon crop farming altogether. Livestock introduce more options to cut losses if crops fail during the latter part of the growing season, as fields with insufficient yield or quality for harvesting can still be used for cattle feed. Furthermore, since cattle operations are less sensitive to minor fluctuations in temperature and moisture, these operations are better able to weather extreme events.

Cattle producers, however, also rely on a suite of management strategies. Ranchers require secure stock water supplies. Traditionally, these needs are met via dugouts, and adaptation strategies involve constructing more or larger dugouts and fencing off existing ones, and using on-demand pumps to maintain water quality as supplies run low. In the Hanna area, a number of producers have started digging shallow pipelines for dugout recharge from secure sources (such as pipeline tap-offs or good wells). In the worst-case scenario in both Hanna and Outlook, ranchers have resorted to hauling water using bulk tankers to bring their cattle through extreme dry periods.

Institutional programs such as crop and hail insurance are, in theory, designed to buffer the insured against the adverse effects of an extreme event or a poor season. However, farmers in both Saskatchewan and Alberta expressed a range of frustrations with the current arrangements, citing issues such as increasing premiums following claims and the removal of spot hail insurance. In the latter case, farmers are unable to claim losses for an individual field, as the loss has to be averaged over the entire operation. Given farm sizes in the hundreds and thousands of acres coupled with the extremely localized impact of hail, these losses are difficult to recoup.

Producers who have access to irrigation are better able to survive moisture variability; however, reliance on irrigation introduces greater sensitivity to late season temperature fluctuations. In addition, access to irrigation generally results in a crop mix with greater moisture requirements. In the Taber area, higher moisture crops, including corn and sugar beets, are common. In the event that irrigation water runs short, the loss is even greater.

Many of the routine climate-related exposure-sensitivities outlined above have associated management strategies. However, all of these strategies come at considerable financial cost, be it increased input costs, higher crop insurance premiums, more fuel costs for pumping of irrigation water, or simply accepting a lower yield during a bad year. Farm economics are based not only on yields but also on input and commodity prices, and thus a "bad year" in terms of crop production can be a "good year" for the bottom line if commodity prices are high. Similarly, a good crop year or one with excellent weight gain in cattle can be a very poor year in the final economic analysis if fuel and input prices are extraordinarily high or commodity prices are low.

The ability to adapt to climate change, thus, is not related only to climate. To a large extent, the economic resources available to an agricultural producer are a determinant of his adaptive capacity. Consequently, non-climatic exposure-sensitivities and adaptive strategies must be considered for a better picture of the vulnerability of the South Saskatchewan River Basin.

COMPOUNDING (NON-CLIMATIC) EXPOSURES-SENSITIVITIES AND ADAPTIVE CAPACITY

Rural communities in the South Saskatchewan River Basin are heavily dependent on agriculture. Agriculture serves many functions, but is, at the end of the day, an economic activity which generates profit. Although quality of life in rural communities arguably encompasses more than income, active agriculture ensures the maintenance of rural population, which is a basic requirement for healthy communities.

The farming and ranching operations of the SSRB, like all agricultural production, is sensitive to global commodity prices. Agriculture is heavily reliant on petrochemical inputs, including fuel for machinery, fertilizer, pesticides, and transportation costs. In recent years, fuel requirements have increased along with farm sizes as grain elevators consolidated. The closest elevator facilities to

Hanna are in Trochu, a road distance of over 100 km. Farmers in Outlook must travel 70 km to the nearest elevator (Pittman, 2008: 14). Furthermore, the removal of the Crow's Nest Freight Rate during the 1990s meant even higher rail costs for shipping grain. The interior Canadian Prairies are particularly sensitive to transportation costs as they are located roughly equidistant from the ports of Thunder Bay and Vancouver.

High petroleum prices are translated into higher production costs for farmers. However, these are offset for some commodities, such as corn, due to competition for biofuels and low world supply for certain grains and oilseeds, leading to higher prices. Although prices started softening both for the products as well as for farm inputs with the world economic slowdown in late 2008, oil prices had begun to increase during mid-2009, and thus indicate a likely return to high commodity and input prices.

Global commodity prices fluctuate based on many factors, including but not limited to production costs, and relative supply and demand. However, farmers in the SSRB have little control over macro-scale economics, and thus operate within this environment of uncertainty. Generally, though, the cost-price squeeze has been heavily felt, and many agricultural producers have started to diversify incomes beyond the farm. Farm spouses and operators working off the farm have become increasingly common in the study area. While this pluri-activity is not generally prompted by climate change itself, income diversification leads to higher adaptive capacity to absorb stresses, including climate-related stresses, and thus represents a major adaptation strategy.

Adaptation to insecure farm economics is particularly facilitated in the SSRB by the presence of oil and gas. Producers do not own sub-surface minerals, but are compensated for oil and gas industry access to both owned and leased land. In this way, farmers in oil and gas-rich parts of the Hanna area sometimes have far higher oil-lease revenues than gross farm receipts. This influence of oil and gas money is increasingly being felt in Saskatchewan as well. Furthermore, the oil and gas industry is a source of high-paying employment, which in turn can supplement farm income.

The lure of high-income non-farm jobs, however, has negative implications for rural communities in the SSRB. Agricultural producers in all three study communities expressed frustration over the lack of available farm labour. In all cases, the average age of operators is increasing, with fewer young people entering the industry. Rural areas are declining in population, and in the case of Outlook, this is reflected in the community's total population. With fewer people, services such as local schools and health care become more difficult to provide.

CONCLUSION

The rural population of the South Saskatchewan River Basin is heavily reliant on agriculture, and agriculture and climate are inextricably linked. The IPCC expects a three-degree warming in this area in the coming century, and current exposure-sensitivities will only be exacerbated with the attendant warmer winters, higher moisture requirements, and possibly decreased snowpack and thus surface water availability.

Currently, agricultural operations in the SSRB are well adapted to climate change, but these adaptive strategies generally come at an economic cost. Thus, there is a direct relationship between the frequency at which some costly adaptive strategies are employed and the overall viability of farm enterprises. In this way, adaptive capacity is a function of available capital resources. In light of the current economic boom, management of problematic conditions has been achieved without widespread farm bankruptcy as there was during the dry years of the early 20th century. However, if oil and gas revenues dry up or climate change increases the frequency of employing adaptive strategies, the currently high adaptive capacity cannot be maintained. Furthermore, a substantial reduction in overall moisture, particularly in the rivers in the southern portion of the basin, may be beyond the capacity of the system to adapt regardless of the buffer of capital. Finally, it should be noted that an adaptive capacity that is strongly related to the infusion of non-agricultural capital introduces differential vulnerability among producers, particularly the "haves," with oil leases, and the "have-nots."

ENDNOTES

1. Funding for this project was provided by the Social Sciences and Humanities Research Council of Canada's (SSHRC) Major Collaborative Research Initiatives (MCRI).

REFERENCES

Adger, N., S. Agrawala, C. Conde, K. O'Brien, J. Pulhin, R. Pulwarty, B. Smit, B. and K. Takahashi. 2007. "Assessment of Adaptation Practices, Options, Constraints and Capacity." Pp. 717–43 in M.L. Parry, O.F. Canziani, J.P. Palutikof, P.J. van der Linden and c.e. Hanson (eds.). *Climate Change 2007: Impacts, Adaptation and Vulnerability. Contribution of Working Group II to the Fourth Assessment Report of the Intergovernmental Panel on Climate Change.* Cambridge University Press: Cambridge.

Bruneau, J., D.R. Corkal, E. Pietroniro, B. Toth and G. Van der Kamp. 2009. "Human Activities and Water Use in the South Saskatchewan River Basin." *Prairie Forum* 34, no. 1: 129–52.

Füssel, H.M. 2007. "Adaptation Planning for Climate Change: Concepts, Assessment Approaches and Key Lessons." *Sustainability Science* 2: 265–75.

Hammond A., S. Murray, J. Abramovitz and C. Revenga. 1998. *Watersheds of the World: Ecological Value and Vulnerability.* World Resources Institute, 178: Washington.

Hurlbert, M. 2009. "Comparative Water Governance in the Four Western Provinces." *Prairie Forum* 34, no. 1: 45–78.

Klein, R.J.T., S.E.H. Eriksen, L.O. Næss, A. Hammill, T.M. Tanner, C. Robledo and K.L. O'Brien. 2007. "Portfolio Screening to Support the Mainstreaming of Adaptation to Climate Change into Development Assistance." *Climatic Change* 84: 23–44.

Kok, M.T.J. and H.C. de Coninck. 2007. "Widening the Scope of Policies to Address Climate Change: Directions for Mainstreaming." *Environmental Science & Policy* 10, nos. 7–8: 587–99.

Marchildon, G.P. 2007. "Institutional Adaptation to Drought and the Special Areas of Alberta, 1909–1939." *Prairie Forum* 32, no. 2: 251–72.

McGee, D. 2007. Personal Communication (September). Edmonton.

O'Brien, K., S. Erikson, A. Schjolden and L. Nygaard. 2004. "What's in a Word? Conflicting Interpretations of Vulnerability in Climate Change Research." CICERO Working Paper: Oslo.

Pittman, J. 2008. *Report on the Community Vulnerability of Outlook* [online]. IACC Working Paper: Regina (accessed May 2009). Available at: *http://www.parc.ca/mcri/iacc063.php*

Prado, S. 2008. *Report on the Community Vulnerability of Taber* [online]. IACC Working Paper: Regina (accessed May 2009). Available at: *http://www.parc.ca/mcri/iacc066.php*

Rush, R., J. Ivey, R. de Loë and R. Kreutzwiser. 2004. *Adapting to Climate Change in the Oldman River Watershed: A Discussion Paper for Watershed Stakeholders.* Department of Geography: Guelph.

Schneider, S.H., S. Semenov, A. Patwardhan, I. Burton, C.H.D. Magadza, M. Oppenheimer, A.B. Pittock, A. Rahman, J.B. Smith, A. Suarez and F. Yamin. 2007. "Assessing Key Vulnerabilities and the Risk from Climate Change." Pp. 779–810 in M.L. Parry, O.F. Canziani, J.P. Palutikof, P.J. van der Linden and c.e. Hanson (eds.). *Climate Change 2007: Impacts, Adaptation and Vulnerability. Contribution of Working Group II to the Fourth Assessment Report of the Intergovernmental Panel on Climate Change.* Cambridge University Press: Cambridge.

Smit, B. and J. Wandel. 2006. "Adaptation, Adaptive Capacity and Vulnerability." *Global Environmental Change* 16: 282–92.

Toth, B., D.R. Corkal, D. Sauchyn, G. Van der Kamp and E. Pietroniro. 2009. "The Natural Characteristics of the South Saskatchwan River Basin: Climate, Geography and Hydrology." *Prairie Forum* 34, no. 1: 95–128.

Wandel, J., E. Wall and B. Smit. 2007. "Process-Based Approach to Climate Change Adaptation." Pp. 42–50 in E. Wall, B. Smit and J. Wandel (eds.). *Farming in a Changing Climate: Agricultural Adaptation in Canada.* UBC Press: Vancouver.

Wandel, J., G. Young and B. Smit. 2009. "Vulnerability and Adaptation to Climate Change: the Case of the 2001–2002 Drought in Alberta's Special Areas." *Prairie Forum* 34, no. 1: 211–34.

Weiterman G. and F. Thauberger. 2006. *Sustainable Irrigation Development* [online] [accessed May 2009]. Available t: *http://www.irrigationsaskatchewan.com/ICDC/content/Sustainable%20Irrigation%20Development.pdf*

Young, G. and J. Wandel. 2007. *Community Vulnerability in the South Saskatchewan River Basin: A Case Study of Hanna, Alberta* [online]. IACC Working Paper: Regina (accessed May 2009). Available at: *http://www.parc.ca/mcri/iacc060.php*

CHAPTER 14

THE ST. MARY RIVER

Ryan MacDonald, James Byrne and Stefan Kienzle

INTRODUCTION

Water supply on the western Prairies of Canada is highly dependent on snowmelt from the eastern slopes of the Rocky Mountains. Under expected future atmospheric warming, it is anticipated that mountain snow accumulations will decline (Hamlet and Lettenmaier, 1999), resulting in a reduction of available water from snowpack in mountainous regions (Barnett, Adam and Lettenmaier, 2005; Lapp et al., 2005). Spring snowmelt is relied upon to provide up to 80% of the annual flow volume in snow-dominated watersheds (Stewart, Cayan and Dettinger, 2004), and has been the most predictable source of water for human use (Stewart, Cayan and Dettinger, 2005). This study investigates the effects of climate change on mountain snowpack in the St. Mary River watershed, which supplies water for approximately 200,000 hectares of irrigation farming in southern Alberta, Canada.

To assess potential changes in mountain snowpack in the St. Mary watershed, a fine-scale hydro-meteorological model is applied and driven by GCM (general circulation model)-derived scenarios of future climate. A number of GCM scenarios are used to perturb the model and test the sensitivity of winter snow hydrology to a range of possible future climates. Changes in snowpack in St. Mary River watershed are assessed for the 2020s (2010–39), 2050s (2040–69), and 2080s (2070–99).

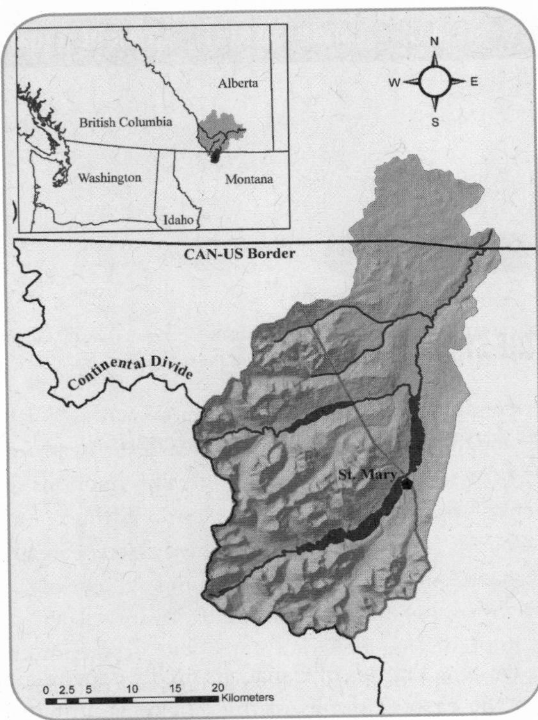

Figure 1. St. Mary River watershed, Montana

STUDY AREA

The headwaters of the St. Mary watershed lie on the eastern slopes of the Rocky Mountains, with the majority of the upper watershed residing within Glacier National Park, Montana. This watershed is physiographically representative of many of the watersheds on the eastern slopes of the Rocky Mountains, especially those in southern Alberta. The St. Mary River flows from the continental divide, through the upper and lower St. Mary lakes, and ends in southern Alberta, where it meets the Oldman River (Figure 1).

RESULTS AND DISCUSSION

This modelling effort has demonstrated that snow hydrology in the St. Mary River watershed is highly susceptible to changes in temperature and, to a lesser degree, precipitation. Table 1 shows the predicted 30-year mean decrease in snow water equivalent (SWE) for three climate change scenarios. This consistent reduction in SWE suggests water supply from snowpack is likely to be reduced in the future. This presents a significant problem for the St. Mary River due to the fact that water supply from this river is already over-allocated.

Each of these future scenarios represents some degree of global greenhouse gas (GHG) emission reduction. Nakicenovic et al. (2000) describe how each scenario represents future political, economic, and environmental change. These scenarios demonstrate the need for adaptation, as even with GHG reduction, it is likely that water supply from snowpack on the eastern slopes of the Rocky Mountains will be reduced in the future. This poses important questions about water resources in the future.

One of the most significant impacts from climate change is likely to be changes in ecosystems. Trends towards an earlier onset of spring have already been

	Historical	CGCM2 (A2)			CSIRO—Mk2 (A1)			NCAR-PCM (B2)		
Watershed Average (mm SWE)	213	192	166	77	200	162	138	192	197	208
Percent Change relative to historical		-2%	-22%	-63%	-1%	-12%	-17%	-2%	-8%	-3%

Table 1. Comparison of mean annual snowpack (mm of SWE) and percent change in mean annual snowpack for the entire St. Mary River watershed relative to the 1961–90 historical period.

recorded in the historical record in numerous mountainous regions across North America (Mote et al., 2005). Using the date of maximum SWE as a surrogate for the onset of spring, this study shows that even under significant reductions in GHG emissions, the advancement of spring is likely to occur. Here, only the NCAR-PCM (B2) and CGCM2 (A2) scenarios are shown, as they represent, respectively, the least and greatest predicted changes in the timing of maximum SWE. The mean date of maximum snowpack over the watershed is April 8 for the historical period. Both scenarios are consistent in predicting an earlier mean date of maximum snowpack over the 2080 period, with the mean date of maximum SWE occurring, by the 2080s, on April 2 in Scenario 3 and March 6 in Scenario 1 (Figure 2).

An earlier onset of snowmelt has important implications for water supply, and could result in a significant impact on water resources for human and ecosystem use (Field et al., 2007). An earlier onset of melt will disrupt current management practices by modifying the predictability of seasonal deliveries of streamflow (Regonda et al., 2005). Reservoir operations and flood management will have to adapt to changes in the onset of peak runoff, while maintaining adequate water supply for late season demand.

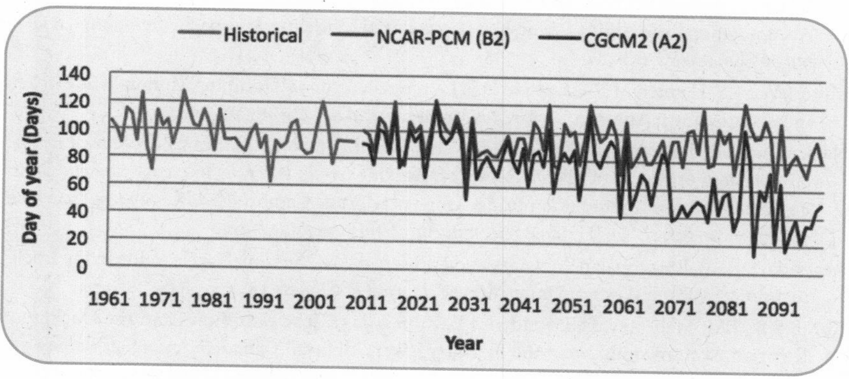

Figure 2. Predicted date of maximum SWE over the St. Mary River watershed.

Shifts in peak runoff to earlier in the spring or late winter lead to decreased water availability later in the summer months, when water demand is highest (Barnett, Adam and Lettenmaier, 2005). With melt occurring earlier, soil moisture may be depleted sooner, drought potential would be accentuated, and summer streamflow volumes could decrease. Further stresses may be imposed as glacier contributions to late season streamflow decline as a result of diminishing glacial ice mass. Reduced late summer flows will create management problems for human uses, and will likely have adverse impacts on aquatic ecosystems (Schindler, 2001). Reduced late season flow can result in warmer water temperatures, forcing native cold-water species like the bull trout (*Salvelinus confluentus*) into isolated headwater portions of mountain streams (Schindler, 2001). Riparian ecosystems could be impacted by late season low flows and associated low soil water, leading to greater stress on riparian forests, and more frequent die-off of young trees (Rood et al., 1995). These potential impacts of climate change demonstrate the importance of adopting management strategies that account for decreased water supply and increased water demand for both human and ecosystem function.

REFERENCES

Barnett, T.P., J.C. Adam and D.P. Lettenmaier. 2005. "Potential Impacts of a Warming Climate on Water Availability in Snow-dominated Regions." *Nature* 438: 1–7.

Field, C.B., L.D. Mortsch, M. Brklacich, D.L. Forbes, P. Kovacs, J.A. Patz, S.W. Running and M.J. Scott. 2007. "North America." Pp. 617–52 in M.L. Parry, O.F. Canziani, J.P. Palutikof, P.J. van der Linden and C.E. Hanson (eds.). *Climate Change 2007: Impacts, Adaptation and Vulnerability. Contribution of Working Group II to the Fourth Assessment Report of the Intergovernmental Panel on Climate Change*. Cambridge University Press: Cambridge, UK

Hamlet, A.F. and D.P. Lettenmaier. 1999. "Effects of Climate Change on Hydrology and Water Resources in the Columbia Basin." *Journal of the American Water Resources Association* 35: 1597–1623.

Lapp, S., J. Byrne, S.W. Kienzle and I. Townshend. 2005. "Climate Warming Impacts on Snowpack Accumulation in an Alpine Watershed: A GIS Based Modeling Approach." *International Journal of Climatology* 25: 521–26.

Mote, P.W., A.F. Hamlet, M.P. Clark and D.P. Lettenmaier. 2005. "Declining Mountain Snowpack in Western North America." *American Meteorological Society* 86: 39–49.

Nakicenovic, N., J. Alcamo, G. Davis, B. de Vries, J. Fenhann and S. Gaffin. 2000. *Special Report on Emissions Scenarios: A Special Report of Working Group III of the Intergovernmental Panel on Climate Change* [online]. Cambridge University Press: Cambridge, UK. 599 pgs. Accessed July 7, 2009: *http://www.grida.no/climate/ipcc/emission/index.htm.*

Regonda, S.K., B. Rajagopalan, M. Clark and J. Pitlick. 2005. "Seasonal Cycle Shifts in Hydroclimatology over the Western United States." *Journal of Climate* 18: 372–84.

Rood, S.B., J.M. Mahoney, D.E. Reid and L. Zilm. 1995. "Instream Flows and the Decline of Riparian Cottonwoods along the St. Mary River, Alberta." *Canadian Journal of Botany* 73: 1250–60.

Schindler, D.W. 2001. "The Cumulative Effects of Climate Warming and Other Human Stresses on Canadian Freshwaters in the New Millennium." *Canadian Journal of Fisheries and Aquatic Sciences* 58: 18–29.

Stewart, I.T., D.R. Cayan and M.D. Dettinger. 2004. "Changes in Snowmelt Runoff Timing in Western North America under a 'Business as Usual' Climate Change Scenario." *Climatic Change* 62: 217–32.

——. 2005. "Changes Toward Earlier Streamflow Timing Across Western North America." *Journal of Climate* 18: 1136–55.

CHAPTER 15

IMPACTS AND ADAPTATIONS TO EXTREME CLIMATIC EVENTS IN AN ABORIGINAL COMMUNITY: A CASE STUDY OF THE KAINAI BLOOD INDIAN RESERVE

Suren Kulshreshtha, Virginia Wittrock, Lorenzo Magzul, and Elaine Wheaton

INTRODUCTION

One of the well-recognized impacts of climate change is increased frequency of extreme events. According to Stern (2007: 68), climate change is likely to increase the costs imposed by extreme weather, both by shifting the probability distribution upwards (more heat waves, but fewer cold snaps) and by intensifying the water cycle, so that severe floods, droughts and storms occur more often. Increasing occurrences of extreme events have the potential to bring a high cost to society, especially in those areas that are already under stress with current climatic and socio-economic conditions. The latter set of factors also affects a community's ability to make proper adaptation, partly on account of low adaptive capacity, low ability to implement adaptation, and/or considerable property damage. For Indigenous peoples, climate change and extreme events bring different kinds of risks and opportunities, threaten cultural survival, and negatively impact Indigenous human rights (IWGIA, 2008).

In this study, biophysical and socio-economic impacts of climate-related extreme events on an Aboriginal community are studied along with their exposures, sensitivities and methods of adapting to these events. Based on this information, we assess their future vulnerability. The study of past and present experience with coping mechanisms and adaptation measures adapted to these changing climate regimes is needed for designing future climate change programs and policies. This study[1] provides information on the current physical and social vulnerabilities related to water resource scarcity and surplus, and assesses the technical and social adaptive capacity of various institutions in the Kainai Blood Indian Reserve[2] study region.

CHARACTERISTICS OF THE KAINAI BLOOD INDIAN RESERVE

This is a study of an Aboriginal community called the Kainai Blood Indian Reserve No. 148 (KBIR)[3]. The KBIR, with a land base of 1414.03 square kilometres (equivalent to 356,755 acres), is located in the southwest corner of Alberta, west of the St. Mary River, east of the Belly River, and south of the Oldman River (Figure 1). Its capital is Standoff, with some members residing in six other communities (Moses Lake, Levern, Old Agency, Fish Creek, Fort Whoop Up, and Bullhorn) on the reserve.

This reserve is located within the Palliser Triangle,[4] the driest part of the Canadian Prairies. This area is characterized by its aridity and has a mixed grass-

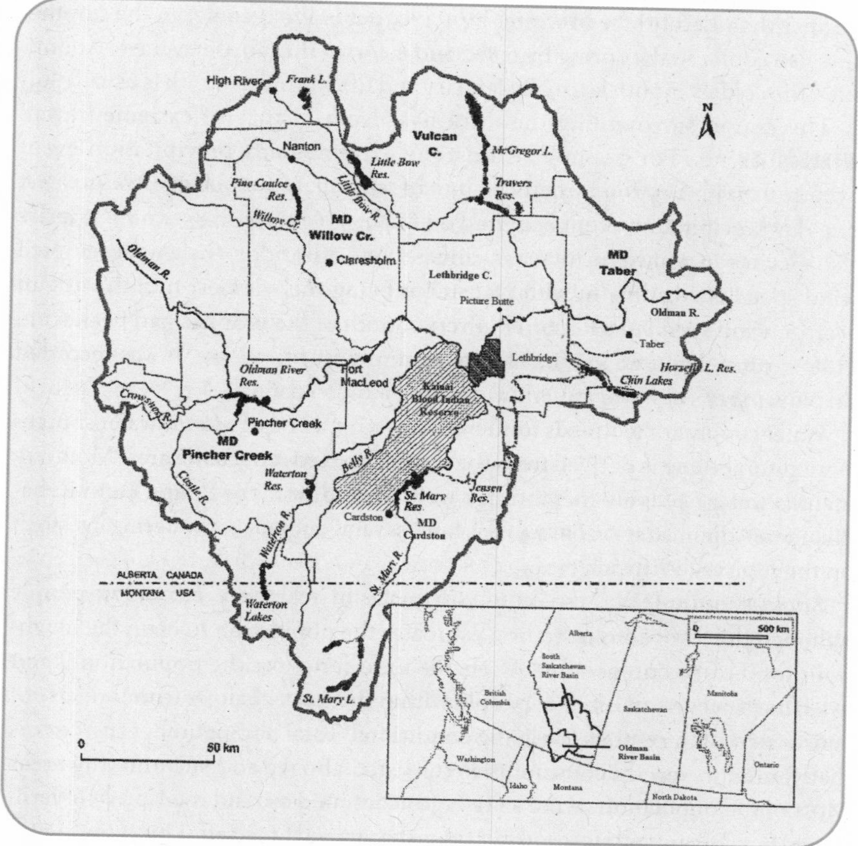

Figure.1. The Location of the Kainai Blood Indian Reserve and Surrounding Areas (adapted from Stratton, 2005).

land ecoregion (Nemanishen, 1998). As indicated in the chapter on future climate change scenarios, the frequency of drought and severe flood events are expected to increase in the area. The recently released Fourth Assessment Report from the Intergovernmental Panel on Climate Change (IPCC) stated that future increases[5] in the area affected by drought are likely (i.e. 66% probability of occurrence) (IPCC, 2007a). The area of very dry regions on a global basis has doubled since the 1970s and much of the Canadian Prairie Provinces has exhibited drying since the 1970s (Dai, Trenberth and Qian, 2004). Global warming enhances the hydrological cycle, so floods as well as droughts, could increase. The IPCC (2007b) estimates that the frequency of heavy precipitation events will very likely increase in the future (90–99% probability).

CLIMATE AND SOCIO-ECONOMICS

The KBIR does not house any long-term weather stations, but two long-term stations, Cardston and Lethbridge, are located relatively close to it. Cardston is warmer than Lethbridge in winter by 1.1°C, but is 1.3°C cooler in the summer, and also cooler in the spring by 0.8°C and 0.2°C in the fall, on average. At both sites, the coldest month is usually January and the warmest month is usually July.

The region surrounding the KBIR has also a history of extreme/severe weather events. For example, Cardston's extreme daily precipitation event occurred on June 6, 1995 when 106 mm of rain fell. In Lethbridge, the greatest one-day precipitation event occurred on May 23, 1980, with 85.4 mm. Wind is also a factor in southern Alberta's climate. At Lethbridge, the average annual wind speed is 18.2 km/h, with December being the windiest month with an average wind speed of 21.2 km/h. Every month of the year has had peak wind gusts greater than 120 km/h. The maximum gust of 171 km/h was recorded on November 19, 1962, at Lethbridge (for 1955 to 2008).

Water supply at the KBIR is from both surface water and groundwater sources. Agricultural water is derived from the three rivers existing in the area. Most residential water needs are met through wells. In addition, the Prairie Farm Rehabilitation Administration (PFRA) has built several dugouts for watering livestock on the reserve (Wittrock et al., 2008).

Socio-economically, the KBIR is a place of residence for 4,177 people, although the KBIR administration estimates the population to be in the neighbourhood of 10,000 people.[6] A reliable time series for the population is not available since the number of people within the KBIR changes from census-to-census period in response to living conditions. Total population of the KBIR is distributed in several communities (as listed above) and surrounding areas. Most of the population of the KBIR is concentrated around the Local Government (Band Council) Headquarter townsite, named Standoff. This is the centre of the reserve and the hub of activity.

The most significant economic activity in the KBIR is agriculture. Of the total land base of the reserve of 356,756 acres (144,436 ha), 50% is cultivated and another 15.6% is grassland. At the present time, all of these cultivated areas and grasslands are leased to neighbouring (non-KBIR member) farmers. Title to the KBIR lands is held by Her Majesty in right of Canada, although some members have historical 'occupancy' rights for the use of that land only (Magzul, 2007). Occupants, however, cannot sell their land or use it as collateral against any loans from a financial institution.

One of the largest single tracts of irrigation in Alberta is on the KBIR lands, with a target irrigation area estimated at 10,000 ha.[7] These lands, along with dryland area, are a major source of income to the KBIR residents through cash-leasing these lands to non-native farmers. Irrigated area commands a rent of $370/ha/year, while the dryland area is rented for $76/ha/year. Water for irrigating these lands is provided by the Blood Tribe Agricultural Project (BTAP).

According to Armstrong (2001), overall socio-economic well-being of the KBIR is below average Canadian conditions[8] for Aboriginal communities. This well-being was examined in terms of education, employment, income, and housing. In general, Aboriginal (First Nations) people endure ill health, run-down and overcrowded housing, polluted water, inadequate schools, poverty, and family breakdown at rates found more often in developing countries than in Canada (Minister of Supply and Services, 1996). Housing availability is a major issue on the reserve. As a result, many of the band members have to find residences outside the KBIR (such as in the bordering towns of Cardston and Lethbridge).

Compared to the educational attainment of a typical Albertan, Kainai Nation adults over the age of 15 years have lower attainment. Almost half of these people have no high school level (or beyond) of education. Similarly, for 2006, only 8% of the KBIR labour force had university degrees compared to 21% for the province of Alberta.

As a result of lack of employment opportunities on the KBIR, and lower educational attainment of workers, the unemployment rate is very high. Statistics Canada (2008) estimated that in 2006, 23.6% of the total population was unemployed.[9] The unemployment rate for youths (persons aged 15–24 years) is higher than for adults (persons aged 25 and over).

VULNERABILITY TO EXTREME EVENTS

Climate related extreme events, such as droughts or floods, generally result in biophysical and socio-economic impacts. During 2001 and 2002, most of southern (which included the KBIR) and central Alberta experienced a major drought, and in some areas, multi-year droughts. Longer droughts are more difficult to deal with because impacts become more severe and adaptation tends to use more resources, including time for recovery. Severe floods were also experienced at the KBIR in 1995, 2002 and 2005. Estimation of the impact (particularly the socio-economic impact) of these events was difficult since precise and sufficient information was not available.

IMPACT OF THE 2001–2002 DROUGHT

Droughts in the region affected water availability—both surface water and soil moisture as well as groundwater. On the KBIR, surface water is used for irrigation purposes. The source of this water is the Waterton, Belly and St. Mary rivers through a canal system. Water level in the St. Mary Reservoir was low during 1999 to 2001 due to a combination of drought conditions and dam maintenance. There was no reported impact on the level of water available for irrigation as a result.

Most of the residential water needs on the KBIR are met using groundwater. Unfortunately, the reserve does not have any groundwater observation wells. Using data from a nearby observation well (located near Lethbridge), it was noted that the water level declined slightly in late 2001 to early 2002, but the level rebounded quickly by mid-2002. Another observation well, located 9 km northwest of the reserve, recorded lower water levels during the drought years. The KBIR administration reported that, during the drought years, some wells went dry, and the water quality deteriorated.[10]

Dugouts are a primary source of water for livestock in the Canadian Prairies. As of 1993, the KBIR had 63 dugouts[11] (Wittrock et al., 2008). By 2000, dugouts on the reserve were one-quarter full to dry. These low water levels continued until 2002 when the early June precipitation filled them.

The economic cost of the droughts could not be estimated precisely on account of inadequate information collected and/or reported. Based on information provided by officials of the KBIR, some of these impacts were:

- Since all of the agricultural lands are cash-leased, and since these lease rates are not affected by drought conditions, there was no direct economic effect[12] of the drought on gross revenue received by the KBIR from crop production.
- Since BTAP provides water for irrigation under the lease arrangements, it is conceivable that higher pumping costs may have been incurred on account of the droughts. However, actual estimates of such costs were not available. This would have reduced the net operating margin of the BTAP (thus, of the KBIR).
- Exact cost of the drought impacts on KBIR livestock production could not be estimated, however, based on discussions with the KBIR officials, the Blood Ranch had to reduce cattle from 700 before the drought, to 200 to 300 head during the drought period. Although this measure would have yielded higher gross revenue for the drought period, it would have a nega-

tive effect on gross income over the next few years due to reduced marketing of cattle. However, the exact economic cost of this measure could not be ascertained.

- During the drought period, several grass fires were reported which required additional expenses for controlling them.
- Some of the wells went dry, so the Blood Tribe Works Department had to deliver water to more than the usual number of homes. The exact cost of this operation was not available.
- During 2001, some homes had cracked cisterns and/or septic tanks. These had to be repaired and/or replaced.
- More boil water advisories were issued during the drought period. However, no major health issues were reported because of the droughts.

On account of the inadequate information available from the KBIR officials, the exact economic and social cost of the droughts could not be estimated. This perhaps is one of the major weaknesses of the institutions that govern the KBIR.

IMPACT OF THE FLOODS

Since the reserve is surrounded by three rivers—making it a natural flood zone—flooding of a part of the KBIR lands is a recurring problem. As noted above, major floods were experienced during 1995, 2002 and 2005, and the last flood had the worst impacts. Flooding was caused by an intense rainfall during a short period. Most of the flood damage was in the rural areas. Some of the major impacts/damages of the flood in 2005 were:

- Road washouts, including culverts, and other road damages;
- Flooded basements;
- Damaged personal belongings;
- People were evacuated from their houses, and the expenses related to lodging in hotels or other suitable accommodation during the flooding and recovery period;
- Increased stress levels among those affected;
- Treatment of the well water at a cost of $4,000 per treatment[13];
- Impacts on the cropland, and
- Reduced fishing in the river due to high river level.

However, on the positive side, although flooding is costly to the Blood Tribe administration and the people affected, it does increase the number of people temporarily employed on the reserve; therefore, providing income to the people. In addition, contractors undertaking flood damage repairs or selling carpets do better business due to these events.

During the flood of 2005 many homes were affected. According to Blood Tribe Housing Department,[14] almost half of the homes (600 homes of a total of 1,218) were damaged by this flood. Major damages resulted from flooded basements and sewer backups. Since these homes are not covered under insurance policies, much of this cost had to be borne by the occupants of the homes or by the KBIR Band Council, although some assistance is provided by Indian and Northern Affairs Canada for such repairs or reconstruction of homes. Lack of insurance appeared to be a source of vulnerability to flood impacts on housing. In contrast, floods, droughts, and other impacts to cropland are not sources of vulnerability because of the cash-lease arrangements with farmers living off the reserve.

Drinking water was contaminated and several water advisories were issued. According to Magzul (2007), the cost of the 2002 and 2005 floods to the KBIR administration was $8 million and $6.5 million, respectively. In addition to the above cost of the floods, crop production aspects under the management of the BTAP were also affected. Some fields were under water, and seeding was delayed or could not be cultivated. In the 1995 flood, it was reported that two pump sites used for irrigation were also flooded.

IMPACT OF OTHER EXTREME EVENTS

Besides droughts and floods, the strong winds that characterize the region are another type of extreme weather event that causes damages. They mostly affect homes and other buildings by causing damages to the roofs and outside structures. Some parts of the reserve are more vulnerable to wind damage than others. Rural homes, or those located in open areas, perhaps belong to this category. Severe windstorms can also cause other problems such as crop and soil damage. Assessment of the impacts of other extreme weather and climate events (such as heat waves, hail storms, dust storms) was beyond the scope of this work.

NATURE OF ADAPTATION UNDERTAKEN

From the above discussion of impacts of extreme events, it appears that the KBIR is sensitive to extreme events, and can be highly vulnerable. In addition, high unemployment (resulting in lack of skills and training), lack of business opportunities, and lack of capital resources results in low adaptive capacity, which makes the reserve more vulnerable to natural extreme events.

The KBIR adapted to previous droughts, specifically the 1988 drought, by implementing a leasing policy. This policy allows reserve land to be leased on a cash basis to non-reserve farmers, and results in stable yearly incomes for people with occupancy rights, and decreases susceptibility to changing climatic and/or economic conditions. In 2001 and 2002, BTAP was the major agricultural organization on the reserve, and since no major impacts of the drought were experienced due to the cash-leasing of agricultural land to non-native farmers, no adaptation measures were reported by the reserve. As noted above, this arrangement tends to reduce vulnerability to droughts, floods, and other sources of crop loss. However, it was felt that if there are more droughts in the future, the irrigated land may have to revert back to dry-land agriculture[15] due to possible lack of irrigation water (Magzul, 2005). The exposure risk to drought on the KBIR may increase in the future because the KBIR administration is reducing the amount of leased land, and opting to farm the land themselves.[16] This increased risk may be a result of lack of experience with droughts in the past, lack of resource (particularly financial), and need for better information for the managers of this land.

The droughts affected management of the livestock herd because of decreased feed supplies and water supply. The result was a large number of cattle were culled, feed was purchased, and water was brought to the cattle via pipeline from one of the rivers (Magzul, 2005). The droughts also resulted in some residents' wells going dry. This resulted in water being hauled to supply water to these residents or new wells were established (Magzul, 2005).

FUTURE VULNERABILITY TO CLIMATE CHANGE

It is commonly understood that the vulnerability of a community (and the residents) to climate change will be greater where and when stresses from other sources, such as poverty, unequal access to resources, food insecurity, environmental degradation, and risks from natural hazards, are high. The KBIR is not shielded from these latter sources of vulnerability. The present-day plan for the BTAP is to abolish the leasing arrangement and take over cultivation of lands under the direct control of the Band Council.[17] This would make the KBIR more vulnerable to future climate extreme events that impact agricultural activities.

The Blood Tribe Works Department, in association with the Blood Tribe Housing Department, has begun to survey the reserve in order to locate homes in less flood-prone areas. The same surveys will also be used for designing and locating roads to various homes on the reserve. It is reported that no new houses would be permitted to be built in the flood plain, and each new house must have proper drainage (Magzul, 2005). These measures should reduce vulnerability to flooding in the future. However, some Kainai Nation members have occupancy rights on land that is only on the flood-prone land and may not have the option of locating their homes in areas with less risk of flooding.

In summary, the many impediments to implementing adaptation measures to flooding include lack of funds and housing, and inability to move houses due to occupancy rights (Magzul, 2005). According to the Royal Commission on Aboriginal Peoples (see Minister of Supply and Services, 1996), distributing land and resources will greatly improve their chances for jobs and a reasonable income. After that, the tools most urgently needed are: capital for investment in business, and enhanced technical, management and professional skills to realize new opportunities.

Over the past two decades, Indian and Northern Affairs Canada has transferred decision-making powers to the KBIR tribal government over matters such as education, health, policing, and social services. Regaining control over these services may lead to regaining capacity to develop their human and natural resources, and create a more stable economy in the community.

In conclusion, the KBIR is in a region that is already affected by drought and floods, and these hazards and others are expected to increase in the future. Their adaptive capacities are constrained by many socio-economic factors; however, they are currently buffered from drought, and sometimes flood effects, by factors including their use of groundwater, irrigation, and by their cash-lease system for agricultural land. In general, others in the Prairies have less access to these adaptation measures. Key questions regarding their vulnerability to climate change and water scarcity include the sustainability of those measures, and the complex interaction of socio-economic and biophysical stresses with future climate extreme events.

ENDNOTES

1. This study was funded by the University of Regina's Institutional Adaptation to Climate Change project, supported by the Social Sciences and Humanities Research Council of Canada. This material is heavily borrowed from Wittrock et al. (2008) and Magzul (2005 and 2007).
2. In subsequent discussion, the 'Kainai Blood Indian Reserve' is also referred to as "reserve."
3. The members of the reserve are called "Kainai Nation Members." The official administrative units at the KBIR are called Blood Tribe Departments. The Statistics Canada refers to the KBIR as the Blood 148 community.
4. The Palliser Triangle is in the "rain-shadow" of the Rocky Mountains. It is north of the American border, bounded by Cartwright, Manitoba; Lloydminster, Saskatchewan; and Calgary and Cardston, in Alberta. It was reported by Captain John Palliser that the land was not suitable for agricultural settlement (Nemanishen 1998).
5. No forecasts of drought frequency were available specifically for the region where the Kainai Nation is located. These observations apply to the entire arid region of the Prairies.
6. The reason for the discrepancy in the two estimates is due to use of different definitions. The Statistics Canada estimate is based on the definition of people living within the KBIR boundaries, while the KBIR definition is based on band members regardless of their place of residence.
7. This target is estimated in light of water rights assigned to the KBIR. However, at this time only 8,354 ha are irrigated.
8. According to this study, all Alberta First Nations communities were more than one standard deviation below the overall Canadian average for Aboriginal communities.
9. Given that the participation rate of the KBIR was 47.3% of the total population, this would make an unemployment rate for the KBIR of almost 50%.
10. This statement is based on information collected from the KBIR officials in July 2008.

11. It is not known whether more dugouts were constructed since that date.
12. This is not to suggest that the non-reserve producers did not incur a loss on account of the drought. However, since the focus of this study was on the KBIR, these costs have been excluded.
13. Information on the nature and number of these treatments was not available.
14. This is based on information collected through informal interviews of KBIR officials in July 2008.
15. Although the amount of water allocated to the KBIR is adequate to irrigate 25,000 acres of cropland, problems may arise in the future during drought periods, particularly if droughts tend to be back-to-back or longer in duration. Furthermore, during these periods, the province may also impose restrictions on amount of water that can be used. Both of these factors may suggest that some areas in the future may have to revert back to dry-land production.
16. This is based on information gathered during the 2007 visit to the KBIR.
17. Based on discussions with the Blood Tribe Agriculture Project officials in July 2008.

REFERENCES

Armstrong, R. 2001. *The Geographical Patterns of Socio-Economic Well-being of First Nations Communities in Canada*. Catalogue No. 21–601-MIE01046. Statistics Canada: Ottawa.

Dai, A., K.E. Trenberth and T. Qian. 2004. "A Global Dataset of Palmer Drought Severity Index for 1870–2002: Relationship with Soil Moisture and Effect of Surface Warming." *Journal of Hydrometeorology* 5: 1117–29.

IPCC (Intergovernmental Panel on Climate Change). 2007a. *Climate Change 2007: Impacts, Adaptation and Vulnerability. Contribution of Working Group II to the Fourth Assessment Report of the Intergovernmental Panel on Climate Change.* M.L. Parry, O.F. Canziani, J.P. Palutikof, P.J. van der Linden and C.E. Hanson, (eds.). Cambridge University Press: Cambridge, UK. 976 pgs.

——. 2007b. "Summary for Policymakers." Pp. 7–22 in M.L. Parry, O.F. Canziani, J.P. Palutikof, P.J. van der Linden and C.E. Hanson (eds.). *Climate Change 2007: Impacts, Adaptation and Vulnerability. Contribution of Working Group II to the Fourth Assessment Report of the Intergovernmental Panel on Climate Change.* Cambridge University Press: Cambridge, UK.

IWGIA (International Work Group for Indigenous Affairs). 2008. "Conference on Indigenous Peoples and Climate Change—Conference Report." International Work Group for Indigenous Affairs, Copenhagen. February 21–22, 2008. 12 pgs.

Magzul, L. 2005. *Fieldnotes*. Community Vulnerability Assessment of the Blood Indian Reserve, Institutional Adaptations to Climate Change Project. University of Regina: Regina.

——. 2007. *Report on the Blood Tribe (Kainai Nation): Community Vulnerability.* Report of the Institutional Adaptation to Climate Change Project. University of Regina: Regina.

Minister of Supply and Services. 1996. *Highlights from the Report of the Royal Commission on Aboriginal Peoples: People to People, Nation to Nation.* [online]. [accessed February 12, 2009]. Available at: *www.aunc-inac.gc.ca/ap/pubs/rpt/rpt.eng.asp*

Nemanishen, W. 1998. *Drought in the Palliser Triangle.* Agriculture and Agri-food Canada. 58 pgs.

Statistics Canada. 2008. *2006 Community Profiles* [online] [accessed May 3, 2008]. Available at: *http://www12.statcan.ca/english/census06/data/profiles/community/*

Stern, N. 2007. *The Economics of Climate Change—The Stern Review.* Cambridge University Press: Cambridge, MA. 692 pgs.

Stratton, E.K. 2005. "Local Involvement in Water management and Adaptive Capacity in the Oldman River Basin, Alberta." Master's thesis. University of Guelph, Guelph.

Wittrock, V., S. Kulshreshtha, L. Magzul and E. Wheaton. 2008. *Adapting to Impacts of Climatic Extremes: Case Study of the Kainai Blood Indian Reserve, Alberta*. SRC Publication No. 11899-6E08. Saskatchewan Research Council: Saskatoon. 94 pgs.

CHAPTER 16

ASSESSMENT OF THE 2001 AND 2002 DROUGHT IMPACTS IN THE PRAIRIE PROVINCES, CANADA

Elaine Wheaton, Suren Kulshreshtha, and Virginia Wittrock

A warmer climate, with increased climate variability, will increase the risk of both floods and droughts (Wetherald and Manabe, 2002). Future trends of increased heavy precipitation events over most areas globally are assessed to be very likely (greater than 90% likelihood). Likewise, it is considered likely (greater than 66% likelihood) that the area affected by droughts will increase (IPCC, 2007). Although parts of the eastern Prairies can face greater flood than drought risks, most of the Prairie Provinces, from time to time, are prone to drought conditions—moderate to severe. In more recent periods, major droughts have been experienced in the region in 1931, 1961, 1988, 2001, and 2002 (Wheaton et al., 2008).

Drought is one of the world's most significant natural hazards. Droughts have major impacts on the economy, environment, health, and society. In this chapter, a review of the impacts of the 2001 and 2002 droughts is presented. Although these droughts were experienced throughout Canada and our original assessment was Canada-wide (Wheaton, Kulshreshtha and Wittrock, 2005 a, b; Wheaton et al., 2008), this discussion is focused on the Prairie Provinces. In this discussion, both impacts on biological and physical conditions as well as socio-economic impacts are discussed. In addition, some of the adaptation options undertaken by farmers, communities and governments are also described.

DROUGHT TYPOLOGY

Droughts are much more complex and difficult to define and quantify than other hazardous weather events (Wheaton, Kulshreshtha and Wittrock, 2005a: 2). A comprehensive definition of a drought is: "a prolonged period of abnormally dry weather that depletes water resources for human and environmental needs" (AES Drought Study Group, 1986). According to Kundzewicz et al. (2007), the term *drought* may refer to several types of droughts: meteorological drought (amount of precipitation well below average), hydrological drought (low river flows and water levels in rivers, lakes, and groundwater), agricultural drought (low soil moisture during the crop growing season), and environmental drought (a combination of the above). Any one of these droughts may result in socio-economic impacts on the region through the interaction between natural conditions and human factors, such as changes in land use and land cover, and water demand and use. The 2001 and 2002 droughts were a combination of agricultural and hydrological droughts.

OVERVIEW OF THE DROUGHTS OF 2001 TO 2002

In general, droughts in Canada affect only one or two regions, are relatively short-lived (one or two seasons), and only impact a smaller number of sectors of the economy. However, the droughts of 2001 and 2002 in Canada were unique in that they covered massive areas, were long-lasting, and brought conditions unseen for at least a hundred years in some regions. The years 2001 and 2002 may have brought the first coast-to-coast drought on record, and were rare as they struck areas that are less accustomed to dealing with droughts. The drought was concentrated in the west, with Saskatchewan and Alberta the hardest hit provinces (Wheaton, Kulshreshtha and Wittrock, 2005a, b; Wheaton et al., 2008). In addition, they brought devastating impacts to many sectors of the economy, posed considerable adaptation challenges, and made history. Although agriculture was the major sector affected, other sectors were also affected either directly or through indirect linkages that exist in an economy.

Drought conditions were observed for all three Prairie Provinces, but more severe conditions were observed for east-central and southern Alberta, and for the Peace River regions. In Saskatchewan, the most affected regions were the southwest and central parts of the province. In Manitoba, drought conditions were more prevalent in the west-central part of the province (Wheaton, Kulshreshtha and Wittrock, 2005a, b).

BIOPHYSICAL AND SOCIO-ECONOMIC IMPACTS OF THE DROUGHTS

The 2001–2002 period in the Prairies was a culmination of dry to drought conditions in previous years. For example, well below normal precipitation was reported in parts of Alberta and Saskatchewan for consecutive seasons for more

Province	Particulars	Amount in Million Dollars	
		2001	2002
Alberta	Change in Farm Cash Receipts	-$271	-$1,334
	Change in Net Farm Income	-$267	-$1,312
Saskatchewan	Change in Farm Cash Receipts	-$652	-$968
	Change in Net Farm Income	-$652	-$953
Manitoba	Change in Farm Cash Receipts	-$7	-$14
	Change in Net Farm Income	-$7	-$14
Prairie Provinces	Change in Farm Cash Receipts	-$930	-$2,316
	Change in Net Farm Income	-$926	-$2,279

Table 1. Summary of Crop Production-Related Economic Impacts of Droughts in the Prairie Provinces, 2001 and 2002. Source: Kulshreshtha and Marleau (2005a)

than four years, extending from fall of 1999 to 2002. The Palmer Drought Severity Index[1] (PDSI) in various regions was less than -4, indicating an extreme drought. Values of this index below -5 were observed for Saskatchewan in 2001. In Manitoba, drought conditions persisted only during 2001; by the summer of 2002, higher PDSI values were observed (Wheaton et al., 2008).

CROP PRODUCTION

In general, crop production was very low across the Prairie Provinces. The worst-hit region in both 2001 and 2002 was Saskatchewan, although Alberta was a close second. Crop production and related economic impacts are summarized in Table 1. Impacts on crop production for Manitoba were fairly small. For the Prairies as a whole, the 2001 drought led to a loss of $930 million in sales of crop products. In 2002, this loss doubled to $2,316 million. On account of payments from various safety nets (particularly the crop insurance payments), the resulting net farm income from crop products was reduced by only $926 million in 2001 and by $2,279 million in 2002 (Kulshreshtha and Marleau, 2005a).

LIVESTOCK PRODUCTION

Estimated impacts of the 2001 and 2002 droughts on livestock production in the Prairie Provinces are summarized in Table 2. Based on the available evidence, the 2001 drought did not have any adverse impacts on livestock production in

Province	Impact in 2001	Impact in 2002	Distribution of 2002 Impact (% of Western Canada)
Alberta	0	-$66.4	58.6
Saskatchewan	0	-$33.2	29.3
Manitoba	0	-$13.8	12.1
Prairie Provinces	0	-$113.4	100.0

Table 2. Estimated Impacts of Droughts on the Livestock Sector, Prairie Provinces. Source: Kulshreshtha and Marleau (2005b)

any of the Prairie Provinces. Producers did cull their cow herd in anticipation of reduced forage and feed grain availability. This increases farm income in the current period, but has implications for the future production of livestock enterprise. In contrast, the 2002 drought had a larger impact, particularly in Alberta. The total impact of the 2002 drought on livestock production was estimated at $113.4 million, 58% of which was in Alberta. The impact was smallest in Manitoba, partly because of the fact that the drought region in this province was smaller and more localized in the west-central region. In contrast, in Alberta, the area affected by the drought was larger, and in some regions, back-to-back droughts were experienced (Kulshreshtha and Marleau, 2005b).

FORESTRY SECTOR

The forestry sector in the Prairie Provinces was affected by the drought through area burnt and the cost of fire management. For example, the area burnt in Alberta in 2002 was about five times higher than the previous 10-year average. For Saskatchewan, the area burnt was twice the previous 10-year average. However, burnt areas or level of cost of fire fighting were not reported for Manitoba.

POWER GENERATION

In Saskatchewan, hydroelectric power generation was reduced in 2001 to 66% of the previous four-year average. This shortfall was made up by purchases from other parts of North America. For Manitoba, such an impact of the drought was not reported. Since Alberta is more dependent on thermal power generation, no direct effects of the drought were reported.

OTHER ECONOMIC SECTORS

In addition to the economic sectors listed above, various non-agricultural economic activities were also affected by the droughts. Many food processing industries found it more challenging to procure the raw material for processing. In some cases, the issue was that of reduced quantity, which had to be made up through imports. In other cases, the quality of the feedstock was also an issue. However, these industries did not report any impact on sales or employment.

Recreational and tourism activities were also affected by the droughts. Reports indicate that on account of the droughts, there were fire restrictions at campgrounds, which may have deterred attendance at these recreational sites. Some recreational sites were closed for a period of time on account of low water levels in the water bodies adjacent to these sites.

SECONDARY ECONOMIC IMPACTS OF THE DROUGHTS

In addition to direct impacts reported above, the prairie economy also faced another type of economic impact—i.e., secondary impacts, which are divided into indirect and induced effects caused by direct changes generated by the droughts. The indirect impacts are generated by change in the purchasing of inputs by the industries affected by droughts, while the induced impacts are through changes in the purchasing power of consumers. Since data for impacts of the droughts on non-agricultural industries were not amenable to further analysis, only agricultural data were used.

The prairie agricultural production was reduced by an estimated $3.6 billion for 2001–2002, including a greater reduction of $2.2 billion caused by the 2002 drought. These direct changes were used in an input-output model of Canada to estimate secondary impacts. Estimates indicate that the Prairie Provinces lost a total of $4.5 billion in terms of gross domestic product, and 27,886 person-years of employment[2] (Kulshreshtha, 2005).

ADAPTATION TO THE DROUGHTS

Generally speaking, producers adjust their cultural and crop decision practices in response to an imminent drought. However, this requires that the information regarding the occurrence of drought be available to them before the start of the growing season or before sensitive decision-making and operations. Unfortunately, for the 2001 and 2002 droughts, this was generally not the case.[3] Thus, most producers did not alter their production practices significantly. Some producers reported reducing their seeded area, and/or increased summerfallow slightly. On account of reduced forage availability, some of the poorly developed crops were used as forage for livestock.

Rural communities facing water shortages also adopted measures to either improve water supply or reduce water use. In some cases, new sources of water were obtained, while, in other cases, rationing was practiced to reduce water use.

The various levels of government assisted producers through crop insurance and safety net programs. In addition to financial assistance, help was available to producers accessing new water sources, in offsetting the cost of producing a crop, and in deferring taxable income because producers had to cull their herd due to lack of feed.

SUMMARY

Repercussions of the 2001–2002 droughts were severe and far-reaching; some of these were:

- Agricultural production levels, through crop production losses, were devastating for a wide variety of crops across the Prairie Provinces, particularly in 2001. Total value of production dropped an estimated $3 billion for the 2001 and 2002 drought years, with the largest loss in 2002 at more than $2.2 billion;
- The Gross Domestic Product fell some $4.5 billion for 2001 and 2002, again with the larger loss in 2002 at more than $3.1 billion;
- Employment losses exceeded 27,883 jobs, including nearly 17,803 jobs in 2002;
- Net farm income was negative or zero for several provinces for the first time in 25 years;
- Livestock production was especially difficult due to the widespread scarcity of feed and water;
- Water supplies that were previously reliable were negatively affected, and several failed to meet the requirements;
- Multi-sector effects were associated with the 2001–2002 droughts, unlike many previous droughts that affected single to relatively few sectors. Impacts were felt in areas as wide-ranging as agricultural production and processing, water supplies, recreation, tourism, health, hydroelectric production, transportation, and forestry;
- Long-lasting impacts included soil and other damage caused by wind erosion, deterioration of grasslands, and herd reductions; and
- Expenditures of various levels of governments increased in response to the drought conditions, providing payments to producers under various safety net programs. These programs partially offset negative socio-economic impacts of the 2001–2002 drought years.

Several lessons were learned from the assessment of the 2001–2002 droughts. Several adaptation measures were suggested and used; however, many were costly and disruptive. Many adaptations proved insufficient to deal with such an intense, large-area, and persistent drought, underlining Canada's vulnerability to such events. Drought causal factors are not well understood. The large-area atmospheric and oceanic patterns suspected to cause previous major droughts were distinctly different than those associated with these recent droughts (Bonsal and Wheaton, 2005). Evidence indicates that droughts may become worse as a result of climate change, requiring a far greater adaptive capacity in all areas. Drought monitoring and assessment of causes, impacts, adaptation and vulnerability research requires additional coordination, resources and expertise.

ENDNOTES

1. The Palmer Drought Index, according to Felch (1978), serves as a means of quantitatively evaluating drought events. Using persistently normal weather in terms of temperature and precipitation, an index value of zero is estimated. An extended period of dryness can produce an index of -6. It is a commonly used drought index in North America.
2. For Canada as a whole, these impacts were estimated at $5.7 billion for gross domestic production and a loss of 41,414 person-years of employment.
3. Although such information was available in scientific reports (such as Wittrock, 2002), it was not effectively received by decision makers, including producers.

REFERENCES

AES (Atmospheric Environment Service) Drought Study Group. 1985. *An Applied Climatology of Drought in the Prairie Provinces.* Canadian Climate Centre Report No. 86–4. Atmospheric Environment Service: Downsview, ON.

Bonsal, B. and E. Wheaton. 2005. "Atmospheric Circulation Comparison between the 2001 and 2002 and the 1961 and 1988 Canadian Prairie Droughts." *Atmosphere-Ocean* 43, no. 2: 63–72.

Felch, R.E. 1978. "Drought: Characteristics and Assessment." Pp. 25–42 in N.J. Rosenberg (ed.). *North American Droughts.* Westview Press: Boulder, CO.

IPCC. 2007. "Summary for Policymakers." In S. Solomon, D. Quin, M. Manning, Z. Chen, M. Marquis, K.B. Averyt, M. Tignor and H.L. Millers (eds.). *Climate Change 2007: The Physical Science Basis. Contribution of Working Group I to the Fourth Assessment Report of the Intergovernmental Panel on Climate Change* (IPCC). Cambridge University Press: Cambridge, UK and New York, USA.

Kulshreshtha, S.N. 2005. *Canadian Droughts of 2001 and 2002: Secondary Economic Impacts of the Agricultural Droughts in Canada.* SRC Publication No. 11602–37E03. Saskatchewan Research Council: Saskatoon, SK.

Kulshreshtha, S.N. and R. Marleau. 2005a. *Canadian Droughts of 2001 and 2002: Impact of the Droughts on Crop Production in Western Canada.* SRC Publication No. 11602–34E03. Saskatchewan Research Council: Saskatoon, SK.

——. 2005b. *Canadian Droughts of 2001 and 2002: Drought Impact on Livestock Production in Western Canada.* SRC Publication No. 11602–32E03. Saskatchewan Research Council: Saskatoon, SK.

Kundzewicz, Z., L. Jose Marta, N. Arnell, P. Kabat, B. Jimenez, K. Miller, T. Oki, Z. Sen and I. Shiklomanov (with contribution from J. Asnuma, R. Betts, S. Cohen, C. Milly, M. Nearing, C. Prudhome, R. Pulwarty, R. Schultze, R. Thayyen, N. van de Giesen, H. van Schaik, T. Wilbank and R. Wilby). 2007. "Freshwater Resources and their Management." Pp. 173–210 in M. Parry, O. Canziani, J. Palutikof, P. van der Lindeen and C. Hanson (eds.). *Climate Change 2007: Impacts, Adaptation and Vulnerability.* Report of the Working Group II Contributions to the Fourth Assessment. Report of the Intergovernmental Panel on Climate

Change. Cambridge University Press: Cambridge: UK.

Wetherald, R.T. and S. Manabe. 2002. "Simulation of Hydrologic Changes Associated with Global Warming." *J. Geophysic. Res.* 107 (D19), 4379, dou:10.1029/2001JD001195.

Wheaton, E., S.Kulshreshtha and V. Wittrock (eds). 2005a. *Canadian Droughts of 2001 and 2002: Climatology, Impacts, and Adaptations.* Volumes I and II. SRC Publication No. 11602–1E03: Saskatoon, SK. 1323 pgs.

——. 2005b. *Canadian Droughts of 23001 and 2002: Climatology, Impacts, and Adaptations.* Volume II. SRC Publication No. 11602–1E03: Saskatoon, SK.

Wheaton, E.W., S. Kulshreshtha, V. Wittrock and G. Koshida. 2008. "Dry Times: Hard Lessons from the Canadian Droughts of 2001 and 2002." *The Canadian Geographer* 52, no. 2: 241–62.

Wittrock, V. 2002. *Preliminary description of the 2001 Drought in Saskatchewan.* Publication No. 11501–1E02. Prepared for PFRA, AAFC. Saskatchewan Research Council: Saskatoon, SK.

CHAPTER 17

AGRICULTURAL ADAPTATION THROUGH IRRIGATION

Dave Sauchyn and Suren Kulshreshtha

Irrigation is an adaptation measure for agriculture in dry environments. It reduces the impacts of drought and farm risks, supports higher crop diversity, increases profit margins, and improves the long-term sustainability of smaller farm units. This case study is a review of irrigation activity in the Prairie Provinces. It encompasses the extent of irrigation in the Prairie Provinces, its biophysical and socio-economic impacts, and water use issues, including improving water use efficiency.

NEED FOR IRRIGATION ON THE PRAIRIES

Irrigation is such a natural part of agriculture that it has no recorded beginning (Fjeld, 1986). Plants are the major source of food world over, but they need a proper climate and three elements—soil, water, and air. Among these three, water is in short supply. Irrigation is the act of artificially applying water to crops. In the semi-arid and more drought-prone region of the Prairies, agricultural production has to be optimized; therefore, application of water to crops artificially (i.e. irrigation) is a necessity.

Irrigation also improves product quality. For example, potatoes grown under irrigation are of more uniform size, which are preferred by potato processors. This provides an added incentive to producers to adopt irrigation.[1]

FUNCTIONS OF IRRIGATION

Groenfeldt (2006) suggests agricultural water (most of which is for irrigation) has four types of associated functions and values: (1) economic and productive, (2) environmental, (3) socio-cultural, and (4) rural development. The economic value of water for agricultural production arises from the fact that use of water (as irrigation) increases a farmer's income, and thus has a positive value[2]. Improvement in the physical productivity of irrigated lands has also been significant, with implications for food security and income distribution, particularly in developing countries (Hussain, 2007). Environmental value of irrigation water comes from groundwater recharge, flood prevention (that can also have economic values), habitat and eco-tourism. One set of socio-cultural values includes aesthetics and landscape improvement.

Regional economic development is also enhanced under irrigated conditions; the irrigation crop mix includes specialty crops that can be processed. These industries are located near the irrigated areas, given the cost and time sensitivity of transporting raw food products. Irrigated farms also employ more labour and other farm inputs, which are sold in the region. Frequently this leads to new businesses being established in the irrigated region. The combined effects of irrigation and agricultural processing industries through increased labour demand are increased size of the rural communities and community stability over time.

CURRENT STATE OF IRRIGATION

The distribution of Canada's irrigated land is shown in Figure 1. The Prairies have about 75%, a total of 632,290 ha (Statistics Canada, 2006). Alberta has the largest proportion of this total—at 85%, followed by Saskatchewan at 11%, and Manitoba at 4%. In fact, Alberta had 64% of the total Canadian irrigated area in 2006. Most of the irrigated areas in Alberta and Saskatchewan are located in the southern semi-arid region, known as the 'Palliser's Triangle.[3] In this part of the region, irrigation is a virtual necessity, as water deficit is in excess of 100 mm (Shady, 1989). Without irrigation, crop production is very risky due to low yields and high variability.

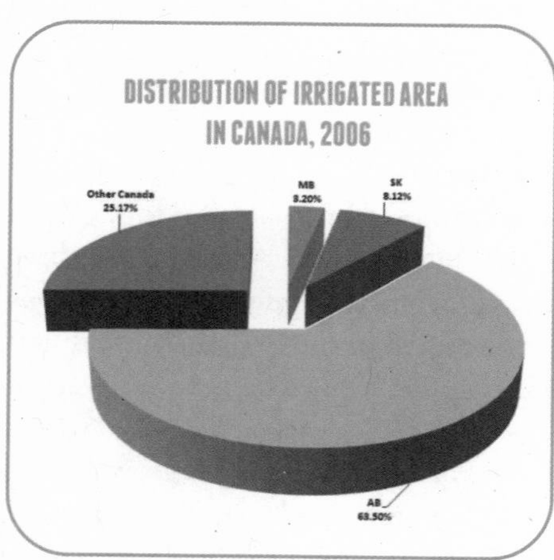

Figure 1. Distribution of Canadian Irrigated Area by Regions, 2006

WATER USE

Irrigated agriculture is by far the largest water user on the Prairies. Beaulieu, Fric and Soulard (2007) estimated that, in 2001, Canadian irrigation water use constituted on average 92% of total agricultural water use. However, this proportion is higher for the Prairie Provinces (estimated at 96%). Irrigation water use is lower for Manitoba (at 50% of the total) and highest for Alberta (97% of the total agricultural water use). Irrigation in the Prairie Provinces used about 3.43 million dam^3 of water.[4]

Irrigation water requirements within the prairie region vary according to level of potential evapotranspiration. In parts of Manitoba, where precipitation is higher, only 1.1 dam^3 of water[5] is applied per hectare as supplementary irrigation. In contrast, in Alberta, 5.4 dam^3 of water is required per hectare. Similar requirements for Saskatchewan were estimated to be 7.3 dam^3.

Irrigation water use is determined by the technology for delivering water to the farm gate, and for applying water on crops. Use of pipelines, as done in parts of the Lake Diefenbaker Development Area (LLDA), is a more efficient method of delivering water to the farm gate. Open canals and similar infrastructure lead to large losses through seepage, percolation, and evaporation. The most efficient method of applying water to plants/crops involves water conservation technologies, such as drip irrigation. According to Wade (1986), the efficiency of surface irrigation systems in delivering water to plants is 60–70%, as against 80% for sprinkler systems.

Small improvements in irrigation efficiency save considerable amounts of water. Advances in centre-pivot systems, including the irrigation of field corners and low-pressure application devices, have significantly improved the efficiency and effectiveness of irrigation. With labour savings and the ability to irrigate rolling land, the irrigated area in Alberta has more than doubled since 1970. In 2006, the irrigation infrastructure in the province consisted of 7,796 km of conveyance works (canals and pipelines) and 49 reservoirs. Off-stream reservoirs accommodate seasonal variations in supply and demand, but are not as effective as the on-stream reservoirs in meeting in-stream flow needs and apportionment. Capital costs are considerable for water distribution. For example, a new irrigation development in the Qu'Appelle South Irrigation Project in Saskatchewan would cost $6,301 to $6,619 per acre.[6] This project would cost a total of $805 million and would develop an additional 122,350 acres of irrigation.

BIOPHYSICAL IMPACTS OF IRRIGATION

Generally speaking, most cereals and oilseeds yields are consistently higher under irrigation than under dryland production systems. For example, under irrigation, spring wheat yield is estimated, on average, at 65 bushels, a little over two times higher than that under dryland farming (Table 1). Crop mix, under

Crop	Unit	Dryland Yield*	Irrigated Yield**	% Increase in Irrigated Yield over Dryland
Spring Wheat	Bu/ac	27.90	65.00	133%
Durum Wheat	Bu/ac	28.70	70.00	144%
Barley	Bu/ac	42.90	90.00	110%
Canola	Bu/ac	23.60	50.00	112%
Flax	Bu/ac	17.20	40.00	133%
Dry peas	lb/ac	1,870.00	5,500.00	194%
Lentils	lb/ac	1,640.00	3,610.00	120%
Alfalfa	tonnes/ac	2.00	4.00	100%

* Ten-year average ** Yield for 2007

Table 1. Comparison of Average Dryland and Irrigated Production of Selected Crops. Source: Clifton Associates Ltd. (2008b).

irrigation, is significantly different from that under dryland production. Major differences are in terms of area under specialty crops and forages. Higher yield and crop diversity lead to further value-added activities and regional economic development, as described in a section below.

SOCIO-ECONOMIC IMPACTS OF IRRIGATION

The result of these higher and more stable crops yields is two major benefits to crop producers:

1. Farm income levels are higher under irrigation. This implies that under average conditions, irrigation has returns above those received under dryland production systems, even after paying for all new capital investment at the farm level. Samarawickrema and Kulshreshtha (2008a) estimated irrigation to generate $17 to $25 more than dryland crop production per dam³ of water. Similar estimates for Saskatchewan have ranged from $10 to $62 per dam³.
2. There are additional benefits during droughts, given the sustained yields of crops during these dry periods from applying appropriate quantities of water. For the South Saskatchewan River Basin of Alberta, additional benefits of $40.70 per dam³ during a drought year have been reported (Samarawickrema and Kulshreshtha, 2008b).

Irrigation crop mix has created opportunities for further economic development opportunities in the region. First, major food processing industries have evolved in southern Alberta, where the production of specialty crops (potatoes, beans, sugar beets) is enabled by the longer growing season, high heat units, and relatively secure water supply derived mainly from snowmelt in the Rocky Mountains. Second, inclusion of forages in the crop mix has provided some competitive advantage to beef cattle enterprises. Processing of beef cattle, therefore, has been attracted to the region providing further economic development. In southern Alberta, 4% of the cultivated irrigated land in various irrigation districts produces 18.4% of Alberta's agri-food gross domestic product, exceeding the productivity of dryland farming by 250–300% (Irrigation Water Management Study Committee, 2002). For every dollar of farm income, personal incomes elsewhere increase almost six times (Kulshreshtha et al., 1985). For Manitoba, this ratio was estimated to be eight-fold (Kulshreshtha and Grant, 2003), and six-fold for Saskatchewan (Kulshreshtha and Russell, 1988).

ISSUES SURROUNDING IRRIGATION

Veeman (1986) suggested that the major issues surrounding irrigation are improvement of water use efficiency and cropping intensity with irrigation systems. These issues still persist. One of the major economic concerns of irrigation in Saskatchewan and Alberta is that it is focused mostly on production of low-value crops. Although some high-value specialty crops are produced, the amount is relatively small, relative to cereals and forage (Figure 2). The issue of improving water use efficiency is very relevant in the context of climate change and projected lower water supplies. As available water is already a constraint for further development of irrigation, improving water use efficiency may be the only avenue to increase irrigation area.

Another option might be the construction of large reservoirs; however, in recent years, a more cautious attitude to large dams has been adopted world

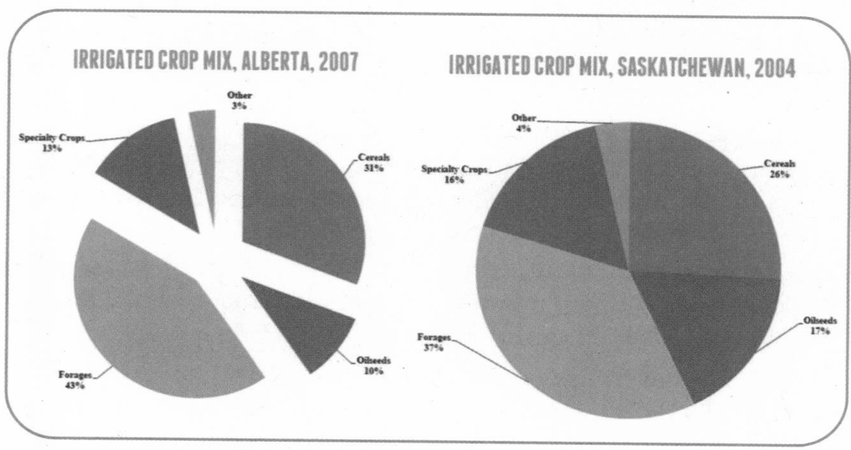

Figure 2. Irrigated Crop Mix in Alberta and Saskatchewan

over, partly because of the cost of such structures, but also because of concern about their stability and safety, and risks of social and environmental disruptions (Carruthers and Clark, 1981).

FUTURE IRRIGATION NEEDS UNDER CLIMATE CHANGE

Climate change affects both water supply for irrigation and the amount needed to cope with changed precipitation and evapotranspiration (Peterson and Keller, 1990). The most direct effects of changing climate on the hydrological regime will be through variations in precipitation amount, type, timing, and intensity, and through changes in evaporation rate, which is closely related to temperature (Ripley, 1987). With the possibility of more precipitation coming as rain (versus snow), water availability may not coincide with water requirements of crops. Further research into crop varieties more suitable under these conditions may be necessary. Long-term strategies are also needed, including formulation of policies and strategies to cope with scarce water conditions (Rydzewski and Abdullah, 1992). To adequately fund irrigation development, Small and Carruthers (1991) suggested user fees implemented by a financially autonomous irrigation agency. This would address the increasing shortage of funds to develop, operate and maintain irrigation projects, and poor overall performance relative to expectations. Both of these problems are related.

SUMMARY

A study of irrigation requirements and opportunities was initiated in 1996 by the Alberta Irrigation Projects Association (AIPA), representing the 13 irrigation districts in Alberta, the Irrigation Branch of Alberta Agriculture, Food and Rural Development (AAFRD), and the Prairie Farm Rehabilitation Administration (PFRA) of Agriculture and Agri-Food Canada. The project report, *Irrigation in the 21st Century* (see Irrigation Water Management Study Committee, 2002), includes the following key findings:

- A move towards increased forage production to support the livestock industry, and an increased area of specialty crops for value-added processing, will result in slightly higher future water requirements than those of the current crop mix.
- On-farm application efficiency—the ratio between the amount of irrigation water applied and retained within the active root zone and the total amount of irrigation water delivered into the on-farm system—increased from approximately 60% in 1990 to about 71% by 1999. Efficiencies could approach 78% with new technologies, although a 75% on-farm application efficiency is considered to be a reasonable target for planning purposes for the foreseeable future.

- Properly levelled and designed surface irrigation systems can have efficiencies of up to 75%, whereas poorly designed and managed surface irrigation systems may have efficiencies less than 60%. Low-pressure, down-spray sprinklers also improve water use efficiency.

Under climate change, and for drought-proofing the farm economy, irrigation offers potential adaptation options. Stability in crop production ensures stable farm income, although, depending on price of products, the producers will face some variability. However, through diversification of farm business (e.g. with the introduction of livestock), further farm-level and community/regional-level stability is also attained. Since development of irrigation infrastructure is costly, and future (under climate change) water supplies are going to be constrained, proper irrigation policies would need to be developed.

ENDNOTES

1. A case in point is the recent addition to French fry processing. In order to attract a new plant in Manitoba, the processor insisted on expansion of irrigated area in the region.
2. This aspect is further discussed later in this chapter.
3. For more details on the Palliser Triangle, see Lemmen and Dale-Burnett (1999).
4. A cubic decametre (dam^3) is 1,000 m^3, which is equivalent to 1 million litres.
5. These estimates are based on provincial water use for irrigation as provided by Beaulieu et al. (2007) and Statistics Canada (2006) estimates of irrigated area. Caution is advised since these values are not based on data collected from a single source.
6. Based on Clifton Associates Ltd. (2008a). The initial technical feasibility report was prepared by UMA Ltd.

REFERENCES

Beaulieu, M., C. Fric and F. Soulard. 2007. *Estimation of Water Use in Canadian Agriculture.* Catalogue No. 21-601-MIE, No. 087. Statistics Canada: Ottawa.

Clifton and Associates Ltd. 2008a. *A Time to Irrigate—The Economic, Social and Environmental Benefits of Expanding Irrigation in the Lake Diefenbaker Region.* Volume I. Report Prepared for the Saskatchewan Irrigation projects Association. Regina. 163 pgs.

——. 2008b. *A Time to Irrigate—The Economic, Social and Environmental Benefits of Expanding Irrigation in Saskatchewan.* Volume II. Report Prepared for the Saskatchewan Irrigation projects Association. Regina. 163 pgs.

Carruthers, I. and C. Clark. 1981. *The Economics of Irrigation.* Liverpool University Press: Liverpool. 300 pgs.

Fjeld, H. 1986. "Irrigation on the Prairies—1985." Pp. 42–49 in *Irrigation in the Prairies.* Fourth Annual Western Provincial Conference—Rationalization of Water and Soil Research and Management. Saskatchewan Water Corporation: Moose Jaw.

Hussain, I. 2007. "Pre-Poor Intervention Strategies in Irrigated Agriculture in Asia: Issues, Lessons, Options, and Guidelines." *Irrigation and Drainage* 56: 119–26.

Groenfeldt, D. 2006. "Multifunctionality of Agricultural Water: Looking Beyond Food Production and Ecosystem Services." *Irrigation and Drainage* 55: 73–83.

Irrigation Water Management Study Committee. 2002. *South Saskatchewan River Basin: Irrigation in the 21st Century. Volume 1: Summary Report.* Alberta Irrigation Projects Association: Lethbridge, AB. 163 pgs.

Kulshreshtha, S. and C. Grant. 2003. "Economic Impact Assessment of Irrigation Development and Related Activities in Manitoba." *Canadian Water Resources Journal* 28, no. 1: 53–67.

Kulshreshtha, S.N. and K.D. Russell. 1988. "An Ex Post Evaluation of the Contributions of Irrigation Development in Alberta: A Case Study." *Review of Regional Studies* 18, no. 2: 10-22.

Kulshreshtha, S.N., K.D. Russell, G. Ayers and b.c. Palmer. 1985. "Economic Impacts of Irrigation Development in Alberta upon Provincial and Canadian Economy." *Canadian Water Resources Journal* 1, no. 2: 1–10.

Lemmen, D. and L. Dale-Burnett. 1999. "The Palliser Triangle." P. 41 in K. Fung (ed.). *Atlas of Saskatchewan.* University of Saskatchewan: Saskatoon.

Peterson, D.F. and A.A. Keller. 1990. "Irrigation." Pp. 269–306 in P.E. Waggoner (ed.). *Climate Change and U. S. Water Resources.* John Wiley and Sons: New York.

Ripley, E. 1987. "Climatic Change and the Hydrological Regime." Pp. 137–78 in M. C. Healy and R. Wallace (eds.). *Canadian Aquatic Resources.* Department of Fisheries and Oceans: Ottawa.

Rydzewski, J.R. and S. Abdullah. 1992. "Water for Sustainable Food and Agriculture Production." Pp. 4.1–4.25 in *International Conference on Water and the Environment: Development Issues for the 21st Century.* Keynote Papers: Dublin, Ireland.

Samarawickrema, A. and S. Kulshreshtha. 2008a. "Value of Irrigation Water for Crop Production in the South Saskatchewan River Basin." *Canadian Water Resources Journal* 33, no. 3: 257–72.

——. 2008b. "Value of Irrigation Water for Drought Proffing in the South Saskatchewan River Basin (Alberta)." *Canadian Water Resources Journal* 33, no. 3: 273–82.

Shady, A. 1989. *Irrigation Drainage and Flood Control in Canada.* Canadian International Development Agency: Ottawa. 307 pgs.

Small, L. and I. Carruthers. 1991. *Farmer-Financed Irrigation—The Economics of Reform.* Cambridge University Press: Cambridge, New York. 233 pgs.

Statistics Canada. 2006. *Census of Agriculture* [online]. Statistics Canada: Ottawa [accessed January 19, 2009]. Available at: *www.statcan.ca/*

Veeman, T. 1986. "The Economics of Water Use and Development in Agriculture on the Prairies." Pp. 212–24 in *Irrigation in the Prairies.* Fourth Annual Western Provincial Conference—Rationalization of Water and Soil Research and Management. Saskatchewan Water Corporation: Moose Jaw.

Wade, J.C. 1986. "Efficiency and Optimization in Irrigation Analysis." Pp. 73–100 in N.K. Whittlesey (ed.). *Energy and Water Management in Western Irrigated Agriculture.* Westview Press: Boulder, CO.

CHAPTER 18

GOVERNMENT INSTITUTIONS AND WATER POLICY

Margot Hurlbert, Darrell R. Corkal, and Harry Diaz

INTRODUCTION

Climate change on the Prairies is expected to significantly affect regional water resources (as detailed in Chapter 5). Intuitively, impacts on communities and economies on the Prairies will be significant; communities will need to develop adaptation responses to cope with these impacts.

An important determinant in the ability of a community to adapt to current climate variability and future climate change impacts is its institutional setting and the degree to which this setting either facilitates or hinders the community's adaptive capacity (O'Riordan and Jager, 1996; Willems and Baumert, 2003). Also important are water policies that determine the management of the community's resources. The chapter focuses on the institutional setting established by government institutions and their water policies in Alberta and Saskatchewan, as they apply to a changing climate. Particular attention will be paid in this chapter to water policy and drought planning, which supplements the information on climate change and adaptation policies referenced in Chapter 12, "Government Institutions and Climate Change Policy."

This chapter is based on a review of secondary sources relating to water governance, as well as primary information gathered from community vulnerability assessments and interviews with stakeholders and government representatives. Community vulnerability assessments were carried out during the period from 2005 to 2008 by researchers who established residence and conducted interviews in several rural communities in Alberta and Saskatchewan. In addition, approximately 60 interviews were conducted with representatives from water user associations, watershed and environmental groups, and individuals from local, provincial and federal government agencies. Several focus groups with water users, irrigators, rural communities, and all orders of government were also held. From these research activities, data was gathered respecting past, current and inferred future water practices, policy and legislation. Specific emphasis was placed on governance institutions and their respective roles or mandates relating to water resources.[1]

GOVERNMENT INSTITUTIONS

The government departments in the Prairies most affected by the expected impacts of climate change on water resources are those provincial agencies responsible for water management. All will have to factor anticipated changes to water resources into their strategies and plan in order to reorganize their programs to secure the sustainability of water resources, and reduce the vulnerability of those social and economic sectors directly impacted by the expected changes.

Canadian water management is essentially the mandate of the provinces, which share jurisdictional roles with the senior federal government (e.g. transboundary flow, environmental protection) and delegate some of their functions to local municipal governments (e.g. drinking water, environmental protection). The myriad of government institutions at these levels will be identified.

THE FEDERAL LEVEL—THE GOVERNMENT OF CANADA

Both Environment Canada and Natural Resources Canada have significant roles relating to water because of their mandates pertaining to climate change impacts and adaptation as outlined in this volume's chapter on climate policy. No single department has a monopoly in relation to water issues because of the overlapping mandates of environment, industry and agriculture departments, whose activities impact water. There are 19 departments which have some responsibilities relating to water, spending in the order of $750 million annually on water-related activities (Banks and Cochrane, 2005). The principal departments responsible for water are the 5NR (five natural resource departments, Morrison and Gee, 2001). Environment Canada is responsible for environmental protection of water resources, and conducts critical water research activities. Natural Resources Canada has a significant responsibility in groundwater resource mapping. Health Canada, although without specific jurisdiction over natural resources, has indirect influence through its health mandate; it has many responsibilities relating to public safety, safe drinking water, and establishing the Guidelines for Canadian Drinking Water Quality in partnership with federal-provincial-territorial governments. Fisheries and Oceans Canada is responsible for fisheries and the protection and restoration of fish habitat.

Because of agriculture's use and dependence on water, climate and land resources, Agriculture and Agri-Food Canada (AAFC) plays a significant role in relation to water. AAFC conducts extensive agricultural and agri-environmental research, and promotes adoption of agricultural best (beneficial) management practices to protect the environment from agricultural pollution. The Prairie Farm Rehabilitation Administration, which is a part of AAFC, was created as a special agency in 1935, in response to severe climate impacts on the Prairies caused by the multi-year droughts of the 1920s and 1930s. Many practices and programs promoted by PFRA are targeted at land and water stewardship. These programs have been instrumental adaptations for helping the agricultural sector cope with climate impacts for the water-scarce prairie region.[2]

Many other federal departments have mandates indirectly relating to water. A few examples are presented. Indian and Northern Affairs Canada has shared responsibilities with First Nations Band Councils for water on First Nations reserve communities, and is responsible for many water acts in northern Canada. Industry Canada has a mandate related to industry and development (whose industrial activities may risk polluting water). Parks Canada has a mandate for water resource management and protection within national parks. Transport Canada is concerned with waterways used for transportation. The Department of Foreign Affairs and International Trade becomes involved with water export issues or policies. The International Joint Commission includes membership from the Canadian and United States federal governments, and makes recommendations to each of the governments in respect of international transboundary waters.

THE PROVINCIAL LEVEL IN THE PRAIRIE PROVINCES

The provincial governments of two Prairie Provinces face a similar challenge within the bureaucratic structure in respect of water as the federal government does, with many agencies and departments involved in water because of its ubiquitous nature.

In Alberta, Alberta Environment is tasked with the bulk of the responsibilities relating to water which includes surface and groundwater allocation, flow regulation, water supply and flood forecasting, pollution control, regulation of municipal potable water systems, and developing watershed management plans in conjunction with local groups. Alberta Environment also establishes and enforces drinking water and waste water objectives and legislation. Other Alberta government departments also have impacts on water policy, including the Alberta Ministry of Sustainable Resource Development, the Ministry of Agriculture and Rural Development (Alberta Agriculture encourages adoption of best practices, plays a significant role in sustainable rural water management, and helps rural citizens with their on-farm water supplies), Infrastructure and Transportation, and Regional Health Authorities (funded and controlled by

Alberta Health and Wellness). Although the multiplicity of government departments would appear to hinder the development of overarching water policy, Alberta has developed a successful "Water for Life" policy which is outlined below in the section on "Water Strategy" in this chapter.

Saskatchewan manages its water resources by the separate Crown corporation, the Saskatchewan Watershed Authority. The Saskatchewan Watershed Authority was created to manage water more holistically, and was established in 2002 as a direct response to the North Battleford drinking water disease outbreak in 2001. Saskatchewan Watershed Authority is responsible for protecting source water, water management, water allocations and diversions, and developing watershed and source water protection plans in partnership with watershed groups. Again, other departments have key water-related roles, including the Ministry of the Environment (environmental protection, municipal drinking regulation and treatment standards), Ministry of Agriculture (agricultural operations to protect water), the Ministry of Health (public health protection, monitoring of safe drinking water and enforcement of corrective actions), and Sask Water (Crown corporation with water treatment and distribution systems). When political support exists, Saskatchewan has also coordinated all of these ministries in relation to water. An ongoing effort exists to incorporate an integrated water management approach involving not only these provincial entities, but also the federal government (Swanson et al., 2005: 12).

FEDERAL AND PROVINCIAL SHARED JURISDICTIONS

Federal and provincial ministries appear to overlap or duplicate water mandates and services. However, in practice, much collaboration and cooperation exists between the various provincial and federal departments. One of the most notable shared arrangements is the Prairie Provinces Water Board (PPWB), created in 1948 (Prairie Provinces Water Board, 2007). The PPWB administers the 1969 Master Agreement on Apportionment that ensures equitable water sharing for streams flowing from Alberta through Saskatchewan and onto Manitoba. The board includes federal and provincial government representatives from Environment Canada, Agriculture and Agri-Food Canada's Prairie Farm Rehabilitation Administration branch, Alberta Environment, the Saskatchewan Watershed Authority, and Manitoba Water Stewardship. The PPWB is considered a model for inter-governmental shared water management by those interviewed in this study.

WATER POLICIES AND STRATEGIC PLANNING

Government institutions in Alberta and Saskatchewan have managed the water resource through various pieces of legislation and policy for many decades. Both have policies and programs that relate to coping with the variability of water resources which were not designed with climate change as a driver, but rather drought or water management, which are key issues for the semi-arid prairie region. Alberta has the most experience with competing water demands and water shortages in respect of water licenses. This is confirmed by an examination of Alberta's legislated provisions dealing with these issues. However, Saskatchewan also has significant experience with drought, intensive agriculture demands, and drinking water issues unique to the region's natural resources, geography and land use practices. Because of the breadth and depth of water-related issues, this section will focus on water policy and strategy in relation to drinking water, overarching water strategy, drought planning, and the involvement of local watershed planning groups (because of their relative recent emergence in the water governance institutional setting).

DRINKING WATER

Recently, water management in Canada has been dramatically affected by drinking water disease outbreaks in: Walkerton, Ontario, in 2000; North Battleford, Saskatchewan, in 2001; and Kasheshewan First Nation, Ontario, in 2005 (Corkal, Adkins and Inch, 2007). The Walkerton and North Battleford cases each resulted in costly and extensive provincial inquiries, and the Kasheshewan case resulted in a broad panel review of water on First Nations across Canada (Polaris Institute, 2008). Virtually all provinces and the federal government have subsequently made changes to water management mandates or roles, and most orders of government are now placing an increasing emphasis on implementing integrated water resource management strategies, citizen engagement in participatory planning, source water protection, and increased research and public awareness.

Alberta developed its Water for Life Strategy in 2003 to protect drinking water, safeguard healthy ecosystems, and ensure water's role in a sustainable economy (Alberta Environment, 2003). The strategy (which is more fully discussed in the next section) was renewed in 2008, to update current water management principles in the face of rapid growth and climate change, and to accelerate actions to improve water management and water policies. Central to Alberta's approach is citizen engagement and the participation of watershed groups (Alberta Environment, 2008a).

Saskatchewan responded with "Saskatchewan's Safe Drinking Water Strategy" and a Water Management Framework (Government of Saskatchewan, 2009), which outlines a clear, targeted strategy for achieving safe drinking water. Accountability of municipal service providers and increased testing and reporting requirements have resulted through legislative vehicles. A "State of the Watershed" report is released annually by the provincial government and is available on-line (SaskH20, 2009).

WATER STRATEGY

Because the impacts of climate change on water are expected to be significant, and because water is essential to many human activities, many government departments are now considerering anticipated changes to the water resource, which will affect their programs, policies and mandates. Alberta has one overarching water policy and Saskatchewan has a combination of targeted strategy documents.

The Alberta Water Act required the Minister of the Environment to set a provincial water management planning framework within three years (it was to include a strategy for protecting the aquatic environment and could include matters relating to the integration of water management with land and other resources). As a result of this legislated mandate, Alberta's "Water for Life" strategy was developed in 2003 and renewed in 2008 (Alberta Environment, 2003, 2005, 2008a). This strategy envisioned a partnership approach to managing water and led to the formation of three groups: a Provincial Water Advisory Council, Watershed Planning and Advisory Councils, and Watershed Stewardship Groups. The goals have been identified as:

- Healthy, sustainable ecosystems;
- A safe, secure drinking water supply;
- Reliable, quality water supplies for a sustainable economy; and
- The knowledge necessary to make effective water management decisions

(ALBERTA ENVIRONMENT, 2008A).

The strategy is aimed at promoting conservation, and will complete an evaluation and make recommendations on the merit of economic instruments to meet water conservation and productivity objectives (Alberta Environment, 2008a: 15). The strategy envisions and details specific actions to achieve these goals over the next several years. All of these measures will be achieved through partnership and cooperation, not legislation and regulation. Again, although not specifically labeled a climate change adaptation plan, responding to the changing Alberta water resource issues is significant in relation to climate change preparedness due to the strong relationship between climate and water interactions.

Saskatchewan's strategies in relation to water appear in several documents such as the Saskatchewan Safe Drinking Water Strategy. Many of the activities in relation to water are coordinated by Saskatchewan Environment and the mandates of the Saskatchewan Watershed Authority and Sask Water Corporation (Swanson et al., 2005). Saskatchewan's Safe Drinking Water Strategy (Saskatchewan Watershed Authority, n.d.) seeks to keep source water clean, ensure treatment process make drinking water safe, and prove that the drinking water is safe. This strategy provides goals of protecting watershed, knowing risks to source water, having a clear and efficient regulatory system, ensuring waterworks systems have trained staff, and providing access to information by citizens. This strategy also details actions to achieve these goals. The Saskatchewan Water Conservation Plan complements this strategy with detailed conservation measures (Saskatchewan Watershed Authority, 2006).

Recognizing the interdependence of the Western provinces in relation to water, the provinces of British Columbia, Alberta, Manitoba, and Saskatchewan have agreed to a comprehensive water resource and conservation plan which would entail conservation measures from low-flush toilets and municipal "leak audits," to new technology to reduce the water used in oil sands projects (Wood, 2008). In May 2008, the Western provinces created the Western Watershed Stewardship Council to ensure safe water supplies for each jurisdiction. The Council is hoping to look at and make recommendations to governments regarding the western region's needs for water monitoring and assessments, water demand management, aquatic ecosystem health, education, outreach, watershed planning and governance, and regional drought preparedness (Government of Saskatchewan, 2008).

The Government of Canada's Federal Water Policy was established in 1987 (Environment Canada, 1987). The 1987 Federal Water Policy was visionary for its day, and makes reference to issues that remain current by today's standards: the need for better understanding on the value of water, science leadership, integrated planning, legislation, and public awareness. The policy summarizes many issues, including environmental protection, irrigation expansion, protection of aquatic environments, transboundary flows, and interestingly, drought, flooding, climate change, and climate variability. This strategy was the outcome of *Currents of Change, Reporting the Findings from the 1985 Inquiry on Federal Water Policy*. The inquiry had been established "in response to a growing environmental consciousness and concern about the management of Canada's freshwater resources." (Pearse, Bertrand and MacLaren, 1985).

Looking back in time, during a review of this strategy done in 1994, Pearse commented that the 1987 Federal Water Policy was not fully implemented and was essentially stalled in 1994, because of changing priorities, unfulfilled commitments, and changes in bureaucracy such as reorganization and staff reassignments or retirements (Pearse, 1994). Interestingly, most of the drinking water issues identified in the drinking water inquiries at the beginning of the 21st century had essentially been identified in the 1987 Federal Water Policy. It appears, however, that the recent drinking water disease outbreaks were the real catalysts to effect recent changes.

More recently, there have been calls by stakeholders for a national water strategy (de Loë, 2008; Barlow, 2007). Benefits cited are: more consistent and effective responses to concerns with national dimensions such as water exports (Barlow, 2007: 207) and climate change; increased accountability; enhanced environmental protection; stronger national capacity to respond to threats and crises; and better positioning to meet growing international expectations and obligations as well as public support (de Loë, 2008: iii).

A growing volume of water governance literature has been published in Canada since 1985, from diverse sources, including academia, government and non-government agencies, concerned scientists, and citizens. Essentially, the common theme in this body of literature is a call for improved clarity in Canadian water governance and better implementation of integrated water resource management for water stakeholders and all orders of government in Canada (Hurlbert, Diaz and Corkal, 2008).

DROUGHT PLANNING

Drought response and adaptation has been a constant for the people and provincial governments of the Prairie Provinces since their inception. The region exhibits one of Canada's most widely variable ranges of natural climate (from extreme heat to extreme cold) and hydrological resources (frequent droughts and floods). One of the most significant impacts of climate change on the Prairies is an increase in the intensity and frequency of drought (Sauchyn and Kulshreshtha, 2008). This will force the region to bring an increasing focus on adaptation policies that address drought and extreme event preparedness. The current government programs in this area include programs applied for by individual farmers based on their personal economic circumstances, in order to assist in the mitigation of the short- and long-term effects of drought:

(a) The federal government's programs include AgriInsurance, which is a form of crop insurance for weather, pests and disease; AgriInvest, which is a savings account that provides coverage for small income declines to help mitigate risks or improve market income; AgriStability, which provides support when a producer experiences declines of more than 15% of a producer's average income from previous years (replacing the Canadian Agricultural Income Stabilization Program known previously as the Net Income Stabilization Account); and AgriRecovery, which allows for special federal-provincial responses to disasters not covered by other programs (Government of Canada, 2009);
(b) Alberta has a combination of Crop Insurance, a Farm Income Disaster Program and a Disaster Assistance Loan Program, and specific water facility funding programs;
(c) Saskatchewan has Crop Insurance, a Livestock Drought Assistance Program, and a Livestock Drought Loan Program. (Wittrock and Koshida, 2005).

The federal government established rural water programming to address drought in the prairie region in 1935, following devastating multi-year droughts in the 1920s and 1930s. The Rural Water Development Program existed from 1935 to 2004, and provided funding to help develop secure on-farm water supplies in Alberta, Saskatchewan, Manitoba, and the Peace River region of British Columbia. For the period from 1980 to 2004, when group and community projects were added, the program expended an estimated total of $154 million (Kerster, 2009). This rural water program was managed by PFRA, and also provided direct assistance to water management projects commencing in 2001 (Government of Canada, 2002). The success of this regional program was extended nationally with the creation of the federal National Water Supply Expansion Program, with significant funds commencing in 2002. The goals of the national program included expanding water supply and water management measures (e.g. rural regional water distribution pipelines, investigating schemes for increasing existing water supplies, conducting studies to investigate solutions for rural water supply problems, conducting research into rural water quality and rural water availability issues). The national rural water program existed from 2002 to 2009, and expended approximately $102 million across Canada (Kerster, 2009). Of this amount, roughly $68 million was expended on the Prairies (largely due to its greater exposure to water shortages). Monies were also expended for research and investigation into research in the National Land and Water Information Service (Wittrock and Koshida, 2005: 9). These federal programs were most often shared programs. The participating provinces and local groups or project proponents provided additional funding, support and contributions towards these rural water projects, which were designed to enhance resiliency to drought and develop greater water security for the agri-food sector.

On the provincial front, Alberta Agriculture has developed an *Agricultural Risk Drought Management Plan 2007* focused on drought preparedness, reporting and drought response; and an Agricultural Drought Management Committee decides upon areas in the province to target assistance based on science-based indices (Alberta Agriculture, 2007). The 2002 draft, *Drought Risk Management Plan for Saskatchewan,* was designed to help government agencies develop a coordinated response to assist stakeholders (agriculture, forestry, municipalities) prepare for, mitigate and respond to drought (Agriculture and Agri-Food Canada, 2002). The existence of these overarching federal and provincial policies are a significant step in responding and planning to climate change, and may define—if finalized—a common vision and set of goals for all organizations.

The City of Regina responded to the 1988 drought by developing contingency plans, including water conservation programs and the expansion of water treatment and delivery capacity (Cecil et al., 2005). Although other cities on the Prairies do not currently have such contingency plans in place (Wittrock, Wheaton and Beaulieu, 2001), the issue is no doubt germane to some.

LOCAL WATERSHED PLANNING GROUPS

Alberta and Saskatchewan have adopted watershed level planning and management through engagement with municipal and local stakeholders. Each province has utilized a slightly different model. Most ministries within the Government of Canada now emphasize a "source to tap" holistic protection, management and use of water resources. Canada's recent drinking water disease outbreaks have generated renewed interest in water planning and policy at this level. These incidents have increased the effort to include civil society in water resource management, and the need to be able to plan for, and to adapt to the unknown or the uncertain.

Alberta is now managing water in accordance with river basin councils and watershed groups. Alberta has both local watershed stewardship groups and, also, regional watershed planning and advisory councils (which existed as basin councils prior to the Water for Life strategy). Many of the local watershed stewardship groups also existed prior to the Water for Life strategy, as they had emerged to address local needs. Their key role focuses on watershed stewardship activities such as riparian health assessment, water quality and quantity monitoring. (Alberta Environment, 2005). The regional watershed planning and advisory councils prepare "state of the watershed" indicators and watershed management plans consistent with the Water for Life policy direction (Alberta Environment, 2002).

The Saskatchewan Watershed Authority was created to manage water on a watershed and aquifer basis in partnership with watershed groups. It has worked with several sub-watersheds to produce watershed plans in accordance with a Watershed and Aquifer Planning Model for Saskatchewan (Saskatchewan Watershed Authority, 2003). Staff of the Saskatchewan Watershed Authority facilitate and support local committees (which may contain municipal, First Nation, and irrigation district representatives) with planning facilitation and technical support. The primary focus is on source water supply with additional issues of interest to the local residents.

ANALYSIS

Developing strategy and policy in relation to water is as challenging an area as climate change. Similar challenges such as the nature of the Canadian electoral system (with potential for policy changes resulting from changing short-term priorities), the ubiquitous nature of water (as it has important implications for many agencies and departments within government), climate change and environmental impacts, and the uncertainty of future climate change impacts and social responses, exist and must be accepted in relation to water. The study of water policy illustrates similar themes to those in relation to climate change. These themes will be discussed in relation to water policies under the headings of integration and long-term planning, attention to climate impacts, focus on adaptation, civic engagement, and flexible and responsive policy.

INTEGRATION AND LONG-TERM PLANNING

Although both Alberta and Saskatchewan have either an overarching water policy or a water policy contained in several instruments, a current comprehensive water strategy integrating federal, provincial and local water strategies is lacking (Diaz et al., 2009: 52, 54). Water resources do not recognize provincial or administrative boundaries. The provincial governments have the mandate for resource management in Canada, including water management. However, it is clear that water management activities are often shared with the federal government and other provinces. An example of this is the federal-provincial Prairie Provinces Water Board, which has been a successful regional agency. The PPWB is uniquely placed as an institution currently overseeing certain water-related functions and duties, and has the potential to play an increasing role for interprovincial coordination in relation to water and anticipated climate-induced water stress. The current conservation strategy and the creation of the Western Watershed Stewardship Council by the Western provinces are also good initiatives in relation to regional integrated water resource management approaches, albeit with a limited mandate to date. Expanding the breadth of issues for which there is common understanding and vision between provinces to such things as water quality, extreme weather events and consequent disaster planning, and potential drainage or drought issues may prevent future conflict.

Better integration of water policy and planning would also address two other shortcomings identified in the Institutional Adaptation to Climate Change Project. First, many water data collection issues were reported. Identified gaps in the data pool (water quality, quantity and use, and climate data) were identified. Uncertainty exists about what data is available, what can be accessed by whom, and who is responsible for collecting and sharing (Diaz et al., 2009: 53). Second, better integration is needed to address the complexity of water governance arrangements and operational challenges. Rural communities and their residents are often frustrated by the need to deal with a large number of agencies, and are often unsure which agencies are responsible for various aspects of water policy (Hurlbert, Corkal and Diaz, 2009: 196).

A joint commitment by all orders of government to implement a baseline plan with a long-term vision has the potential to strengthen coping capacities in the face of critical water management issues, including climate-induced water stress. The 1987 Federal Water Policy was an example of a long-term vision.

Had Canada's 1987 Federal Water Policy been fully implemented, better integrated water resource management may have been more fully achieved by all watershed stakeholders at an earlier point in time. In recent years, after Canada's waterborne disease outbreaks, these principles and their application have been strengthened in practice. Water stakeholders and all orders of government have adopted a more vigilant and holistic "source to tap" water management approach. This is a positive step that will help protect water users, the environment and economic activities. Designing and implementing effective long-term climate, environment and water policies may help build resilience to avert other such potential crises, particularly when science and good governance arrangements can foresee or anticipate potential impacts and possible solutions. Some improvements are listed below.

ATTENTION TO CLIMATE IMPACTS ON WATER RESOURCES

More work is needed on long-term policy responses to drought and flood events (Diaz et al., 2009: 54). These are typically viewed as "extreme events," yet it is known that the Canadian Prairies have one of the most variable climate and hydrological regimes in Canada. In the last 100 years, at least 40 droughts have occurred in the Canadian Prairies (Marchildon et al., 2008). Proxy data using tree rings, lake sediments, and geological and biological archives indicate that drought (including multi-decadal drought) and flood events are recurrent in the history of the region (Sauchyn and Kulshreshtha, 2008). If we understand that droughts and floods are a natural feature of climate variability in the Prairies, might it not be possible to design long-term and flexible policies and programs (e.g. modifications to crop insurance) to improve responses to the realities of droughts and floods as a natural feature of our climate, rather than as an *ad hoc* response to an "uncertain recurrent extreme event"? If we anticipate a greater impact by a changing climate regime (increased intensity of droughts, floods, storms), how can society better prepare for the impact that forecasted climate scenarios may have on water resources? Can our institutions and economic sectors build resilience and capacity by creating more flexible and effective responses for future water stress? These types of questions can become instrumental in guiding water management approaches that consider present and future climate impacts.

FOCUS ON ADAPTATION TO CLIMATE CHANGE AND NATURAL CLIMATE VARIABILITY

A review of the current water programs and policies makes it abundantly clear that the full impacts of anticipated climate change listed by Byrne, Kienzle and Sauchyn in Chapter 5 have not been considered in these policies. It would be wise for all orders of government to review the water policies mentioned herein and reassess them in consideration of potential climate change impacts and hence, potential economic and social impacts. Such an exercise would undoubtedly require historic and economic evaluation, and risk-based analyses. The emphasis on adaptation to climate impacts would prove valuable in identifying future adaptation plans, and may help address current vulnerabilities caused by natural climate variability (e.g. better regional drought preparedness, one of the desired activities envisioned by the Western Watershed Stewardship Council).

Policy responses will require flexibility to address more extreme climate events. For instance, crop insurance currently provides an economic adjustment for farmers in relation to one crop season. The use of these economic instruments and their appropriateness for extended multi-year droughts is an open issue (yet there is evidence that multi-year droughts are part of the natural prairie climate regime, and may very well occur in the future, whether or not global warming and climate change impacts occur).

CIVIC ENGAGEMENT

Increasingly, it is being recognized in the literature that responding to increasing water demands such as climate change, especially in relation to water governance and development decisions in order to foster sustainability, will require significant engagement of civil society (World Bank, 2003; RIFWP, 2007). This engagement should occur at the overarching climate change policy development stage. Alberta has consulted widely on its climate change policy, and has done so now on at least two formal occasions. In addition, Alberta has implemented civil society engagement in water conservation, environmental decisions, and development in its water management framework for watershed groups and watershed councils. For example, significant citizen engagement occurred in watershed planning with the Industrial Heartland and Capital Region, which is comprised of a 470-square-kilometre area northeast of Edmonton (Alberta Environment, 2008b). The federal government is also supporting integrated water resource management (e.g. technical support for watershed protection plans; promoting the adoption of environmental farm planning and agricultural beneficial management practices) which would engage civil society in development decisions.

The governments of Alberta and Saskatchewan are managing watersheds in a more holistic fashion, and are increasingly engaging citizens for involvement in source water protection plans, as outlined above, in relation to local watershed planning partnerships. Challenges remain in financing watershed groups, incorporating these groups into the formal institutional arrangements of water governance, and funding watershed activities, but the involvement of civil society in water management is a positive development (Diaz et al., 2009: 53).

Such engagement of citizens has not been broadly attempted in relationship to an overarching integrated federal-provincial-local water strategy. The risk of the provinces and federal government going different directions to address climate and water issues, coupled with limited citizen engagement, is problematic for developing adaptation measures and planning for sustainability, particularly when regional approaches are desired. Perhaps this has been the same challenge facing water strategies across Canada since 1987, as evidenced by the extensive literature which calls for more clarity in the roles of all orders of government. Meaningful citizen engagement has begun with leadership from watershed groups. Formally incorporating contributions from citizen and stakeholder groups would facilitate the formulation of policies implementing publicly compatible solutions that will be beneficial in the event of water shortage and the requirement to resolve competing priority uses. In this manner, the values and priorities of the public will be incorporated into policy. The values and priorities of the public may relate to economic savings and solutions, but also to environmental protection and conservation (e.g. protection of aquatic ecosystems). Expenditures for adaptation measures that sustain environmental health, for example, may be embraced by the public, as the benefits are perceived, understood, valued, and seen as useful to protect both lifestyle and the environment.

FLEXIBLE AND RESPONSIBLE WATER POLICY

Responding to the changing water resource as a result of impacts caused by natural climate variability and climate change requires more than a myriad of unrelated adaptation measures, but rather a structured response (for which the tool of policy is ideal) allowing for identification, prevention and resolution of problems that are created (Sauchyn and Kulshreshtha, 2008). All water policies, including the ones listed specifically herein, should be reviewed simultaneously, and thought given to addressing gaps, increasing efficiency and investigating further research and implementation activities.

Responsive policy requires constant feedback loops, revisiting and review by government (IISD, 2006) together with stakeholders to ensure its currency, accuracy, and appropriateness. Although from time to time governments are revising their water-related policies, mechanisms making the review of policy automatic, planned and transparent (or participatory for all of civil society) could be improved. The source water protection plans developed by local watershed groups require not only full implementation and review to ensure adaptation to expected climate change impacts, but also formal mechanisms for regular meaningful review.

CONCLUSION

This case study of water policy in relation to Alberta and Saskatchewan's wide range of natural climate variability and their changing climate illustrates many of the specific comments that are equally made in relation to climate change mitigation and adaptation policy. More focus on the long-term policy in relation to water is required, with specific thought to an integrated local-provincial-federal policy approach to address the region's needs. Significant consultation on what this might encompass, and what specific provisions might be, would be required. More attention on adaptation to climate change issues is required, with built-in flexible policy responses to reduce vulnerability and risk to climate change.

Promising developments in relation to civic engagement and water policy are occurring in Alberta and Saskatchewan. This process of forming watershed advisory groups should be supported and made a permanent water governance practice for all affected institutions. Incorporating this civic engagement would help strengthen responsive policy by allowing constantly informed public and stakeholder contributions to ongoing water issues, leading to better decisions incorporating local community values. These promising developments should receive continuous support and critical assessment to ensure that present and future adaptations to climate-induced water stress will build social and economic resilience, and achieve the maximum benefits possible.

ENDNOTES

1. These activities were carried out as part of the research project "Institutional Adaptations to Climate Change" (www.parc.ca/mcri). The IACC project has been supported by the Social Sciences and Humanities Research Council of Canada.
2. In spring 2009, PFRA evolved from a prairie-focused agency to become the Agri-Environment Services Branch, expanding its agri-environmental focus nationally.

REFERENCES

Agriculture and Agri-Food Canada. 2002. *Drought Risk Management Plan for Saskatchewan, 2002.* In Canada-Saskatchewan MOU on Water Committee Minutes, June 24, 2002; Agriculture and Agri-Food Canada—PFRA, Regina.

Alberta Agriculture. 2007. *Drought Action Plan* [online] [accessed 14 December, 2007]. Available from World Wide Web: *http://www1.agric.gov.ab.ca/$Department/deptdocs.nsf/all/ppe9026*

Alberta Environment. 2002. *Framework for Water Management Planning*, Alberta Environment: Edmonton. 37 pgs.

——. 2003. *Water for LABife: Alberta's Strategy for Sustainability* [online]. Edmonton, Alberta, 32 pgs. [accessed May 26, 2008]. Available at: *http://www.waterforlife.gov.ab.ca/html/background2.html*

——. 2005. *Enabling Partnerships: A Framework in Support of Water for Life*. Alberta Environment: Edmonton, AB.

——. 2008a. *Water for Life: A renewal* [on-line]. Alberta Environment: Edmonton. 20 pgs. [accessedd May 28, 2009]. Available at: *http://environment.gov.ab.ca/info/library/8035.pdf*

——. 2008b. *The Water Management Framework for the Industrial Heartland and Capital Region.*[online] [accessed May 28, 2008]. Available at: *http://environment.gov.ab.ca/info/library/7864.pdf*

Barlow, Maude. 2007. *Blue Covenant*. McLelland and Stewart Ltd.: Toronto.

Banks, Tommy and E. Cochrane. 2005. *Water in the West: Under Pressure* [online]. Standing Senate Committee on Energy, the Environment and Natural Resources, Honourable Tommy Banks, Chair [accessed May 28, 2008]. Available at: *http://www.parl.gc.ca/38/1/parlbus/commbus/senate/com-e/enrg-e/rep-e/rep13nov05-e.pdf.* p. 15.

Cecil, B., H. Diaz, D. Gauthier and D. Sauchyn. 2005. *Social Dimensions of the Impact of Climate Change on Water Supply and Use in the City of Regina.* Report prepared by the Social Dimensions of Climate Change Working Group for the Canadian Plains Research Center, University of Regina: Regina, Saskatchewan. 54 pgs.

Corkal, D.R., P.E. Adkins and B. Inch. 2007. *The Case of Canada—Institutions and Water in the South Saskatchewan River Basin* [online]. Working Paper for Institutional Adaptations to Climate Change [accessed May 30, 2009]. Available at: *http://www.parc.ca/mcri/pdfs/papers/iacc045.pdf*

de Loë, Rob. 2008. *Toward a Canadian National Water Strategy Final Report.* Prepared for Canadian Water Resources Association: Guelph, ON, Rob de Loë Consulting Services.

Diaz, H., M. Hurlbert, J. Warren and D. Corkal. 2009. *Saskatchewan Water Governance Assessment Final Report* [online]. Unit 1E Institutional Adaptation to Climate Change Project [accesed September 4, 2009]. 304 pgs. Available at: *http://www.parc.ca/mcri/pdfs/papers/gov01.pdf*

Environment Canada. 1987. *Federal Water Policy, 1987* [online] [accessed May 26, 2008]. Available at: *http://www.ec.gc.ca/Water/en/info/pubs/fedpol/e_fedpol.htm* and *http://www.ec.gc.ca/Water/en/info/pubs/fedpol/e_fedpol.pdf*

Government of Canada. 2002. *Backgrounder on Drought Measures. Farm Financial Assistance Programs.* Government of Canada: Ottawa.

——. 2009. *The New Business Risk Management Suite* [online] [accessed June 26, 2009]. Available at: *http://www4.agr.gc.ca/AAFC-AAC/display-afficher.do?id=1200408916804&lang=eng*

Government of Saskatchewan. 2008. *Communiqué—May 30, 2008—Premiers Establish Western Watershed Stewardship Council* [on-line]. Western Premiers Conference, Prince Albert, Saskatchewan [accessed May 29, 2009]. Available at: *http://www.wpc.gov.sk.ca/Default.aspx?DN=2f83f100-bfac-460b-bd58-a605aa60da23*

——. 2009. *Government of Saskatchewan Safe Drinking Water Strategy* [online]. Document ID 11347, Queen's Printer: Regina, SK [accessed February 27, 2009]. Available at: *http://www.swa.ca/Publications/Documents/SafeDrinkingWaterStrategy.pdf*

Hurlbert, M., D.R. Corkal and H. Diaz. 2009. "Government and Civil Society: Adaptive Water Management in the South Saskatchewan River Basin." *Prairie Forum* 34, no. 1: 181–210.

Hulbert, M., H. Diaz and D.R. Corkal. 2008. *Government, Water Governance and Adaptive Capacity in the South Saskatchewan River Basin* [online]. Climate 2008, Hamburg University of Applied Sciences [accessed May 29, 2009]. Available at: *http://www.climate2008.net/?a1=pap&cat=2&e=43*

IISD (International Institute for Sustainable Development). 2006. *Designing Policies in a World of Uncertainty, Change and Surprise, Adaptive Policy-Making for Agriculture and Water Resources in the Face of Climate Change, Phase 1 Research Report* [online] [accessed July 24, 2008]. Available at: *http://www.iisd.org/PUBLICATIONS/pub.aspx?id=840*

Kerster, G. 2009. Personal communication with G. Kerster, Manager of National Water Supply Expansion Program. Rural Water Development Program and National Water Supply Expansion Program overview based on data and records collected by Prairie Farm Rehabilitation Administration (PFRA). March 6, 2009.

Marchildon, G.P., S. Kulshreshtha, E. Wheaton and D. Sauchyn. 2008. "Drought and Institutional Adaptation in the Great Plains of Alberta and Saskatchewan, 1914–1939" *Journal of Natural Hazards* 45: 391–411.

Morrison, H.A. and J. Gee. 2001. "DIAGNOSTIQUE: Federal Water Policy and Ontario Region." As reported with updated information in *Federal Government Departments' Roles and Man-*

dates for Water in the Prairie Provinces, A Brief Summary. Federal Prairie Water Quality Workshop November 6–9, 2001.

O'Riordan, T. and J. Jager. 1996. *Politics of Climate Change in Europe: A European Perspective*. Routledge: New York. 416 pgs.

Pearse, P.H. 1994. "Development of Federal Water Policy: One Step Forward, Two Steps Back." Keynote Address at the Canadian Water and Wastewater Association National Management Seminar, Calgary, February 3–4, 1994.

Pearse, P.H., F. Bertrand and J.W. MacLaren. 1985. *Currents of Change, Final Report, Inquiry on Federal Water Policy*. Environment Canada: Ottawa, 222 pgs.

Polaris Institute, in collaboration with Assembly of First Nations and supported by Canadian Labour Congress. 2008. *Boiling Point*. Ottawa, ON.

Prairie Provinces Water Board. 2007. *Overview* [online] [accessed May 26, 2008]. Available at: *http://www.mb.ec.gc.ca/water/fa01/fa01s01.en.html*

RIFWP (Rosenburg International Forum on Water Policy). 2007. *Report of the Rosenburg International Forum on Water Policy to the Ministry of Environment, Province of Alberta* [online]. February, University of California, Division of Agriculture and Natural Resource, University of California, Berkley [accessed May 26, 2008]. Available at: *http://www.waterforlife.gov.ab.ca/docs/Rosenberg_Report.pdf*

Saskatchewan Watershed Authority. 2003. *A Watershed and Aquifer Planning Model for Saskatchewan*. Moose Jaw, 12 pgs.

——. N.d., *Saskatchewan's Safe Drinking Water Strategy* [online] [accessed September 4, 2009]. Available at: *http://www.swa.ca/Publications/Documents/SafeDrinkingWaterStrategy.pdf*

——. 2006. *Saskatchewan Water Conservation Plan* [online] [accessed February 26, 2009]. Available at: *http://www.swa.ca/WhatsNew/Advisories.asp*

SaskH2O. 2009. SaskH2o Web Site [online] [accessed February 26, 2009]. Available at: *http://saskh2o.ca/*

Sauchyn, D. and S. Kulshreshtha. 2008. "Prairies [online]." Pp. 1–54 in D.S. Lemmen, F.J. Warren, J. Lacroix and E. Bush (eds.). *From Impacts to Adaptation Canada in a Changing Climate* Government of Canada, Ottawa, ON [accessed May 26, 2008]. Available at: *http://www.adaptation.nrcan.gc.ca/assess/2007/index_e.php*

Swanson, Darren, Stephan Barg, Henry David Venema and Bryan Oborne. 2005. *Prairie Water Strategies, Innovations and Challenges in Strategic and Coordinated Action at the Provincial Level* [online]. International Institute for Sustainable Development, Winnipeg, Manitoba. Available from the World Wide Web: *http://www.iisd.org*

Willems, S. and K. Baumert. 2003. *Institutional capacity and climate actions: Organization for Economic Co-operation and Development*, Environment Directorate, International Energy Agency, Publication COM/ENV/EPOC/IEA/SLT(200305). 50 pgs.

Wittrock, V. and G. Koshida. 2005. *Canadian Droughts of 2001 and 2002: Government Response and Safety Net Programs—Agriculture Sector*. Saskatchewan Research Council, SRC Publication Nol. 1162–2E03: Saskatoon.

Wittrock, V., E.E. Wheaton and C.R. Beaulieu. 2001. *Adaptability of Prairie Cities: the Role of Climate, Saskatchewan Research Council, Current and Future Impacts and Adaptation Strategies*. Environment Branch, Publication 11296–1E01. 230 pgs.

Wood, James. 2008. "Climate Change, Provinces to Combine on Water Conservation." Saskatchewan News Network, *The Leader-Post*, January 30, A4.

World Bank. 2003. *World Development Report 2003, Sustainable Development in a Dynamic World, Transforming Institutions, Growth and Quality of Life*, A co-publication of the World Bank and Oxford University Press: Washington, DC.

CHAPTER 19

CLIMATE CHANGE IMPACTS AND MANAGEMENT OPTIONS FOR ISOLATED NORTHERN GREAT PLAINS FORESTS[1]

Norman Henderson, Elaine Barrow, Brett Dolter and Edward Hogg

INTRODUCTION

In the midst of the Great Plains, scattered from central Alberta to Texas, are island forests, refugia of trees and tree-dependent species isolated in a sea of grass (for a location overview see Figures 1 and 2). This chapter summarizes the impacts climate change will have on a subset of these unique and valuable forests and outlines the options, including recommendations, for the management of these impacts. Henderson et al. (2002) provide more detail on each individual island forest's post-glacial and modern management history, and each forest's current vegetation status, than is provided here.

Plains island forests are of four basic types: highland, sand dune, riparian and scarp. Highland forests owe their existence to increased effective soil moisture levels resulting from an orographic precipitation effect and from cooler high-elevation temperatures (which reduce evapotranspiration). Sand dune forests owe their existence to high water tables and low near-surface soil moisture levels that result from the rapid infiltration of moisture down through the sand. This infiltration shifts the competitive advantage away from grasses to deeper-rooted shrubs and trees. Riparian forests are based on river water whose origin typically lies in a distant, more humid, environment. Sometimes these riparian forests spread out across the nearby Plains; more often, they are confined to river valley slopes and to a valley's tributary ravines. Scarp woodlands, like riparian forests, are typically linear in shape and may stretch in long narrow corridors across the Plains landscape. An escarpment may generate a little more precipitation and may be the site of springs. A dissected scarp face will retain pockets of moisture

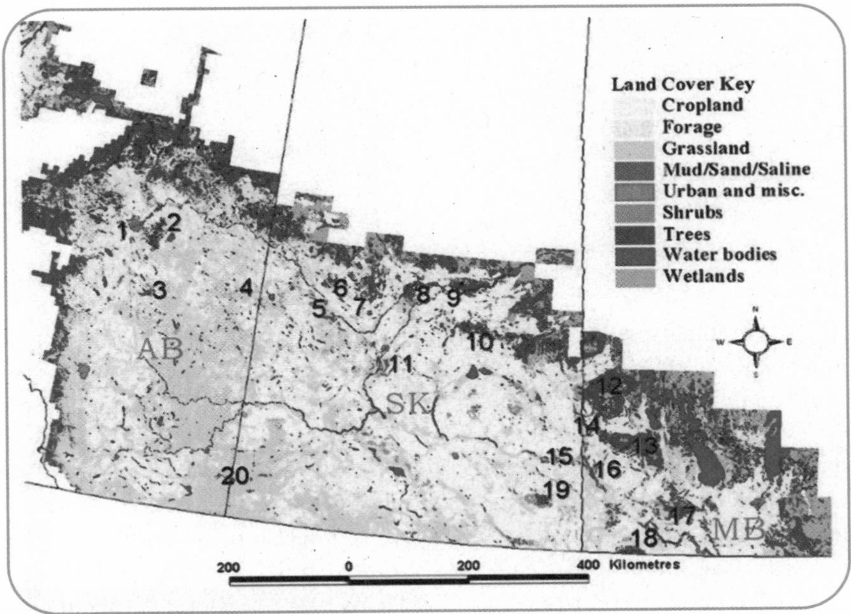

Figure 1. Island Forests of the Canadian Plains[2]

and north or east-facing slopes may be a little cooler. Escarpments also often act as effective firebreaks, which encourages the development of some types of woodland (Wells, 1970).

We selected five island forests as climate change impacts study sites. All five forests are relatively close to the United States–Canada international boundary. All are highland forests, except for Spruce Woods in southwestern Manitoba, which is a combination riparian and sand dune forest. Turtle Mountain forest straddles the North Dakota–Manitoba international boundary. Cypress Hills forest straddles the Sask-atchewan-Alberta interprovincial border.

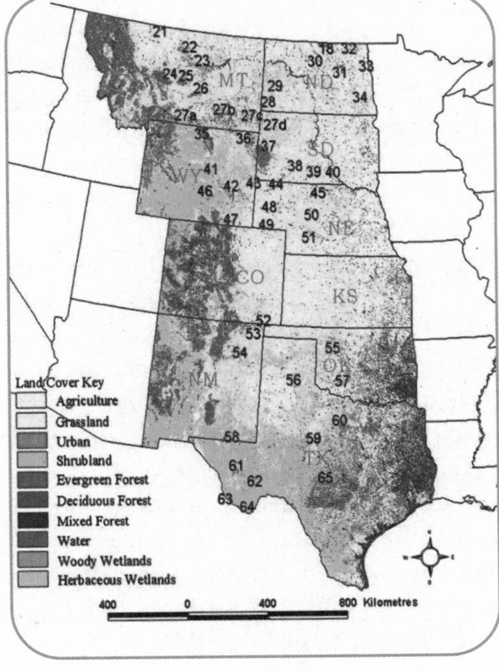

Figure 2. Island Forests of the American Plains[3]

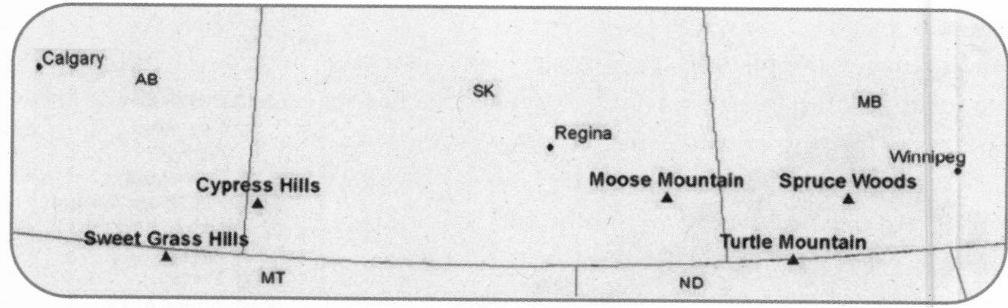

Figure 3. Location of the Island Forest Study Sites.

Moose Mountain forest is in southeastern Saskatchewan. Sweet Grass Hills forest is in north-central Montana. The study site locations are shown in Figure 3.

Although it presented special difficulties (even weather instrumentation varies between Canada and the United States), our selection of study sites from five jurisdictions in Canada and the United States was deliberate. As climate change affects the entire northern Great Plains, it makes sense to consider climate, ecology and policy on both sides of the international boundary and in various states and provinces.

There is good reason to suppose that Plains island forests are at significant risk from climate change. They are marginal or ecotone systems, borderline between grassland and forest ecosystems, and therefore sensitive to relatively small changes in environmental conditions. Saporta, Malcolm and Martell's (1998) review of the literature confirmed the common sense expectation that climate change can be expected to affect species populations at their range limits most profoundly. With regard to the boreal forest at its southern boundary with the Plains, IPCC Working Group II (1996) noted that a warming of average annual temperature of as little as 1°C (in the absence of increased moisture) could be enough to shift the forest northward, while the southern limit of the boreal forest converts to grassland or other non-boreal species (Wheaton, 1997).

By focusing on trees as an indicator for forest ecosystems, we adopted a "keystone" species approach. The assumption is that within a given ecosystem, whilst all species and individual organisms are ultimately interrelated by a network of structures and processes such as food webs or energy flows, a few particular species may play a critical role in supporting the existence of many others (Paine, 1980); one is therefore justified in paying particular attention to these species' habitat needs. Anderson et al. (1998), for example, noted the importance of keystone species in climate change modelling. Trees are clearly keystone species in island forests and, therefore, it is logical, where possible, to model them. This does not, however, preclude that other species may have key roles affected by climate change in these island forests, or that systems-wide impacts may occur.

MANAGEMENT ISSUES

As a management response to climate change, Webb (1992) and Dudley, Jeanreaud and Markham (1996) agree that we need protected natural areas to allow and assist the migration of species to areas of more suitable habitat. Yet, even in large contiguous forests, natural migration rates may not be nearly fast enough to cope with current and predicted climate change. Scott and Suffling (2000), for example, suggest white spruce can only move at the rate of 3 to 4 km every 40 years. In any case, ecological rights-of-ways for trees through grassland from island forest to island forest are not feasible (except perhaps for the very light seed of some trees in the poplar family), and a key characteristic of island forests is that, by definition, the trees within them have nowhere else to migrate to. Further, as they are relatively small ecosystems, island forests may not be as ecologically diverse or as resilient to climate change as larger systems. They may exhibit reduced genetic diversity and greater vulnerability to severe disturbance such as wildfire, insect attack or severe drought. Should a virulent pest or invader reach an isolated forest, it may be some time, if ever, before an appropriate natural control also reaches the forest. Periodic major disturbances are the norm, not the exception, in the greater Plains environment in which the island forests are embedded (Henderson, 2001). All these factors contribute to the vulnerability of these isolated habitats.

While vulnerable, island forests are also valuable landscapes. They typically contain important species and ecosystem outliers at the very edge of their natural range, making them of conservation and scientific importance. As trees are valued on the Plains, all of the study site forests are important for tourism and recreation. Island forests sometimes contain small lakes and ponds that are valuable for a great variety of wildlife, especially waterfowl. They are often of cultural and spiritual importance to First Nations peoples. Highland forests often supply valuable water to the surrounding plains.

Typically, island forests are not managed for timber harvest, but primarily for recreation and nature conservation purposes. These purposes can sometimes be in conflict. A further complication in the island forests is that, normally, there are pre-existing management or use stresses such as grazing, mining, hunting, or oil and gas extraction activities. Government is typically the majority landholder in Plains island forests, and therefore has a responsibility to act as responsible trustee for sustainable forests. But how to manage a forest for nature conservation objectives when it is undergoing significant climate change impacts is unclear. In contrast, climate change adaptation in a forest managed primarily for timber maximization, is more straightforward (Cohen and Miller, 2001). In this situation, the forest manager can continue traditional harvest strategies, perhaps employing new tree species and varieties.

STUDY SITE SELECTION AND RESEARCH QUESTIONS

We selected our specific study sites on the basis of: the existence of vegetation management issues; the presence of species of particular interest; a demonstrated level of government and public interest; site managers' interest in climate change input into their management decisions; the existence of ecological linkages and comparability between sites; site modelability; and on manageability of total study site numbers. Because of some climate, vegetation and physiography similarities, Turtle Mountain and Moose Mountain make a logical pair for comparison of site ecology and management, as do the Cypress Hills and the Sweet Grass Hills. Spruce Woods is an ecologically unique site with numerous management issues and is an interesting example of a riparian and sand dunes island forest.

These five island forests are located in the southern Canadian and northern United States Great Plains within a study region defined by 47.5°N to 51.0°N and 115°W to 95°W. This is a region of cold winters, hot summers, an annual temperature range exceeding 80°C, and large climate variability (Lemmen et al., 1997). Winters are generally dry, while summer (June–July-August) is the wettest season, and June generally the wettest month. A total of 20–30% of annual precipitation falls as snow, while 70–80% falls as rain. The western part of this region is in the rain shadow of the Rocky Mountains, which form a barrier to the maritime influence of the Pacific Ocean. The eastern part of the study region has greater exposure to southerly flows of warm, moist air from the central United States and the Gulf of Mexico. Precipitation amounts, therefore, tend to increase from west to east. Despite summer being the wettest season, strong sunshine, low humidity and drying winds lead to large evaporative water losses in June, July and August. Lowest mean temperatures generally occur in January, with the warmest winter temperatures located in the south-west of the region and the coldest temperatures in the north-east. In July, generally the hottest month, the pattern is similar, although the difference between the cooler north-east and the warmer south-west is not as marked as in winter.

Joyce et al. (2001) noted that annual precipitation has decreased by 10% in eastern Montana and western and central North Dakota over the past 100 years, while Bootsma (1994) noted there is no historical precipitation trend apparent for the Canadian prairie region. Statistically significant warming of 0.9°C from the late 1800s to the 1980s is evident over this region (Lemmen et al., 1998).

Our key research questions included determining the range of probable future climates for these island forests; determining the impacts that future climates may have on the key tree species of the forests; considering the wider implications for nature conservation management of probable climate change impacts; and considering the immediate and long-term management options in response to probable climate change impacts.

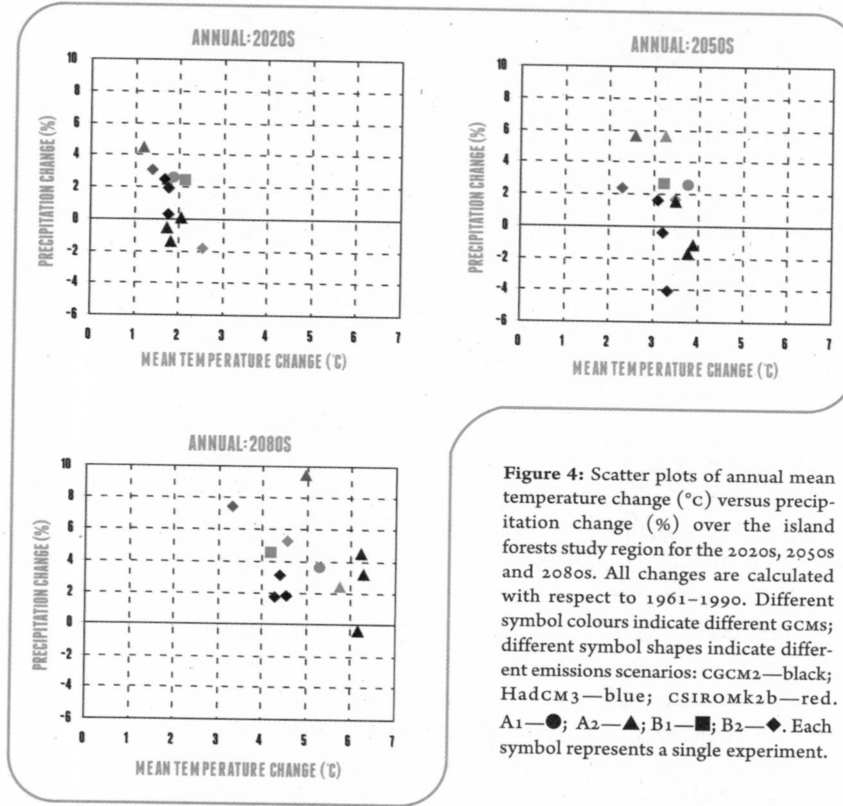

Figure 4: Scatter plots of annual mean temperature change (°C) versus precipitation change (%) over the island forests study region for the 2020s, 2050s and 2080s. All changes are calculated with respect to 1961–1990. Different symbol colours indicate different GCMs; different symbol shapes indicate different emissions scenarios: CGCM2—black; HadCM3—blue; CSIROMk2b—red. A1—●; A2—▲; B1—■; B2—◆. Each symbol represents a single experiment.

SELECTION, CONSTRUCTION AND INTERPRETATION OF CLIMATE SCENARIOS

The first step in determining the range of probable future climate for the study region was the construction of climate change scenarios. We used output from three global climate models incorporating four greenhouse gases and aerosols emissions scenarios representing different world futures with respect to population and economic growth, energy use, and technological development. Using the different emissions scenarios, we constructed a suite of climate change scenarios reflecting the range of probable future climate. The scenario results (valid for the study region defined above) are shown by scatter plots of future temperature and precipitation (Figure 4). All scenarios agree that average temperatures will continue to rise. There may also be slightly more precipitation by the 2080s.

In the northern Plains, a region always susceptible to drought stress, soil moisture levels represent the single most important climate change parameter. Both precipitation and temperature impact on moisture levels (warmer temperatures lead to lower moisture levels because of increased evaporation). The

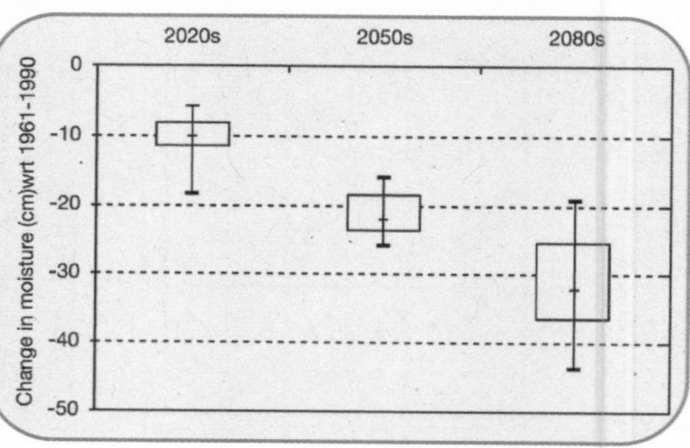

Figure 5: Summary of the projected changes in soil moisture levels (averaged over the five island forest study sites) for the 2020s, 2050s and 2080s. The thin vertical lines in the plot indicate the range of possible future moisture levels compared with the climate of 1961–1990. The boxes indicate the moisture level ranges within which 50% of the scenario projections fall. The horizontal dash within each box indicates the median moisture scenario.

net effect on moisture levels of the foreseen changes in both temperature and precipitation is shown by a box-and-whisker plot (Figure 5).

Without exception, all scenarios project decreases in moisture levels with respect to the baseline climate of 1961–1990. Increased temperatures will have a powerful evaporation effect, such that soil moisture balances will decline substantially. To understand the net impact of precipitation and temperature changes, as illustrated in the box-and-whisker plot, one can imagine that the temperature regime of 1961–1990 is constant, but that annual precipitation declines by 10 cm by the 2020s, 21cm by the 2050s, and 32 cm by the 2080s across the study area. The degree of drying foreseen by these scenarios is sobering, with implications well beyond the issue of island forest survival. It is beyond the focus of this study to examine the many impacts increasingly dry conditions will have in the study region. However, it is obvious, for example, that the climate change impacts on agriculture may be severe.

IMPLICATIONS FOR ISLAND FOREST FUTURES

It must be remembered that in addition to rising temperatures and changing precipitation regimes, there are many other ongoing climate changes that may impact on tree growth and survival in island forests. These include rising CO_2 levels, a lengthening growing season, possibly increasing ground-level ozone, rising UV-b levels, changing diurnal temperature patterns, and changes in the timing and intensity of freeze-thaw events. Climate variability is important too. Some researchers are of the opinion that climate variability is increasing, even as "average" climate shifts (Kharin and Zwiers, 2000; Groisman et al., 2005). This would imply, for example, a greater future frequency of extreme drought and flood events. Thunderstorms and windstorms may also increase in frequency. If there is increasing variability, it will likely further increase stress on trees.

Rising atmospheric CO_2 levels are important, as (unlike most climate change impacts on the forests) they give some cause for hope that increasing aridity may not prove fatal to existing forests—CO_2 fertilization increases the efficiency of water use by some plant species (Long et al., 2004). For example, studies reported a positive CO_2 enrichment effect on the growth of white spruce in southwestern Manitoba (Wang, Chhin and Bauerle, 2006). Johnston and Williamson (2005), with reference to the Saskatchewan boreal forest, found that, even under drought conditions, the CO_2 enrichment effect could result in an increase in productivity. However, a major problem in predicting the impacts of CO_2 fertilization on a specific species is that the enrichment occurs on all vegetation simultaneously. To understand the real-world outcomes, one needs to know the relative growth advantages gained by each vegetation species competing for resources at a specific site (Sauchyn et al., 2008). In their metastudy, Boisvenue and Running (2006: 874) concluded: "there is no clear answer as to whether rising CO_2 concentrations will cause forests to grow faster and store more carbon."

Notwithstanding the influence of other drivers, if the change in the moisture regime is as severe as the scenarios indicate, it seems certain there will be consequences. Hogg and Bernier (2005) suggested increasing vulnerability to drought, insects and fire in Canada's western boreal forest, and it is reasonable to assume these trends also apply to the island forests. One possible response to increasing aridity in the study region forests is slow, cumulative decline such as aspen breakup as aspen passes maturity, is subject to insect and pathogen attack, and experiences conditions too dry for regrowth. Alternatively, the change mechanisms may be spectacular and catastrophic, such as a major fire. Intense fires are increasingly likely as forest stands age and fuel loads increase at all five study sites. Under conditions of climate change, fire or pathogen attack could permanently remove forest cover if conditions have become too dry for new trees to regenerate.

THE OUTLOOK FOR MOOSE MOUNTAIN (SASKATCHEWAN)

Moose Mountain forest is highly vulnerable. Only a few hundred metres of elevation separate Moose Mountain from the surrounding prairie ecosystem. The forest lacks tree species diversity. It also lacks major refugia such as steep coulees or rivers with secure water input from more humid regions. Initially, aspen and balsam poplar will colonize contracting wetland sites, while trees on droughty sites will give way to grassland and bush. Over the long term, the forest will likely change to a very open parkland state, with trees restricted to low hollows and a few northern exposures. Regeneration of existing tree species will prove increasingly difficult. A catastrophic fire is increasingly possible as the forest dries.

Moose Mountain has many relatively small and shallow ponds and lakes. These will shrink in size. Where water remains, it will be warmer in summer, with lower dissolved oxygen and greater susceptibility to algal bloom. Maintaining any sport fishery will likely be impossible. Boating, swimming, and the extraction of water for the existing golf course and waterslide complex will be negatively affected.

THE OUTLOOK FOR TURTLE MOUNTAIN (MANITOBA–NORTH DAKOTA)

The climate change impact outlook for Turtle Mountain is almost identical to that for Moose Mountain. However, at Turtle Mountain, the forest is slightly more diverse in tree species, which may make it somewhat more resilient—bur oak expansion may to a degree compensate for aspen dieback. Bur oak, being relatively drought-hardy, will have the greatest chance of survival amongst the existing trees as moisture levels decline, and an oak savannah landscape may result.

THE OUTLOOK FOR SPRUCE WOODS (MANITOBA)

Aspen is already marginal and short-lived on most sites within the sandhills and can be expected to suffer dieback as moisture levels decline. This could, in turn, negatively impact on spruce regeneration, since in the drier, more open, areas, spruce may get its start under aspen canopy. As at Turtle Mountain, bur oak may prosper in some areas as aspen declines, resulting in an increasingly open savannah landscape. Mature white spruce in the sandhills will likely survive for some years as moisture levels decline, but regeneration will not occur. The long-term ability of trees to persist in the greater sandhills landscape outside the Assiniboine Valley and its tributaries is very doubtful. Trees will, however, very likely persist in sheltered coulees and on valley slopes.

THE OUTLOOK FOR CYPRESS HILLS (ALBERTA–SASKATCHEWAN)

By the 2050s, natural regeneration of aspen, lodgepole pine or white spruce is very unlikely to be possible outside of very localized sites within the Cypress Hills. The future landscape is likely to be one of small patches of stressed woodland persisting only in the most favourable sites—in north-facing high-elevation hollows, or in the riparian bottomlands around Battle Creek, for example. By the 2080s, it is very possible that there will be no regeneration of spruce or lodgepole. Alternatively, a few sheltered sites may remain moist enough to prevent complete extirpation of conifers.

There is already a great and increasing risk of catastrophic fire. It is possible that post-burn forest regeneration would be slow and patchy, even under today's climate conditions, as conditions are already drier than those under which the existing forest developed. Regeneration will be ever more difficult in the future. Lodgepole stands are also becoming increasingly vulnerable to mountain pine beetle attack. It is widely believed that periods of very cold winter weather act as an effective control on mountain pine beetle outbreaks. Periods of sanitizing cold are already less frequent. Spruce budworm attack is also possible on the white spruce.

THE OUTLOOK FOR SWEET GRASS HILLS (MONTANA)

The smallness of the Sweet Grass Hills forests increases their vulnerability to increasing aridity. However, these forests are relatively diverse in conifer species and are, therefore, potentially more resilient than the forests at Cypress Hills. The presence of a forest elevation range of about 700 metres also provides resiliency. However, the rapid pace of climate change may make colonization of higher (i.e. moister) elevations impracticable for some tree species. Those species present, both at the hill summits and far down the slopes, do not need to migrate, but will see their natural range contract up-slope. Tree species presently found only towards the top of the hills, i.e., subalpine fir and whitebark pine, will very likely disappear entirely. Most other tree species will likely persist, either in small sites near the summits or in steep ravine refugia. However, this already small forest will shrink greatly in size. It may decline slowly, with ever-decreasing regeneration. However, the fire risk is increasing and high-impact pathogen disturbances are becoming more likely over time. This means that sudden forest loss is increasingly likely.

MANAGEMENT OPTIONS AND RECOMMENDATIONS

Climate and other environmental influences are only one set of determinants of the island forests' future. Management choices will also co-determine vegetation outcomes. No management option can forestall change entirely. Management that aims simply to retain existing vegetation, or to restore historical vegetation distributions and ecosystems, will fail as the climate moves farther away from recent and current norms. However, active and intrusive management could retain significant areas of forest cover.

We found climate change is not currently considered within the management plans of any of our five island forest study sites. Vandall, Henderson and Thorpe (2006) confirmed this in the case of Moose Mountain and the Saskatchewan Cypress Hills. This omission is serious, and the probable range of climate change impacts (including an analysis of management options to deal with these impacts) should be integrated into future vegetation management strategies for the island forests. However, as trees are rare and valued on the Plains—and likely to become rarer in future—it is a reasonable working assumption that there will be a desire to retain some forest cover at the island forest sites in the face of increasing aridity. To accomplish this, active, anticipatory management will be required.

There are a number of practical measures that could be taken to retain forest cover. Maintaining a diversity of age stands in the forests is one important technique—in all of the study site forests, but particularly at Cypress Hills, Moose Mountain, and Turtle Mountain, there are large stands of old trees that may be less resilient to change agents. Some forest harvest activity would help in creating more age-stand diversity. Unfortunately, the timber in most of the study site forests is generally not economic to harvest, and typically would need to be paid for. Cypress Hills, with higher-value lodgepole pine, is a possible exception, although, even here, the sloped terrain tends to make tree harvest a marginal economic activity at best.

Managing for fire is an imperative in a drying environment. Fuel load build-ups should be avoided to reduce the risk of catastrophic loss. Fire breaks should be maintained and created where necessary. Where safety and other considerations allow it, prescribed fire may be a useful tool to reduce fuel loads or stimulate regeneration. Tree planting may be necessary to aid forest regeneration.

A discovery, provenance and breeding program would be useful. The program should encompass both extant tree species within the island forests and possible new species introductions, with the objective of establishing which varieties and species are best adapted to the range of probable future climates in the island forests. Seed could be collected from dry microsites within and outside the island forests. Plantation site trials could be started within or adjacent to the island forests, or at dry sites outside the island forests, where such sites might serve as analogues for future moisture conditions at the island forest sites.

More aggressive (and controversial) forest retention measures could include countering potentially catastrophic insect or vegetation disturbances by biological, chemical or physical controls, if necessary. Most divergent from current nature conservation policy would be the introduction of non-native tree species into the island forests. The introduction of trees more suited to drier, warmer conditions may, however, ultimately be necessary to retain forest cover as climate change progresses. Henderson et al. (2002) have researched the range suitability of native and non-native trees across the Prairie Provinces under a range of climate scenarios. This research could aid in tree species selection for specific Prairie Provinces sites, including the island forests. Bendzsak (2006) investigated alternative conifers that could be planted in the island forests of central Saskatchewan near Prince Albert and concluded that several non-native conifers could be established with minimal ecological risk. However, there are many issues that need to be carefully considered before introducing a non-native tree into a forest—Thorpe, Henderson and Vandall (2006) discuss them at length. A key question in the decision as to whether or not to introduce a new tree species is whether or not the proposed new species would add resiliency to the forest as a whole.

Zoning can be a useful tool in situations of limited knowledge and uncertain outcomes. It will be advisable to try a number of adaptation techniques in different zones of each forest (including complete non-intervention in some areas). Results between zones can be compared and management adjusted accordingly.

It would be useful to compare current climate change impacts monitoring and adaptation planning and measures across all the Great Plains island forests. At our study sites we found that no climate change impacts monitoring programs were underway, and this may well be representative of the situation at other Plains island forests. A trans-Plains island forest climate change monitoring system could be valuable. It could foster knowledge transfer of best practice across all Plains island forests, identify potential seed and species stock for potential inter-island-forest transfer, and help establish whether particular species assemblages or ecosystems in a given island forest will likely become viable in another.

Communications between climate researchers and land managers and policy-makers is a major challenge. Likewise, communications between scientists and managers on the one hand, and the public on the other, is very necessary. The management of climate change impacts is not yet integrated into protected lands planning (Vandall, Henderson and Thorpe, 2006), and the public is largely unaware of the management choices and challenges that are a consequence of climate change. It would help if existing island forest interpretation centres interpreted site-specific climate change scenarios and impacts to the public. There is also the opportunity to survey island forest visitor preferences for management responses. The Turtle Mountain International Peace Garden could be an ideal binational Plains "flagship" climate change interpretation centre, given its high number of visitors from both sides of the international boundary. Managers will need help and input as to how the public feels about some possible controversial interventions, such as the introduction of new tree species. Whatever our management response, the climate scenarios make clear that change is coming to the island forests.

ENDNOTES

1. We are indebted to many for advice and comments on this study. We wish to thank Ron Hopkinson, Fraser Hunter, Daniel Scott, Don Lemmen, Kelly Redmond, Jim Ashby, Lou Hagener, Richard Hopkins, Shannon Iverson, Stanley Jaynes, Brad Sauer, John Thompson, Bruce Reed, George Seielstad, Edward Cloutis, Jay Malcolm, John Pomeroy, Annabel Robinson, Carlos Gracia, Randy Craft, Adam Wellstead, Bill DeGroot, Bill Schroeder, John Kort, Wybo Vanderschuit, Misty Vermulm, Dawn Wickum, Craig Stange, Gregory Yapp, Geri Morris, John Vandall, Ken Lozinsky, Ron Zukowsky, Bruce Martin, Kelvin Kelly, Brad Mason, Gary Neil, Butch Anderson, Marty Halpape, Les Weekes, Wes Mickey, Archie Landals, Raymond Wong, Helios Hernandez, Gerry Becker, Keith Knowles, Ken Schykulski, Patti Ewashko, Cathy Mou, Dave Spittlehouse, Jeff Thorpe, Narinder Dhir, Erik Eneboe, Casey Kellogg, Shawn Morgan, Brian Prince, Gary Olson, Tom Karch, Ray Weed, Greg Schenbeck, Kurt Atkinson, Bill Duemling, Larry Hagen, Ed Korpela, Bill Baker, Marianne Weston, Ron Davis, John McQueen, Gord Howe, Victoria Muzychuk, Ted Morris, Brent Joss, Sam Kennedy, Dave Sauchyn, and Malcolm Wilson.
2. GIS data from: J. Aston, PFRA's Generalized Landcover, version 1, 2001.
3. GIS data from: USGS National Land Cover Data, 1992.

REFERENCES

Anderson, J.C., I. Craine, A.W. Diamond and R. Hansell. 1998. "Impacts of Climate Change and Variability on Unmanaged Ecosystems, Biodiversity, and Wildlife." Pp. 121–88 in G. Koshida and W. Avis (eds.). *The Canada Country Study: Climate Impacts and Adaptation* (Volume 7: National Sectoral Volume). Environment Canada.

Bendzsak, M. 2006. *Evaluation of Conifer Tree Species Alternatives for Island Forest Renewal.* Saskatchewan Forest Centre: Prince Albert, 37 pgs.

Boisvenue, C. and S. Running. 2006. "Impacts of Climate Change on Natural Forest Productivity—Evidence Since the Middle of the 20th Century." *Global Change Biology* 12: 862–82.

Bootsma, A. 1994. "Long Term (100 yr) Climatic Trends for Agriculture at Selected Locations in Canada." *Climatic Change* 26: 65–88.

Cohen, S. and K. Miller (eds). 2001. "North America." Pp. 735–800 in J. McCarthy, O. Canziani, N. Leary, D. Dokken and K. White (eds.). *Climate Change 2001: Impacts, Adaptation, and Vulnerability,* IPCC. Cambridge University Press: Cambridge.

Dudley, N., J-P. Jeanreaud and A. Markham. 1996. "Conservation in Boreal Forests under Conditions of Climate Change." *Silva Fennica* 30: 379-83.

Groisman, P., R. Knight, D. Easterling, T. Karl and V. Razuvaev. 2005. "Trends in Intense Precipitation in the Climate Record." *Journal of Climate* 18: 1326–50.

Henderson, Norman. 2001. *Rediscovering the Great Plains: Journeys by Dog, Canoe and Horse.* Johns Hopkins University Press: Baltimore, London.

Henderson, N., E. Hogg, E. Barrow and B. Dolter. 2002. *Climate Change Impacts on the Island Forests of the Great Plains and the Implications for Nature Conservation Policy.* Prairie Adaptation Research Collaborative: Regina.

Hogg, E.H. and P.Y. Bernier. 2005. "Climate Change Impacts on Drought-Prone Forests in Western Canada." *Forestry Chronicle* 81: 675–82.

IPCC Working Group II. 1996. *Climate Change 1995—Impacts, Adaptations and Mitigation of Climate Change: Scientific-Technical Analyses.* Cambridge University Press: Cambridge.

Johnston, M. and T. Williamson. 2005. "Climate Change Implications for Stand Yields and Soil Expectation Values: A Northern Saskatchewan Case Study." *Forestry Chronicle* 81: 683–90.

Joyce, L.A., D. Ojima, G.A. Seielstad, R. Harriss and J. Lackett. 2001. "Potential Consequences of Climate Variability and Change for the Great Plains." Pp. 191-217 in National Assessment Synthesis Team, *The Potential Consequences of Climate Variability and Change* (Report for the US Global Change Research Programme). Cambridge University Press: Cambridge.

Kharin, V. and F. Zwiers. 2000. "Changes in Extremes in an Ensemble of Transient Climate Simulations with a Coupled Atmosphere-Ocean GCM." *Journal of Climate* 13: 3760–88.

Lemmen, D., R. Vance, I. Campbell, P. David, D. Pennock, D. Sauchyn and S. Wolfe. 1998. *Geomorphic Systems of the Palliser Triangle, Southern Canadian Prairies: Description and Response to Changing Climate.* Geological Survey of Canada (Bulletin 521): Ottawa.

Lemmen, D., R. Vance, S. Wolfe and W. Last. 1997. "Impacts of Future Climate Change on the Southern Canadian Prairies: A Paleoenvironmental Perspective." *Geoscience Canada* 24: 121-33.

Long, S.P., E.A. Ainsworth, A. Rogers and D.R. Ort. 2004. "Rising Atmospheric Carbon Dioxide: Plants FACE the Future." *Annual Review of Plant Biology* 55: 591–628.

Paine, R.T. 1980. "Food Webs: Linkage, Interaction Strength and Community Infrastructure." *Journal of Animal Ecology* 49: 667-85.

Saporta, R., J. Malcolm and D. Martell. 1998. "The Impact of Climate Change on Canadian Forests." Pp. 319–82 in G. Koshida and W. Avis (eds.). *The Canada Country Study: Climate Impacts and Adaptation* (Volume 7: National Sectoral Volume), Environment Canada.

Sauchyn, D., S. Kulshreshtha, E. Barrow, D. Blair, J. Byrne, D. Davidson, P. Diaz, N. Henderson, D. Johnson, M. Johnston, S. Kienzle, J. Klaver, J. Thorpe and E. Wheaton. 2008. "Prairies." Pp. 275–328 in D. Lemmen, F. Warren, J. Lacroix, and E. Bush (eds.). *From Impacts to Adaptation: Canada in a Changing Climate 2007.* Government of Canada: Ottawa.

Scott, Daniel and Roger Suffling. 2000. *Climate Change and Canada's National Park System*. Parks Canada, Environment Canada, and the University of Waterloo.

Thorpe, J., N. Henderson and J. Vandall. 2006. *Ecological and Policy Implications of Introducing Exotic Trees for Adaptation to Climate Change in the Western Boreal.* Saskatchewan Research Council, SRC Publication No. 11776–1E06: Saskatoon, Saskatchewan, 111 pgs.

Vandall, J.P., N. Henderson and J. Thorpe. 2006. *Suitability and Adaptability of Current Protected Area Policies under Different Climate Change Scenarios: The Case of the Prairie Ecozone, Saskatchewan*. Saskatchewan Research Council Publication No. 11755–1E06.

Wang, G., S. Chhin and W. Bauerle. 2006. "Effect of Natural Atmospheric CO_2 Fertilization Suggested by Open-Grown White Spruce in a Dry Environment." *Global Change Biology* 12: 601–10.

Webb, T. 1992. "Past Changes in Vegetation and Climate: Lessons for the Future." Pp. 59-75, in R. Peters and T. Lovejoy (eds.). *Global Warming and Biodiversity*, Yale University Press: New Haven.

Wells, Philip. 1970. "Postglacial Vegetational History of the Great Plains." *Science* 167: 1574-82.

Wheaton, Elaine. 1997. "Forest Ecosystems and Climate," Appendix B. Pp. 1–31 in R. Herrington, B. Johnson and F. Hunter (eds.). *Responding to Climate Change in the Prairies: Volume Three of the Canada Country Study: Climate Change and Adaptation*. Environment Canada.

CHAPTER 20

WINTER ROADS IN MANITOBA

Danny Blair and David Sauchyn

The most significant impact of climate change on transportation infrastructure in the northern part of the Prairies is the reduced viability of winter roads. The majority of the region's winter roads are in Manitoba, where some 2,200 km of winter roads are built annually to provide access to 28 communities not serviced by permanent roads (Manitoba Infrastructure and Transportation, 2007; 2008). The population of these communities—about 25,000—is expected to double in the next 20 years (Centre for Indigenous Environmental Resources, 2006).

Winter roads are vital links between northern Aboriginal communities and to other parts of Canada. They are social, cultural and economic lifelines in remote communities, enabling the delivery of such essential goods as: food, fuel, and medical and building supplies (Foster, 1996; Kuryk, 2003; Centre for Indigenous Environmental Resources, 2006). There are also safety-related issues, since many northerners use winter roads and trails for hunting, fishing, and cultural and recreational activities.

In its study of five First Nations in Manitoba (Barren Lands, Bunibonibee Cree, Poplar River, St. Theresa Point and York Factory Cree), the Centre for Indigenous Environmental Resources (2006) reported the following key issues.

RELIABILITY OF WINTER ROADS

According to the northern community members, the two most common causes of poor winter road conditions are: 1) warmer weather (attributed to both natural cycles and human-induced climate change), and 2) high, rapidly fluctuating water levels with strong currents (attributed to flow-control structures and naturally high runoff). Poor conditions include weaker and thinner ice, excess slush, earth patches, potholes, hanging ice and ice pockets on roads, and less direct routes than those that cross water bodies.

With climate change, the average length of the winter road season in Manitoba is expected to decrease by eight days in the 2020s, 15 days in the 2050s and 21 days in the 2080s (Prentice and Thomson, 2003).

WINTER ROAD FAILURE AND EMERGENCY MANAGEMENT

Manitoba Infrastructure and Transportation has reported decreased ice thickness, poor ice texture and density, delayed winter road seasons, problematic muskeg, and decreased load limits. There have been cases of equipment damaged beyond repair from a single trip on the winter road. Emergency responses to winter road failure, including the airlifting of supplies, are costly.

PERSONAL SAFETY ON WINTER ROADS, TRAILS AND FROZEN WATER BODIES

When winter road seasons are short, some community members take additional risks on winter roads, trails and water bodies. A road construction worker from Wasagamack First Nation drowned in 2002 when the grader he was driving broke through the ice.

PERSONAL HEALTH CONCERNS

There are concerns about access to health centres and other medical assistance when winter roads and trails are not available. In addition, high rates of diabetes in Aboriginal communities have been linked to decreased access to affordable healthy foods, whether from stores or from the wild. Stress is another health-related concern linked to shortened winter road seasons, in terms of increased financial pressures and greater social isolation.

HIGH COST OF LIVING

Transportation by winter roads minimizes the cost of fuel, goods and services. The cost of shipping goods by air is two to three times greater than that for ground transportation on winter roads. Lower prices also are available at larger centres accessible via all-weather roads. The cost of food is a significant issue, given that unemployment rates in northern communities are as high as 80–90%. Accessing wild meat and fish allows individuals to offset the high cost of food at the local store, but warmer winters are restricting the gathering of traditional foods.

DECREASED PARTICIPATION IN SOCIAL AND RECREATIONAL ACTIVITIES

Winter roads, access trails and frozen waters play important social and cultural roles in northern communities. They provide access to neighbouring communities and larger centres to shop; visit with friends and family, gather for social events (e.g. marriages, births and funerals); participate in recreational activities (e.g. bingos, festivals and fishing derbies); and visit friends, family or the elderly in hospitals or care facilities.

Furthermore, community members use trails for recreational riding. In recent years, some fishing derbies and winter carnivals have had to be cancelled. Overall, individuals feel more disconnected from their friends and relatives in neighbouring communities when winter road seasons are shorter and less reliable. Helicopter flights are available, but most northern residents cannot afford the high fares.

HINDRANCE OF COMMUNITY OPERATIONS AND ECONOMIC DEVELOPMENT

Much economic activity is related to access provided by frozen ground and water bodies. Thin ice cover and poor winter road conditions have restricted some income-generating activities, including commercial ice fishing and the export of resources (e.g. fish and furs) for sale in larger centres. Winter roads enable communities and businesses to more efficiently acquire goods and supplies required for regular operations, maintenance and repairs. Also, the winter roads provide First Nations with income generated from road construction and maintenance contracts with Manitoba Infrastructure and Transportation. Thus, the length and timing of winter road operations can impact economic development, housing, capital, special projects, and equipment maintenance.

ADAPTATION

The Centre for Indigenous Environmental Resources (2006) recommended a variety of actions at the community and government levels to address issues related to degrading winter roads. These can be summarized as follows:

- Increase the security of winter roads (both levels);
- Develop community climate change action plans (community) and provide support for implementing these plans (government);
- Develop a communication strategy (community) and increase communication with other First Nations (government);
- Increase social and cultural-recreational opportunities (community) and provide support for these opportunities (government);
- Increase consumption of local foods (community) and provide support for production of these foods (government);
- Enhance community safety (both levels); and
- Increase funding opportunities for community operations (both levels).

SUMMARY

Winter (ice) roads, particularly in Manitoba, are vital lifelines for many remote and northern communities. They facilitate the transport of machinery, building supplies, fuel, and other bulk goods that would otherwise be extremely expensive to deliver to these communities. Importantly, they also present the members of these communities with expanded social and cultural opportunities, if only for a short period of the year. Unfortunately, there is already a substantial amount of evidence that the length and quality of the winter roads are being negatively affected by climate change, and there is good reason to think that the problems are going to get worse in the coming decades. Given the near certainty of the continuation and expansion of these problems, it is perhaps not surprising that at least some jurisdictions are already adapting to the change. Most prominently, the Government of Manitoba has recognized that future winter road operations are in jeopardy and have consequently started to investigate and invest in alternative modes of transport. In particular, they have made a commitment to relocate many of its northern ice roads to land-based routes, improving safety, and providing opportunities to convert at least some of the winter roads to all-season roads in the future, even though some of these options are likely to be very expensive. Indeed, Manitoba's response to the threat that climate change presents to its winter roads is an excellent example of the type of long-term planning and adaptation that will likely become more common within the transportation sector, and elsewhere, as climate changes become more evident and expected.

REFERENCES

Centre for Indigenous Environmental Resources. 2006. *Climate Change Impacts on Ice, Winter Roads, Access Trails, and Manitoba First Nations* (final report). November 2006. Natural Resources Canada and Indian and Northern Affairs Canada. 210 pgs.

Foster, R.H. 1996. "Winter Roads." Pp. 175–76 in J. Welsted, J. Everitt and C. Stadel (eds.). *The Geography of Manitoba: Its Land and Its People.* The University of Manitoba Press: Winnipeg, Manitoba.

Kuryk, D. 2003. "Seasonal Transportation to Remote Communities—What If?" Pp. 40–49 in B.E. Prentice, J. Winograd, A. Phillips and B. Harrison (eds.). *Moving Beyond the Roads: Airships to the Arctic Symposium II Proceedings,* October 21–23, 2003, University of Manitoba Transport Institute: Winnipeg, Manitoba.

Manitoba Infrastructure and Transportation. 2007. *Annual Report: 2006–2007* [online]. Manitoba Transportation and Government Services, 129 pgs. [accessed February 1, 2009]. Available at: *http://www.gov.mb.ca/mit/reports/annual/2006annual.pdf*

——. 2008. *Annual Report: 2007–2008* [online]. Manitoba Transportation and Government Services, 132 pgs. Available at: *http://www.gov.mb.ca/mit/reports/annual/2007annual.pdf*

Prentice, B.E. and J. Thomson. 2003. "Airship Fuel Tankers for Northern Resource Development: A Requirement Analysis." Pp. 592–606 in *Proceedings of the 38th Annual Conference of the Canadian Transportation Research Forum: Crossing Borders: Travel Trade, Security and Communication*: Ottawa, Ontario.

CHAPTER 21

THE VULNERABILITIES OF FIRST NATIONS COMMUNITIES: THE CASES OF SHOAL LAKE AND JAMES SMITH

Jeremy Pittman

> This climate change is becoming an issue and there is more and more concern about the impacts it is having on our society as First Nations people because of our cultural links with Mother Earth. [Climate change] has made impacts and we have to live in the environment. [The environment] is a gift. We lived in harmony with this continent for 6000 years. We choose a lifestyle that is harmonious to nature. Over the past 500 years or so, there have been changes to both our lifestyle and our ways of doing things. Yet, that characteristic of our culture is still there, to preserve and protect, and it will always be there over the next few years. I am just starting to understand that there are international activities going on with our cultural people to address the challenges of climate change. It is a serious matter, to the extent that there are gatherings just to comprehend and understand and to prepare, not so much a strategy, but to try and develop an interpretation for not just our immediate generations, but for everyone else. — A BAND COUNSELLOR OF THE JAMES SMITH FIRST NATION

INTRODUCTION

Climate change implications for First Nations communities in Canada are broad in scope and complex in nature. Many facets of First Nations' livelihoods and lifestyles are potentially at risk due to projected changes in climate. To understand how climate change may impact First Nations in Western Canada, vulnerability assessments of two *nêhiyawak* (Cree) communities in northern Saskatchewan, the Shoal Lake and James Smith First Nations, were conducted (Pittman, 2009c). The vulnerability approach was employed, where current vulnerability is used as a starting point to inform an analysis of potential future vulnerabilities (Ford and Smit, 2004). Vulnerability was conceptualized as a function of both the sensitivity of the communities to various climatic stressors (i.e. their exposure) and the ability of the communities to cope with the impacts of these stressors (i.e. their adaptive capacity) (Handmer, Dovers and Downing, 1999; Kelly and Adger, 2000; Kasperson and Kasperson, 2001; Smit and Pilifosova, 2001; Downing and Patwardhan, 2003; Turner et al., 2003; Ford and Smit, 2004; O'Brien et al., 2004; Smit and Wandel, 2006). The results of these assessments follow.

EXPOSURE-SENSITIVITY

Exposure-sensitivity refers to the characteristics of the community in combination with the impacts of a stimulus causing negative implications for community members (Downing, 2003; Smit and Pilifosova, 2003). Exposure-sensitivity includes the biophysical conditions and the social conditions making the biophysical conditions problematic (Downing, 2003; Smit and Pilifosova, 2003). Major exposure-sensitivities anticipated from climate change include vegetation shifts in local lands, higher competition for water resources, and loss of culture.

The James Smith and Shoal Lake First Nations are located in the transition area between grasslands and boreal forest. This area in particular is expected to experience drastic changes due to shifts in vegetation and water resources in response to climate change (Wheaton, 1997; Saporta, Malcolm and Martell, 1998). As forest fires, pests, and increased aridity take their toll on local forests, grasslands are expected to migrate north, pushing the tree-line further and further away from these two reserves (Kulshreshtha, Johnston and Wheaton, 2001; Johnston et al., 2001; Johnston, 2008).

Community members have already experienced changes in the local forests from logging. An Elder from Shoal Lake spoke of the importance of the forest and how changes in the forest affect the community (pers. comm, 2007):

> It was beautiful in those days. You could walk around with the snow shoes, even when I was trapping. There used to be shapes like a moose or a deer had slept under a tree in the night. It was good for the animals to run around or get away from danger. There were a lot of berries growing in open places. It was good for us, Aboriginal Indians. That is where we took our food from. We never went to IGA [grocery store] to get our groceries. We just picked up a rifle and got our own meat. We used to share it, too. Nobody asked you for $10 for a chunk of meat, we just give them away for free. It is a treaty right, even rabbits and stuff. The ones we used for food. We used everything for bows and knives, feathers for

> mattresses, and whatever. Nothing was wasted. Now today there is waste. The bush is so thick it is like a jungle . . . There were only limited number of little trees, like white poplar. Nothing really grew under the spruce trees and there were millions of them. It was just like walking in the park. Now today, it is just like a jungle, you can't even get around. I don't think you can even wear snow shoes. In those days, there used to be 3ft of snow. There was a lot of snow in those days. But today there is hardly any snow for the last two years. There are no bears and stuff like that. We used to gather our medicine and we don't now. Everything is destroyed by us. We were all logged out, that is why you don't see any spruce trees. Those trees help a lot. Now today, it is always windy. There are no spruce trees to stop that wind. Every day we have a lot of wind, almost tornadoes, high winds. We miss those spruce trees.

In addition to changes in the local forests, reduced stream flow down the Saskatchewan River is also anticipated (Pietroniro et al., 2006). The implications of reduced flows are far reaching, leading to diminished water supplies in reservoirs, decreased hydroelectric potential, and, undoubtedly, higher competition for water resources. Competition over water resources has already been realized by the people of James Smith. Unfortunately, traditional and cultural activities often lose out to the industrial and consumptive demands of contemporary Canadian culture. The importance of the Saskatchewan River to the people of James Smith, and the ways in which competition manifests into impacts on the lives of people, are reflected in the thoughts of a local hunter (pers. comm, 2007):

> In the spring time, right now, because we had so much water, the river is staying up. In the last two years, we used to have to lift up our motor two or three times. They shut the dam off up river. This water just drops right down and it is hard to go down river and go hunting. Then all of a sudden, next week, [the river] is real high. [The] same [thing happens] at the lake after Melfort Ferry Road. When Nipawin lets out too much [water], that [lake] just drops. And there is one area there where there is a big sand bar, and if they let out too much water, you can't get to them spruce trees and that is where we go after elk. We are stuck there. So we have to go further up and walk. It is bad when these guys let out too much water or these guys shut off their water. We never used to have that water. We used to use canoes. One time my brother [and I] went on canoe from Melfort Ferry Road.

In combination with climate change, the communities are also at risk of cultural change. Although traditional knowledge and values still exist in both communities (Pittman, 2009a), maladaptation to the compounding effects of cultural exile and coercive tutelage has taken their toll on some local people (Dyck, 1991; McLeod, 2007; Ermine, Sauchyn and Pittman, 2008). Some traditional values and behaviours have been eroded away and replaced with dangerous and unhealthy ones such as alcohol, drugs, violence, and gangs. The effects of alcohol, drug abuse, violence, and gangs can be best understood from the experience of a local young man incapable of finding work because he lacks a high school education, does not have a driver's licence, and has a criminal record. Faced with limited opportunities, this particular young man left the reserve for Prince Albert, SK. The following quote from the young man (pers. comm, 2007) helps to frame the discussion of gangs, violence and alcohol abuse he witnessed: "They're alright when they're sober, but when they get drunk they [have] to prove themselves. They [have] to hurt somebody. People just want to kill, kill, [and] kill!" Further reflections from the young man help to characterize his situation and show that even in the face of adversity, cultural resilience and knowledge can relieve certain stresses caused by drastic social change (pers. comm, 2007):

> [On] my father's side, my grandparents died before I was born. I just remember [my grandparents] on my mom's side. This is also hereditary, this alcohol. They all drank, but that's not part of heritage culture. It's a part of life. We used to camp, ride wagons, have a garden a half mile away from the house. I used to help my dad with a team of horses. What my dad taught me, it helped me through life. Tobacco is part of our heritage culture. Hunting, when we hunt, we take [tobacco]. I don't smoke it. I crack it in half and let it go to the ground. Wherever the sun is, that's where heaven is. Give thanks to the animal. Don't feel sorry for the animal. It was put here for us. Thank the creator and thank the animal. Everybody's not all the same. Long ago, people, our culture's lost. We went off the road somewhere . . . white schooling, English. Not a lot of people know our culture. Everybody has a tribe. It's their ways. Everybody has a leader, a chief, and a provider (hunter) . . . My dad taught me to be independent and not to let the animal know you're afraid.

COPING CAPACITY

Coping capacity refers to the ability of the community to manage impacts and risks resulting from their exposure-sensitivity (Wheaton and MacIver, 1999; Bryant et al., 2000; Yohe and Tol, 2002; Füssel and Klein, 2002; Smit and Pilifosova, 2003). Coping capacity is synonymous with adaptive capacity, or the ability to deal with stress. Major issues relating to the ability of the communities to cope with the projected impacts of climate change were human, social, economic, and institutional capital.

An important resource for climate change adaptation is human capital (Smit and Pilifosova, 2003). In recent times, human capacities in the communities have diminished. An Elder from Shoal Lake reflects on human capacities (pers. comm, 2007):

> My grandfather asked me one morning what I was going to do, and I told him: "Nothing; I don't have anything to do." And he said, "Grandson, look around the log house. You will see something to do and go outside and look around. There are a lot of things you can do outside in the bush. So don't ever say that you have nothing to do." Use the brain, the mind and the movement, and you will strengthen. Use it! Even those Elders in those days used to talk to us. They used to sit outside in the shade, outside the log houses. They used to sit outside from like 10AM until 3PM waiting for somebody to come and visit them. They used to lecture us. What to respect and what not to do. What to be afraid of and stuff like that, how to look after your body. Your body is a gift. You have to look after it and only you can look after it. Now today you see kids walking around at night with no control. In those days, we used to have a curfew. Everybody followed it. We would start going to sleep at 9pm, even the adults. By 10pm there were no lights. We didn't have power, only coal lamps and candles. Everybody was well rested in the morning and ready to work. You didn't feel grouchy at the school. I used to run around quite a bit. There are a lot of things missing.

Social capital is another important determinant of coping capacity, not in terms of measured quantities, but rather in the ways it can be employed to collectively confront challenges (Adger, 2001). Shoal Lake demonstrates the use of social capital to collectively confront the negative impacts of social change by hosting their annual Family Camp (Pittman et al., 2009; Pittman, 2009b). Family Camp is an event where the whole community goes into the forest for a week to learn traditional skills and values. It is intended to aid the youth in their struggles against the many challenges they face in reserve life.

Unfortunately, James Smith currently exhibits poor capabilities for collective action. The community is divided into three bands—James Smith, Peter Chapman and *chakastaypasin*—that were forced to amalgamate onto one reserve due to land claim settlements, in the case of Peter Chapman, and the loss of reserve land following the Northwest Rebellion of 1885, in the case of *chakastaypasin*. Compounding the problem are the residual effects of residential schooling and the shattering effect they had on Indigenous communities. An Elder from James Smith discusses the tension and mistrust between community members on-reserve (pers. comm, 2007):

> Long ago this used to be a close-knit reserve. With myself and my family, there is only two of us now, but there used to be 8. We weren't close. We were not the people that hug and I love you and all that. You didn't get that in residential school. I never lost my language. I didn't lose that. They let us talk our language in one school. At high school, in Lebret, we spoke English, but we spoke Cree anyway. There were a lot of languages there, Chipeweyan, Cree, Lesotho. Most of my friends talked Cree. Some people wanted to teach me Saulteaux. I learned some. There is a lot of effect on us. There is always a tension on this reserve. We are kind of fighting amongst each other. There are certain families that don't like each other because of last name. You don't belong here. We get the jobs. You get the second jobs. I am down on a last-name basis. I am *chakastaypasin*. That is the one close to Fenton. It is not a town anymore. It was close to Birch Hills, by the ferry. That is where we are from. We left that reserve on account of the 1885 rebellion. They were trying to recruit our people to the Métis against the government. The Métis told us that if we didn't join them they would kill us off. That is why we came over here. People went all over the place from over there. That is the history I know from there. That is pretty well how things stand nowadays here. Hopefully it changes. You don't know who is your friend here anymore. Me, I have a good idea who. Would [I] say that [I] have friends from other bands? Ya, I got friends. Even some reserves, I have met some of my relatives from *chakastaypasin*. They are all over the place. They married a woman and stayed. There is another family next reserve that was from there. We finally found them. But that is how it is here. It is pretty hard. Hopefully things change now.

Economic capital is limited. There are few employment opportunities on these reserves. Unemployment rates in James Smith and Shoal Lake are 33.8% and 31.8% respectively, with participation rates of 53.2% and 35.5% respectively (Statistics Canada, 2008a; 2008b). In Saskatchewan, the unemployment rate is 5.6%, with a participation rate of 68.4% (Statistics Canada, 2008c). Income levels on-reserve are also low, as average income for individuals over 15 years of age that work full-time is $26,345 in James Smith and $17,148 in Shoal Lake (Statistics Canada, 2008a; 2008b). For the total population 15 years and older in James Smith and Shoal Lake, median income is only $8,296 and $5,648 respectively, with 64.7% and 59.0%, respectively, of total income on-reserve coming from earnings, while 32.9% and 40.0%, respectively, comes from government transfers (Statistics Canada, 2008a; 2008b). In Saskatchewan, median income for all those over 15 years of age is $23,755, with only 12.8% of this from government transfers (Statistics Canada, 2008c).

There are numerous formal institutions defining resource access for First Nations communities; those specifically named by community members included Indian and Northern Affairs Canada (INAC), Canada Mortgage and Housing Corporation (CMHC), and the Saskatchewan Indian Gaming Association (SIGA). These institutions work towards meeting the needs of individuals within the communities, by providing funding or access to capital for various projects, including those related to housing, education, drinking water, culture, and recreation. Despite their efforts to provide these services, community needs are not always sufficiently met, resulting in poor and, sometimes, unhealthy conditions on-reserve.

CONCLUSIONS

First Nations people rely on the environment for physical and spiritual health. Any climate change impacts to the lands of First Nations people will have implications for their communities. The changes in vegetation and river levels in Saskatchewan anticipated under climate change will have direct and indirect effects on local First Nations people. Cultural relations with the forest may be stressed, as local forests are lost. Fluctuations in the Saskatchewan River will affect hunting, and so too will resource competition as a result of climate change. Capacities to cope with the implications of climate change in two key areas—social and human capital—in some cases, have been diminished (James Smith) in recent years due to the compounding effects of social, political and cultural change, but, in other cases, are functioning to increase the ability of communities (Shoal Lake) to cope with change. Limited economic resources, however, will significantly reduce adaptive capacities to cope with future change. Although already occurring, further efforts should be made to attune external institutions to the needs of communities and address issues currently reducing adaptive capacity. Institutional support to cope with the multifaceted exposures existing on-reserve would significantly increase community level adaptive capacities. In conclusion, adaptation to climate change in First Nations communities is constrained by the wide array of social, cultural, institutional, economic and political challenges facing First Nations people.

REFERENCES

Adger, N. 2001. *Social Capital and Climate Change.* Tyndall Center Working Paper No. 8: Norwich, England.

Bryant, C.R., B. Smit, M. Brklacich, T.R. Johnston, J. Smithers, Q. Chiotti and B. Singh. 2000. "Adaptation in Canadian Agriculture to Climatic Variability and Change." *Climatic Change* 45: 181–201.

Downing, T.E. 2003. "Lessons from Famine Early Warning and Food Security for Understanding Adaptation to Climate Change: Toward a Vulnerability/Adaptation Science?" pp. 71–100 in J.B. Smith, R.J.T Klein and S. Huq (eds.). *Climate Change, Adaptive Capacity and Development.* Imperial College Press: London.

Downing, T.E. and A. Patwardhan. 2003. "Vulnerability Assessment for Climate Adaptation, Adaptation Policy Framework: A Guide for Policies to Facilitate Adaptation to Climate Change." United Nations Developmental Program.

Dyck, N. 1991. *What is the Indian "Problem."* The Institute of Social and Economic Research: St. John's, Newfoundland.

Ermine, W., D. Sauchyn and J. Pittman. 2008. *Report Nikan Oti: The Future—Understanding Adaptation and Capacity in Two First Nations.* Climate Change Impacts and Adaptation Program: Ottawa, ON.

Ford, J.D. and B. Smit. 2004. "A Framework for Accessing the Vulnerability of Communities in the Canadian Arctic to Risks Associated with Climate Change." *Arctic* 57, no. 4: 389–400.

Füssel, H.M. and R.J.T. Klein. 2002. "Assessing Vulnerability and Adaptation to Climate Change: An Evolution of Conceptual Thinking." Paper presented at the UNDP Expert Group Meeting in Integrating Disaster Reduction and Adaptation to Climate Change, June 17–19, 2002: Havana, Cuba.

Handmer, J.M., S. Dovers and T.E Downing. 1999. "Societal Vulnerability to Climate Change and Variability." *Mitigation and Adaptive Strategies for Global Change* 4: 267–81.

Johnston, M. 2008. *Impacts of Climate Change on the Island Forests of Saskatchewan.* Saskatchewan Research Council, Pub. No. 12168–1E08.

Johnston, M., E. Wheaton, S. Kulshreshtha, V. Wittrock and J. Thorpe. 2001. *Forest Ecosystem Vulnerability to Climate: An Assessment of the Western Canadian Boreal Forest.* Saskatchewan Research Council, SRC Publication No. 11341–8E01.

Kasperson, R.E. and J.X. Kasperson. 2001. "Climate Change, Vulnerability and Social Justice." Stockholm Environment Institute: Risk and Vulnerability Programme.

Kelly, P.M. and W.N. Adger. 2000. "Theory and Practice in Assessing Vulnerability to Climate and Facilitating Adaptation." *Climatic Change* 47: 325–52.

Kulshreshtha, S., M. Johnston and E. Wheaton. 2001. "Forest Ecosystem Vulnerability to Climate Change: Conceptual Framework for Economic Analysis." Prepared for the Government of Canada's Climate Change Action Fund, Presentation to Social and Economic Perspectives of Boreal Forest Ecosystem Management Conference. Heriot-Watt University, June 5 to 8, 2001, Edinburgh, Scotland. SRC Publication No. 11341–4D01.University of Saskatchewan: Saskatoon, SK.

McLeod, N. 2007. *Cree Narrative Memory: From Treaties to Contemporary Times.* Purich Publishing, Ltd: Saskatoon.

O'Brien, K., S. Eriksen, A. Schjolden and L. Nygaard. 2004. "What's in a Word? Conflicting Interpretations of Vulnerability in Climate Change Research." Center for International Climate and Environmental Research (CICERO): Oslo, Working Paper.

Pietroniro, A., M. Demuth, P. Dornes, J. Toyra, N. Kouwen, A. Bingeman, C. Hopkins, D. Burn and B. Brua. 2006. "Stream Flow Shifts Resulting from Past and Future Glacier Fluctuations in the Eastern Flowing Basins of the Rocky Mountains." Climate Change Resources Users Group.

Pittman, J. 2009a. "Coming to Know: My Experience with Traditional Knowledge as a Wêmistikôsiw in the Land of the Nêhiyawak." University of Guelph, Invited Talk, March 20, 2009: Guelph, Ontario.

——. 2009b. "The Vulnerability of the James Smith and Shoal Lake First Nations to Climate Change and Variability." Association of American Geographers (AAG) Annual Conference, March 22–28, 2009: Las Vegas, Nevada.

——. 2009c. "The Vulnerability of the James Smith and Shoal Lake First Nations to Climate Change and Variability" [online]. M.A thesis, University of Regina. Available from World Wide Web: *http://www.parc.ca/mcri/pdfs/theses/pittman_thesis.pdf*

Pittman, J., E. Cook, P. Diaz and D. Sauchyn. 2009. "Coping with Change—Family Camp at the Shoal Lake Cree Nation." Pp. 207–14 in J. Oakes, R.Riewe and A. Cogswell (eds.). *Sacred Landscapes*. Aboriginal Issues Press, University of Manitoba: Winnipeg, Manitoba.

Saporta, R., J.R. Malcolm and D.L. Martell. 1998. "The Impact of Climate Change on Canadian Forests." Pp. 319–82 in: G. Koshida, and W. Avis (eds.). *Responding to Global Climate Change: National Sectoral Issue*. Environment Canada, Canada Country Study: Climate Impacts and Adaptation vol. VII.

Smit, B. and O. Pilifosova. 2001. "Adaptation to Climate Change in the Context of Sustainable Development and Equity." Pp. 876–912 in J. McCarthy, O.F. Canziana, N.A Leary, D.J. Dokken and K.S. White (eds.). *Climate Change 2001: Impacts, Adaptation, Vulnerability,* Contribution of Working Group II to the Third Assessment Report of the Intergovernmental Panel on Climate Change, Cambridge University Press: Cambridge, UK.

——. 2003. "From Adaptation to Adaptive Capacity and Vulnerability Reduction." Pp. 9–28 in J.B. Smith, R.J.T Klein and S. Huq (eds.) *Climate Change, Adaptive Capacity and Development.* Imperial College Press: London.

Smit, B. and J. Wandel, 2006, "Adaptation, Adaptive Capacity and Vulnerability." *Global Environmental Change* 16: 282–92.

Statistics Canada. 2008a. *James Smith, Saskatchewan (table), Aboriginal Population Profile, 2006 Census* [online]. Statistics Canada Catalogue no. 92–594-XWE: Ottawa. Released January 15, 2008. [accessed November 18, 2008]. Available at: *http://www12.statcan.ca/english/census06/data/profiles/aboriginal/Index.cfm?Lang=E*

——. 2008b. *Shoal Lake Cree Nation, Saskatchewan (table), Aboriginal Population Profile, 2006 Census* [online]. Statistics Canada Catalogue no. 92–594-XWE: Ottawa. Released January 15, 2008. [accessed November 18, 2008]. Available at: *http://www12.statcan.ca/english/census06/data/profiles/aboriginal/Index.cfm?Lang=E*

——. 2008c. *Outlook, Saskatchewan (table), 2006 Community Profiles, 2006 Census* [online]. Statistics Canada Catalogue no. 92–591-XWE: Ottawa. Released March 13, 2007. [accessed November 18, 2008]. Available at: *http://www12.statcan.ca/english/census06 /data/profiles/community/Index.cfm?Lang=E*

Turner, B., R. E. Kasperson, P. A. Matson, J. McCarthy, R. Corell, L. Christensen, N. Eckley, X. J. Kasperson, A. Luers, M. L. Martello, C. Polsky, A. Pulsipher and A. Schiller. 2003. "A Framework for Vulnerability Analysis in Sustainability Science." *Proceedings of the National Academy of Sciences* 100, no. 14: 8074–79.

Wheaton, E. 1997. "Forest Ecosystems and Climate." Appendix B in R. Herrington, B. Johnson and F. Hunter (eds.). *Responding to Global Climate Change in the Prairies: in Volume III of the Canada Country Study: Climate Impacts and Adaptation.* Environment Canada: Ottawa, Ontario.

Wheaton, E.E. and D.C. MacIver. 1999. "A Framework and Key Questions for Adapting to Climate Variability and Change." *Mitigation and Adaptation Strategies for Global Change* 4: 215–25.

Yohe, G. and R.S.J. Tol. 2002. "Indicators for Social and Economic Coping Capacity—Moving Toward a Working Definition of Adaptive Capacity." *Global Environmental Change* 12: 25–40.

CHAPTER 22

TOURISM AND RECREATION

Mark Johnston

Climate scenarios for the Prairie Provinces suggest the future will bring warmer winters with greater precipitation and earlier springs (Flato et al., 2000; McDonald et al., 2004; Barnett, Adam and Lettenmaier, 2005). While winter precipitation will likely be higher, more of it will fall as rain rather than snow due to higher temperatures. Summers may be somewhat warmer, but will be dryer due to increased evapotranspiration (Laprise et al., 2003; Wang 2005). Southern portions of the region will likely experience increased frequency and duration of droughts (Sauchyn, Stroich and Beriault, 2003). All of these changes will affect land- and water-based recreation and tourism opportunities. This case study includes impacts to individual tourism activities, and impacts of climate change on visitation to national parks in the prairie region.

A study of the potential impacts of climate change on visitation to national parks in the southern boreal forest (e.g. Prince Albert National Park, SK) suggests that visitation would increase by 6–10% in the 2020s, 10–36% in the 2050s, and 14–60% in the 2080s, based on a relationship between temperature and visitor days (Jones and Scott, 2006). The primary impact of climate change was to increase the length of the shoulder seasons (i.e. spring and autumn). Climatic conditions along the southern boundary of the boreal forest will cause a shift in vegetation to more drought-resistant species, especially grasses (Thorpe et al., 2001; Hogg and Bernier, 2005). Loss of stands of trees at some sites, and other vegetation changes, are unavoidable.

This shift may affect visitor experience. Increased rates of forest fires can affect immediate use of wildlands, but also change species composition in the long term. More fire in the boreal forest will probably result in a shift to greater dominance by early successional hardwoods (aspen, birch; Johnston, 1996), and these species may be less desirable than conifer forests for recreation. Insect activity is also expected to be higher (Volney and Hirsch, 2005), and may also negatively affect visitor experiences, especially in park settings.

Changes in vegetation will impact wildlife habitat and change species distributions (Gitay et al., 2002). On the one hand, species of interest may no longer inhabit protected areas, where they have been traditionally viewed or hunted. On the other hand, an increase in forest fire activity under future climate conditions (Flannigan et al., 2005) could provide increased habitat for species such as deer and moose that are dependent on early- to mid-successional forests. Wildlife species important for viewing and hunting will adjust rapidly to changing environmental conditions. However, a major impact on hunting could be a loss in waterfowl habitat as prairie potholes dry up, resulting in a decline of as much as 22% in duck productivity (Scott, 2006). Communities dependent on these activities could experience reduced tourism revenues (Williamson, Parkins and McFarlane, 2005).

Lower lake and stream levels, particularly in mid-to-late summer (see Chapter 5), may reduce opportunities for water-based recreation: swimming, fishing, boating, canoe-tripping, and whitewater activities. Early and rapid spring snowmelt may prevent spring water-based activities due to high or dangerous water conditions. Changes in water temperatures and levels will affect fish species distributions (Xenopoulos et al., 2005).Warmer springs would result in earlier departure of ice from lakes, limit the ice-fishing season, and increase the likelihood of unsafe ice conditions.

In Alberta's mountain parks, climate change has already caused vegetation and associated wildlife species to migrate to higher elevations (Scott, Jones and Konopek, 2008), and this will accelerate under future warming. Scott and Jones (2005) and Scott, Jones and Konopek (2008) examined the potential impacts of climate change on visitation patterns in Banff and Waterton Lakes national parks, respectively, utilizing a number of climate change scenarios. They found that climate change could increase visitation to Banff by 3% in the 2020s and 4–12% in the 2050s, depending on the scenarios used. For Waterton Lakes, increases associated with changing climate were forecast to be 6–10% for the 2020s and 10–36% for the 2050s. In both cases, increases were due mainly to increased temperatures. However, Banff's ski industry may be negatively affected by less snowfall. The skiing season could decline by 50–57% in the 2020s and 66–94% in the 2050s in areas less than 1500 m in elevation, although snowmaking will help reduce these impacts (Scott and Jones, 2005). Higher altitude ski areas would be affected much less. Less snow cover and a shorter season will also affect the timing and amount of opportunities for cross-country skiing, snowshoeing and snowmobiling (Nicholls, 2006).

REFERENCES

Barnett, T.P., J.C. Adam and D.P. Lettenmaier. 2005. "Potential Impacts of a Warming Climate on Water Availability in Snow-Dominated Regions." *Nature* 483: 303–09.

Flannigan, M.D., K.A. Logan, B.D., Amiro, W.R. Skinner and B.J. Stocks. 2005. "Future Area Burned in Canada." *Climatic Change* 72: 1–16.

Flato, G.M., G. J. Boer, W.G. Lee, N.A. McFarlane, D. Ramsden, M.C. Reader and A.J. Weaver. 2000. "The Canadian Centre for Climate Modelling and Analysis Global Coupled Model and its Climate." *Climate Dynamics* 16: 451–67.

Gitay, H., A. Suárez, R.T. Watson and D. Dokken (eds.). 2002. *Climate Change and Biodiversity.* Technical Paper 4, Intergovernmental Panel on Climate Change: Bonn.

Hogg, E.H. and P.Y. Bernier. 2005. "Climate Change Impacts on Drought-Prone Forests in Western Canada." *Forestry Chronicle* 81: 675–82.

Johnston, M.H. 1996. "The Role of Disturbance in Boreal Mixedwood Forests of Ontario." Pp. 33–40 in C.R. Smith and G.W. Crook (compilers). *Advancing Boreal Mixedwood Management in Ontario: Proceedings of a Workshop.* Canadian Forest Service, Great Lakes Forestry Centre: Sault Ste. Marie, Ontario.

Jones, B. and D. Scott. 2006. "Climate Change, Seasonality and Visitation to Canada's National Parks." *Journal of Parks and Recreation Administration* 24, no. 2: 42–62.

Laprise, R., D. Caya, A. Frigon, and D. Paquin. 2003. "Current and Perturbed Climate as Simulated by the Second-Generation Canadian Regional Climate Model (CRCM-II) over Northwestern North America." *Climate Dynamics* 21: 405–21.

McDonald, K.C., J.S. Kimball, E. Njoku, R. Zimmermann and M. Zhao. 2004. *Variability in Springtime Thaw in the Terrestrial High Latitudes: Monitoring a Major Control on the Biospheric Assimilation of Atmospheric CO_2 with Spaceborne Microwave Remote Sensing* [online]. Earth Interactions 8, paper no. 20, [accessed September 17, 2009]. Available at: *http://EarthInteractions.org.*

Nicholls, S. 2006. Implications of Climate Change for Tourism and Outdoor Recreation in Europe. *Managing Leisure: An International Journal* 11, no. 3: 151–63.

Sauchyn, D.J., J. Stroich and A. Beriault. 2003. "A Paleoclimatic Context for the Drought of 1999–2001 in the Northern Great Plains." *The Geographical Journal* 169, no. 2: 158–67.

Scott, D. 2006. "Climate Change and Sustainable Tourism in the 21st Century." Pp. 175–248 in J. Cukier (ed.). *Tourism Research: Policy, Planning, and Prospects.* Department of Geography Publication Series, University of Waterloo: Waterloo.

Scott, D. and B. Jones. 2005. *Climate Change and Banff National Park: Implications for Tourism and Recreation.* Report prepared for the Town of Banff, Alberta. Department of Geography, University of Waterloo: Waterloo, ON.

——. 2007. "A Regional Comparison of the Implications of Climate Change of the Golf Industry in Canada." *The Canadian Geographer* 51, no. 2: pp. 219–32.

Scott, D., G. McBoyle and A. Minogue. 2007. "The Implications of Climate Change for the Québec Ski Industry." *Global Environmental Change* 17: 181–90.

Scott, D., B. Jones and J. Konopek. 2008. "Exploring the Impact of Climate-Induced Environmental Changes on Future Visitation to Canada's Rocky Mountain National Parks." *Tourism Review International* 12: 43–56.

Thorpe, J., S. Wolfe, J. Campbell, J. LeBlanc and R. Molder. 2001. *An Ecoregion Approach for Evaluating Land Use Management and Climate Change Adaptation Strategies on Sand Dune Areas in the Prairie Provinces*. Prairie Adaptation Research Collaborative Project: Regina.

Volney, W.J.A. and K.G. Hirsch. 2005. "Disturbing Forest Disturbances." *Forestry Chronicle* 81: 662–68.

Wang, G. 2005. "Agricultural Drought in a Future Climate: Results From 15 Global Climate Models Participating in the IPCC 4th Assessment." *Climate Dynamics* 25: 739–53.

Williamson, T.B., J.R. Parkins and B.L. McFarlane. 2005. "Perceptions of Climate Change Risk to Forest Ecosystems and Forest-Based Communities." *Forestry Chronicle* 81: 710–16.

Xenopoulos, M.A., D.M. Lodge, J. Alcamo, M. Marker, K. Schulze and D.P. Van Vuuren. 2005. "Scenarios of Freshwater Fish Extinctions from Climate Change and Water Withdrawal." *Global Change Biology* 11: 1557–64

CHAPTER 23

THE SENSITIVITY OF PRAIRIE SOIL LANDSCAPES TO CLIMATE CHANGE

Dave Sauchyn

INTRODUCTION

The Prairie Provinces are dominated physiographically by the Interior Plains. In the southern part of this vast geologic province, weathered sediments and glacial deposits and a grassland bioclimate support deep fertile topsoil and more than 80% of Canada's farmland. Thus, soil is a major element of the natural capital of the Prairie Provinces, and historically the basis for the regional agricultural economy. The use of about 90% of the Prairie Ecozone for farming and ranching has exposed tens of millions of hectares of subhumid soil landscapes to higher than natural rates of erosion and soil loss. Erosion and slope failure occur on managed (farmed) and unmanaged (natural) soil landscapes, where the poorly consolidated sediments of the Interior Plains are exposed to the forces of wind, water and gravity, and where farming or aridity limit the vegetation cover. In western Alberta, the high relief and steep slopes of the Rocky Mountains create a dynamic and hazardous environment, where episodic movement of soil and rock is driven largely by hydroclimatic processes.

UNMANAGED SOIL LANDSCAPES

Among the earth's most active landscapes are mountains and drylands, like the Rocky Mountains of Alberta and the valleys and dune fields of the Prairie Ecozone (Lemmen et al., 1998). Catastrophic and hazardous geophysical processes associated with extreme climate events are common on long, steep mountain slopes. For example, in August 1999, a debris flow at 5-Mile Creek in Banff

National Park blocked the Trans-Canada Highway for several days at the peak of the tourist season (Evans, 2002). Such events are nearly always triggered by excess rainfall or by runoff from the rapid melting of snow or ice. With global climate change, an increased frequency of landslides, debris flows, rock avalanches and outburst floods is probable, given current and projected future trends in hydrology and climate that include increased rainfall, especially in winter, and rapid snow melt and shrinking glaciers (Evans and Clague, 1994; 1997). These more frequent events will affect public safety and the maintenance of infrastructure, especially with increasing recreational activity and residential development in the Rocky Mountains. In the longer term, further warming, drought and the complete wastage of glaciers could cause catastrophic events to taper off, although the decay of permafrost could accelerate slope failures at high elevations for many decades (Evans and Clague, 1997).

Out on the plains, the vegetation cover in dry years can be insufficient in places to prevent runoff from rainfall and snowmelt and protect the soil surface from wind and water erosion. Because water is both an agent of slope failure and erosion, and largely determines the extent of vegetation cover, prairie soil landscapes are sensitive to fluctuations in the soil and surface water balance (Lemmen and Vance, 1999). Prolonged dry and wet spells have a strong influence on the resistance of soil and vegetation to extreme weather events (high winds, intense rain, rapid snow melt). The link between aridity and erosion is well established from paleoenvironmental records (e.g., Wolfe et al., 2001) and from the monitoring of erosional processes and regional sediment yields (Knox, 1984). Most scenarios of the future climate of the Western interior (Chapter 4) include larger departures from average conditions and more frequent extreme weather events. Under this scenario, slopes and stream channels would be exposed to more episodes of intense rainfall and excess soil water and also to severe dry spells during which the protective vegetation cover would be stressed and depleted (Ashmore and Church, 2001; Sauchyn, 1998).

Soil moisture also plays a critical role in the stabilization of sand dunes. There are extensive areas of mostly stable sand dunes throughout the Interior Plains. In the Great Sand Hills of southwestern Saskatchewan, a rhythm of reactivation in the last 50 years was matched by the pattern of droughts (Wolfe, Ollerhead and Lian, 2002). Widespread reactivation of sand dunes about 200 years ago is correlated with tree-ring records of prolonged droughts of the mid-to-late 18th century (Wolfe et al., 2001). Dune stabilization has occurred since 1890. The dunes fields became more stable throughout the 20th century despite the droughts (e.g. 1930s, 1980s) that impacted agricultural land in the region (Vance and Wolfe, 1996). This trend could change, however. The drought and increased aridity forecast by GCMs will most likely result in more widespread wind erosion and sand dune activity (Wolfe and Nickling, 1997).

Natural disturbances like climate variation and change cannot be isolated from the effects of land use in sand dune landscapes. Current sand dune activity in the dry core of the Prairie Ecozone serves as a spatial analogue of the potential response of currently stable dune fields on the currently moister margins of the Prairie Ecozone and southern Boreal Forest (Wolfe, 1997; Wolfe and Nickling, 1997). Climate change impact assessments of future vegetation and soil moisture (Thorpe et al., 2001) suggest that vegetation will shift towards more open grassland, with increased potential for sand dune activity. Climate at the driest sites may exceed thresholds for active sand dune crests. More proactive land use management and stringent enforcement of current guidelines and regulations will be required given this increased potential for sand dune mobility under a drier climate.

AGRICULTURAL LANDSCAPES

In prairie soils, an abundance of organisms contributes to the soil fertility and structure. The soil fauna and flora would be reduced under lower summer rainfall, but probably increased with warmer and wetter spring conditions, because they expand with increasing temperature (Johnson and Wellington, 1980). This could result in more rapid turnover of organic matter and greater fertility in some soils, depending on management practices and other effects of climate change on soil—increased variability of rainfall, in particular. An increase in the frequency and intensity or shortages or excesses of rainfall would have immediate impacts on plant productivity and the more long-lasting effects of soil degradation on production (Wheaton et al., 2005). When cropland is bare and exposed to wind and runoff, centimetres of topsoil can be removed during a single episode of erosion, reversing centuries or millennia of soil formation, and rendering land less productive. With the cultivation of prairie soils and the harvesting of annual crops, soil loss has been about 2–3 orders of magnitude higher than on rangeland (Coote, 1983) and about 50–70% of the pre-settlement stored carbon has been removed (Lal, 2003). Since the 1980s, however, there has been a revolution in prairie soil and crop management protecting crop land from further degradation.

The most plausible climate future for Canadian Prairies (Chapter 3) includes a declining net surface and soil water balance in mid-to-late summer, as cumulative water loss by evapotranspiration potentially exceeds precipitation with increasing frequency. Prolonged droughts—like those that characterized the pre-settlement history of this region (Chapter 3)—are forecasted to occur with global warming (Chapter 4), and are more likely to exceed soil moisture thresholds beyond which landscapes are more vulnerable to disturbance, and potentially desertified. The semiarid to subhumid mixed grassland ecoregion of southwestern Saskatchewan and southeastern Alberta is at risk of desertification by definition: "Land degradation in arid, semi arid and dry/sub-humid areas, resulting from various factors, including climatic variations and human impact" (UNEP, 1994: 4). Desertification is an issue in this region, given: 1) trends in some socioeconomic variables put land at increased risk of degradation (Knutilla, 2003), and 2) climate scenarios of increased aridity and more severe drought. When Sauchyn, Kennedy and Stroich (2005) computed an aridity index (the ratio of precipitation to potential evapotranspiration: P/PET), for 1961–90 and for the 2050s (2040–69), using output from the Canadian GCM2 (emission scenario B2), the area of land at risk of desertification (P/PET < 0.65; Middleton and Thomas, 1992) increased by about 50%. This is not a prediction case; just one possible scenario, and there is considerable uncertainty in the projection of future growing season precipitation and especially evapotranspiration. Sauchyn, Kennedy and Stroich (2005) used an algorithm to compute PET based only on temperature data. Other factors that control PET, specifically wind and humidity, will change with global warming, and the actual ET depends on how much water is left in the soil.

ADAPTATION TO MINIMIZE CLIMATE IMPACTS ON SOIL LANDSCAPES

Soil conservation has been an integral part of the adaptation of farming practices to the dry and variable climate of the Interior Plains. A network of federal experimental farms was established during the 1890s to early 1900s, when Manitoba was the only Prairie Province, to develop dryland farming practices that prevent wind erosion and mitigate the impacts of drought. Major federal government programs to combat land degradation, including the Prairie Farm Rehabilitation Administration (PFRA), were introduced in response to the disastrous experience of the 1930s, when the drought impacts were exacerbated by an almost uniform settlement of farmland, without accounting for variation in the sensitivity of soil landscapes and the capacity of the climate and soil to produce crops.

In the 1980s–90s, soil degradation was a major policy and management issue. The Senate Standing Senate Committee on Agriculture, Fisheries and Forestry held hearings and produced the landmark document "Soils at Risk: Our Eroding Future" (Senate of Canada, 1984). Institutional adaptive responses to the soil degradation crisis included the soils component of the Agricultural Green Plan of 1990 and the National Soil Conservation Program (NSCP) of 1989. In the Prairie Provinces, a major component of the NSCP was the Permanent Cover Program (PCP; Vaisey, Weins and Wettlaufer, 1996). The initial PCP was fully subscribed within a few months, removing 168,000 ha of marginal land (Canada Land Inventory classes 4–6) from annual crop production. PCP II, a 1991 extension to the original program, converted another 354,000 ha. The PCP represents a policy adaptation that has reduced sensitivity to climate over a large area, even though this was not an objective of the program, and that this benefit has not been acknowledged in official documents. Mitigation of climate change is a stated objective of the follow-up to PCP, the Greencover Canada Program. The Environmental Farm Planning program is another institutional mechanism for promoting and implementing adaptive soil and crop management practices that reduce vulnerability to climate change.

In recent decades, prairie farmers have achieved progressively higher and more consistent cereal crop yields, while protecting more land from degradation (Sauchyn et al., 2005). With better soil, water and crop management, the production of annual crops has become less vulnerable to climate variability, although "severe and widespread erosion could still occur during extreme climatic events and especially during a period of years with back-to-back droughts" (PFRA, 2000: 32). Within the last 20 years, different cropping systems and the adoption of soil conservation practices, specifically, reduced tillage and zero-till, have begun to reverse the decline in soil productivity across up to one-third of the annually cropped land of the Prairies. Acton and Gregorich (1995) estimated that the implementation of soil conservation practices resulted in a decrease in the risk of wind erosion by 7% and water erosion by 11% between 1981 and 1991. More recent statistics (McRae, Smith and Gregorich, 2000) indicate a 32% reduction in the risk of wind erosion in the Prairie Provinces between 1981 and 1996.

CONCLUSION

Under the future climate projected for the Canadian Plains, with generally longer drier summers, soil landscapes may respond with local instability and erosion, including the following:

- Erosion and shallow slope failure caused by less frequent, but more intense rainfall;
- More widespread wind erosion and sand dune activity under conditions of increased summer aridity and more frequent severe drought; and
- Soil moisture thresholds below which landscapes are more vulnerable to disturbance and potentially desertified; the risk of desertification could expand over a larger area, as the extent of semiarid to subhumid climate expands beyond the dry mixed grassland.

These changes to the rates and extent of soil erosion would have obvious adverse impacts, especially for agriculture, but also for aquatic ecosystems, human health (dust), and even recreation. Most climate change impact assessments, however, do not account for adaptations that reduce or prevent the impacts. Despite the vast area and relatively sparse population of the rural Prairies, most of the landscape is managed. Because management practices have more immediate influences on rates of erosion than climate change (Jones, 1993), they have the potential to significantly mitigate or exacerbate the influence of climate. Preventing soil loss is beneficial, whether or not the impacts of global warming occur as forecast, and thus soil conservation is a prime example of a 'no regrets' adaptation. Land degradation is largely preventable with appropriate soil conservation policy and best management practices. However, the costs of a transition to better management practices are a significant barrier because they are generally borne by the land manager. "Very severe wind and water erosion . . . events may only happen once during the farming lifetime of an individual farmer, making it difficult to justify the expense and inconvenience of many soil conservation practices" (PFRA, 2000: 33).

REFERENCES

Acton, D.F. and L.J. Gregorich (eds.). 1995. *The Health of Our Soils: Toward Sustainable Agriculture in Canada.* Centre for Land and Biological Resources Research Branch, Agriculture and Agri-Food Canada. Publication 1906/E.

Ashmore, P. and M. Church. 2001. "The Impact of Climate Change on Rivers and River Processes in Canada." *Geological Survey of Canada Bulletin* 555: 1–48.

Coote, D.R. 1983. "The Extent of Soil Erosion in Western Canada." Pp. 34–48 in *Soil Erosion and Land Degradation,* Proceedings Second Annual Western Provincial Conference, Rationalization of Water and Soil Research and Management. Saskatchewan Institute of Pedology: Saskatoon.

Evans, S.G. 2002. *Climate Change and Geomorphological Hazards in the Canadian Cordillera.* Natural Resources Canada, Climate Change Impacts and Adaptation Program, Report A099, 14 pgs.

Evans, S.G. and J.J. Clague. 1994. "Recent Climate Change and Catastrophic Geomorphic Processes in Mountain Environments." *Geomorphology* 10, nos. 1–4: pp. 107–28.

——. 1997. "The Impact of Climate Change on Catastrophic Geomorphic Processes in the Mountains of British Columbia, Yukon and Alberta." Chapter 7, pp. 1–16 in E. Taylor and B. Taylor (eds.). *The Canada Country Study: Climate Impacts and Adaptation,* Volume I: Responding to Global Climate Change in British Columbia and Yukon. British Columbia Ministry of Environment Lands & Parks and Environment Canada: Vancouver.

Jones, D.K.C. 1993. "Global Warming and Geomorphology." *The Geographical Journal* 159, no. 2: 124–30.

Johnson, D.L. and W.G. Wellington. 1980. "Post-Embryonic Growth of the Collembolans Folsomia candida and Xenylla grisea at Three Temperatures." *Canadian Entomologist* 112: 687–95.

Knox, J.C. 1984. "Fluvial Responses to Small Scale Climate Changes." Pp. 318–42 in J.E. Costa and P.J. Fleisher (eds.). *Developments and Applications of Geomorphology.* Springer-Verlag: Berlin.

Knutilla, M. 2003. "Globalization, Economic Development and Canadian Agricultural Policy." Pp. 289–302 in H.P. Diaz, J. Jaffe and R. Stirling (eds.). *Farm Communities at the Crossroads: Challenge and Resistance.* Canadian Plains Research Center: Regina.

Lal, R. 2003. "Soil Erosion and the Global Carbon Budget." *Env. Intl.* 29: 437–50.

Lemmen, D.S. and R. E. Vance. 1999. "An Overview of the Palliser Triangle Global Change Project." Pp. 7–22 in D.S. Lemmen and R.E. Vance (eds.). *Holocene Climate and Environmental Change in the Palliser Triangle: A Geoscientific Context for Evaluating the Impacts of Climate Change on the Southern Canadian Prairies.* Geological Survey of Canada Bulletin 534: Ottawa.

Lemmen, D.S., R.E. Vance, I.A. Campbell, P.P. David, D.J. Pennock, D.J. Sauchyn and S.A.Wolfe. 1998. *Geomorphic Systems of the Palliser Triangle: Description and Response to Changing Climate*. Geological Survey of Canada Bulletin 521: Ottawa.

McRae, T., C.A.S. Smith and L.J. Gregorich (eds.). 2000. *Environmental Sustainability of Canadian Agriculture: Report of the Agri-Environmental Indicator Project. A Summary.* Agriculture and Agri-Food Canada: Ottawa.

Middleton, N. and D.S.G. Thomas. 1992. *World Atlas of Desertification.* United Nations Environment Program, Edward Arnold: London, 69 pgs.

PFRA. 2000. *Prairie Agricultural Landscapes: a Land Resource Review*. Prairie Farm Rehabilitation Administration: Regina, 179 pgs.

Sauchyn, D.J. 1998. "Mass Wasting Processes." Pp. 48–54 in D.S. Lemmen, R.E. Vance, I.A. Campbell, P.P. David, D.J. Pennock, D.J. Sauchyn and S.A.Wolfe (eds.). *Geomorphic Systems of the Palliser Triangle: Description and Response to Changing Climate*. Geological Survey of Canada, Bulletin 521: Ottawa.

Sauchyn, D., S. Kennedy and J. Stroich. 2005. "Drought, Climate Change, and the Risk of Desertification on the Canadian Plains." *Prairie Forum* 30, no. 1: 143–56.

Senate of Canada. 1984. *Soil at Risk—Canada's eroding future.* Ottawa. 129 pgs.

Thorpe, J., S. Wolfe, J. Campbell, J. LeBlanc and R. Molder. 2001. *An Ecoregion Approach for Evaluating Land Use Management and Climate Change Adaptation Strategies on Sand Dune Areas in the Prairie Provinces.* Prairie Adaptation Research Collaborative Project: Regina.

UNEP. 1994. *United Nations Convention to Combat Desertification in Those Countries Experiencing Drought and/or Desertification, Particularly in Africa.* Geneva. 71 pgs.

Vaisey, J.S., T.W. Weins and R.J. Wettlaufer. 1996. *The Permanent Cover Program—Is Twice Enough?* Soil and Water Conservation Policies: Successes and Failures, Prague, Czech Republic. September 17–20. Agriculture and Agri-Food Canada: Ottawa.

Vance, R. and S. Wolfe. 1996. "Geological indicators of water resources in semi-arid environments: southwestern interior of Canada." Pp. 251–63 in A.R. Berger and W.J. Iams (eds.). *Geoindicators: Assessing Rapid Environmental Changes in Earth Systems.* A.A. Balkema: Rotterdam.

Wheaton E., V. Wittrock, S. Kulshreshtha, G. Koshida, C. Grant, A. Chipanshi and B. Bonsal. 2005. *Lessons Learned from the Canadian Droughts Years of 2001 and 2002: Synthesis Report.* SRC Publication No. 11602–46E03. Saskatchewan Research Council: Saskatoon.

Wolfe, S. 1997. "Impact of Increased Aridity on Sand Dune Activity in the Canadian Prairies." *Journal Arid Environments* 36: 421–32.

Wolfe, S.A., D. J. Huntley, P.P. David, J. Ollerhead, D.J. Sauchyn and G. M. Macdonald. 2001. "Late 18th Century Drought-Induced Sand Dune Activity, Great Sand Hills, Southwestern Saskatchewan." *Canadian Journal of Earth Sciences* 38: 105–17.

Wolfe, S.A. and W.G. Nicklin. 1997. *Sensitivity of Eolian Processes to Climate Change in Canada.* Geological Survey of Canada, Bulletin 421, 30 pgs.

Wolfe, S., J. Ollerhead and O. Lian. 2002. Holocene Eolian Activity in South-Central Saskatchewan and the Southern Canadian Prairies." *Géographie Physique et Quaternaire* 56: 215–27.

CHAPTER 24

ISLAND FORESTS IN CENTRAL SASKATCHEWAN

Mark Johnston and Tim Williamson

The island forests represent the southernmost extreme of the boreal forest in central Saskatchewan. These isolated patches of forest occur on sandy deposits formed near the end of the last glacial period, which because of low agricultural suitability have remained forested while the surrounding lands have been cleared and farmed. The island forests are centred on Prince Albert (Figure 1). The Nisbet Forest is largely to the west and south of Prince Albert, straddling the North Saskatchewan River, but portions extend east of the city.

Fort à la Corne, the other large island forest, is mainly along the north side of the Saskatchewan River between Prince Albert and Nipawin. Two small island forests, Canwood and Torch River, are approximately 50 km west of Prince Albert and 20 km north of Nipawin, respectively. The Canwood Forest is close to the Nisbet Forest, while the Torch River Forest is close to the Fort à la Corne Forest. Most of the stands in the island forests are dominated by either jack pine (*Pinus banksiana*) or trembling aspen (*Populus tremuloides*).

The transition from forest to grassland in this region is linked to the climatic moisture balance, and the island forests are close to the threshold at which moisture becomes insufficient to support continuous forest vegetation. Hogg (1994) mapped a climate moisture index (CMI) for the Prairie Provinces, calculated as annual precipitation minus annual potential evapotranspiration. The zero value of this index coincides almost exactly with the southern boundary of the boreal forest across Alberta, Saskatchewan and Manitoba, indicating that positive

Figure 1. Location of the Island Forests in central Saskatchewan

values support forest, while negative values support grassland/aspen parkland vegetation. Maps of average CMI presented by Hogg et al. (2007) showed that the Fort à la Corne Forest is roughly at a CMI of -5 cm, and the Nisbet Forest at -10 cm (Hogg, 1994). This indicates that the island forests are climatically marginal for boreal forest, with the Nisbet Forest slightly drier than the Fort à la Corne Forest. The predominantly sandy soils in these forests allow rapid infiltration of rainwater, favouring deeper-rooted trees over shallow-rooted grasses, and allowing forest to develop in this marginal climate.

The island forests may already be showing signs of climate change impacts, and are likely to be severely affected in the future. The number of days with minimum temperatures less than -39°C has declined in the past three decades, and are expected to decline further with a warming future climate. It is this temperature threshold that limits the reproduction of mountain pine beetle and the parasitic plant dwarf mistletoe, both pests of jack pine. The age of this forest also makes it vulnerable. Nearly 60% of the forest is more than 70 years old, with an additional 24% between 50 and 70 years old. These age classes are the most susceptible to pests such as the mistletoe and mountain pine beetle. The future climate is expected to be drier than at present, making the island forests on sandy soils with poor water-holding capacity highly susceptible to droughts. This will add to the likelihood of forest mortality due to pests, as well as to declining tree growth. Modelling analysis for the island forests has indicated that future moisture availability may become similar to that currently in southern Saskatchewan (e.g. Swift Current), and that tree growth could decline by up to 30%.

For these reasons, the island forests are an excellent example of the "canary in the mine shaft," where the impacts of climate change are likely to occur earlier than in the contiguous boreal forest to the north. This area could form part of a national "early warning" network of intensively monitored sites, where the signs of climate change will emerge first. It is also important to link sites in the island forests to existing monitoring programs such as the CIPHA study (*C*limate *I*mpacts on the *P*roductivity and *H*ealth of *A*spen) being conducted by the Canadian Forest Service's Northern Forestry Centre in Edmonton. While the CIPHA study is currently focused on aspen, additional sites could be added and the network expanded to monitor forest health and productivity in other forest types such as jack pine in the island forests area. Another opportunity is to link climate change monitoring to existing provincial forest monitoring programs. These usually comprise networks of permanent plots to monitor the effects of forest management activities on forest ecosystems. Existing programs in the Prairie Provinces could provide important early information on climate change impacts in these vulnerable areas.

Several current and planned developments will affect the island forests and will interact with the effects of climate change. Demand for recreation activities from growing urban populations, exploration and likely extraction of diamonds, other mining potential, and continued forest harvesting will all have impacts on the island forest ecosystems. The southern margin of the boreal forest is important to the local forest industry, particularly to the parties involved in the First Nations Island Forests Management Inc. An integrated land management approach in which all resource development actors cooperate to minimize their footprint is essential for managing the impacts of development in this area, particularly in light of some of the ecological vulnerabilities identified above.

Dealing directly with the vulnerabilities described above may also be possible. For example, dwarf mistletoe, older forest age classes, and the potential for a MPB outbreak all add to the high fire hazard in the island forests. FireSmart, a program to reduce fire hazard developed by the Canadian Forest Service, has shown success in several provinces, including Saskatchewan (Hirsch et al., 2001). At the local stand level, the focus is on communities that occur in fire-prone forest environments. At the landscape level, FireSmart involves treating forest stands to reduce flammability and lower fire behaviour, for example, by replacing coniferous stands with deciduous stands through timber harvesting, in order to break up highly flammable contiguous stands; cleaning diseased and insect-killed trees; and targeting harvest operations at the most vulnerable older stands to increase diversity of age-classes. These activities may incur extra expense and will take several decades to have an impact, but they should be considered now before the risks increase. By reducing fire hazard at the landscape level, the risks for other impacts (insects, disease) are also reduced. Forest harvesting in the island forests could be scheduled to reduce the insect, disease and fire hazard as much as possible, while still providing forest products.

Forest management also has the potential to help deal with some of these vulnerabilities. Immediate and aggressive regeneration of harvested (and possibly burned) stands will help ensure that forest cover is maintained. Selection of seed from drought-resistant individuals could also help maintain forest cover in the future. Experimental planting and monitoring of exotic species (e.g. red pine, ponderosa pine) may help identify species that will grow better under future conditions. Recent studies by Carr, Weedon and Cloutis (2004) and Thorpe, Henderson and Vandall (2006) explore these alternatives in more detail.

In spite of these opportunities for reducing risk, the island forests may permanently lose forest cover with global warming. Regeneration failure following fire or harvest is likely on some sites. This suggests that management planning needs to include the potential for change to grasslands in some locations so that this can be accommodated with a minimum of disruption.

REFERENCES

Carr, A., P. Weedon and E. Cloutis. 2004. Climate Change Implications in Saskatchewan's Boreal Forest Fringe and Surrounding Agricultural Areas. Final Report submitted to Natural Resources Canada, Climate Change Action Fund [online]. Natural Resources Canada, Climate Change Action Fund: Ottawa. Available at: *http://adaptation.nrcan.gc.ca/projdb/pdf/125_e.pdf*, accessed November 20, 2009.

Hogg, E.H. 1994. "Climate and the Southern Limit of the Western Canadian Boreal Forest." *Canadian Journal of Forest Research* 24: 1835–45.

Hogg, E.H., J.P. Brandt, M. Michaelian, D.T. Price, M. Siltanen, and D.M. McKenney. 2007. "Applications of Climate Moisture Index for Assessing Impacts of the 2001–2002 Drought on Aspen Forests in the Prairie Provinces." Presentation given at the Drought Research Initiative Workshop, January 11–17, 2007, Winnipeg, MB.

Hirsch, K.G., V. Kafka, C. Tymstra, R. McAlpine, B. Hawkes, H. Stegehuis, S. Quintilio, S. Gauthier and K. Peck. 2001. "Fire-Smart Forest Management: A Pragmatic Approach to Sustainable Forest Management in Fire-Dominated Ecosystems." *The Forestry Chronicle* 77: 357–63.

Johnston, M., R. Godwin and J. Thorpe. 2008. *The Impacts of Climate Change on the Island Forests of Saskatchewan.* Final report submitted to the Prince Albert Model Forest by the Saskatchewan Research Council. SRC Publication No. 12168–1E08. Saskatchewan Research Council: Saskatoon.

Thorpe, J., N. Henderson and J. Vandall. 2006. *Ecological and Policy Implications of Introducing Exotic Trees for Adaptation to Climate Change in the Western Boreal Forest.* Final Report submitted to the Prairie Adaptation Research Collaborative. Saskatchewan Research Council Publication No. 11776–1E06. Saskatchewan Research Council: Saskatoon.

4. CONCLUSIONS

CHAPTER 25

CONCLUSION

David Sauchyn, Harry P. Diaz and Suren Kulshreshtha

The risks and opportunities presented by climate change in the Prairie Provinces are related to the dry and variable climate; projected temperature increases that are greater than elsewhere in southern Canada; sensitivity of the water resources, ecosystems and resource economies to seasonal and inter-annual variations in climate; and to large departures (e.g. drought) from normal conditions. Recent rapid economic growth, especially in Alberta, a population shift from rural to urban, and the extent of agricultural and irrigated land (most of Canada's) are also important local factors. The chapters and case studies in this book have spanned the broad of scope of our scientific understanding of regional climate change impacts and adaptation options. From this wealth of information and expert interpretation, this concluding chapter identifies common themes and issues that have emerged.

PROJECTED CLIMATE CHANGES ARE OUTSIDE THE RANGE OF RECENT EXPERIENCE WITH NATURAL VARIABILITY

The Prairie Provinces represent the major region of Canada with the driest and most variable climate. Historical climate records show significant recent warming consistent with the forecasts from global climate models (GCMs). Recent trends in annual and seasonal temperature strongly suggest that Saskatchewan is not getting hotter, but rather "less cold." There has been a greater increase in daily minimum (as opposed to maximum) temperatures and the largest warming has occurred during winter and early spring, resulting in a longer frost-free period and more growing degree days. The climate of the past half century, while variable, did not encompass the range of conditions captured by records of the past millennium and projected for the near future under global warming.

With the exception of a few scenarios for the 2020s, all models forecast climates that are outside the range of natural variability experienced and observed in the 20th century. Across a range of global climate models and greenhouse gas emission scenarios, there is a consistent increase in mean annual temperature throughout the Prairie Provinces. Most climate change scenarios also indicate an increase in annual precipitation. These are more favourable climatic conditions for most activities, and especially agriculture, depending, however, on the timing of the extra heat and water. A shift to warmer, wetter winters is almost certain. Model projections of summer precipitation are much less certain; however, the common scenario is less summer precipitation, falling in fewer and more intense storms, resulting in drier, possibly much drier, conditions in the mid-to-later stages of the longer warmer summers. While these scenarios describe the most probable climate for average years, most of the risk from climate change will be from an increase in the range of departures from average conditions, that is year-to-year variability, and the prospect of drought, and unusually wet years, with greater severity and frequency than in the past.

MOST IMPACTS ARE ADVERSE, OR AT LEAST CHALLENGING

A warmer climate will present new opportunities for revenue, cost savings, recreation, and design; however, most climate changes tend to expose vulnerabilities and challenge our capacity for adaptation to limit risk and avoid negative consequences. Most impacts are adverse because our communities and resource economies are sensitive to fluctuations in the quantity and quality of natural resources, and they are not adapted to the larger range of climate conditions projected under global warming. From the short perspective of our post-settlement history, climate and water seem rather consistent and, thus, resource management practices and policies reflect a perception of abundant water supplies and ecological resources. Future water and ecosystem management will have to abandon the assumption of a stationary environment, as climate change produces shifts in climate variability, biodiversity, disturbance regimes, and distribution of water resources and ecological services.

THE MAJOR CLIMATE CHANGE IMPACTS ARE SHIFTS IN THE DISTRIBUTION OF WATER RESOURCES AND ECOSYSTEMS

One of the most certain projections about future hydroclimate is that extra water will be available in winter and spring, while summers will generally be drier as the result of earlier spring runoff, and a longer, warmer summer season of water loss by evapotranspiration. The most noticeable and challenging consequences for the Prairie Provinces are shifts in the distribution of water resources and, in turn, ecosystems. The best and most recent scientific information suggests that rainfall will be more concentrated with larger amounts in fewer storms. As a result, we can expect some unusually wet conditions, but also long dry spells between the rainstorms. These more extreme conditions and a wider range of water levels and moisture conditions likely will determine much of the impact of climate change in Saskatchewan. Droughts, and flooding to a lesser extent, could limit opportunities provided by a warmer climate and will challenge our capacity to adapt to changing conditions. The most challenging scenario would be a prolonged drought like the 1930s, or even longer droughts such as the ones that occurred in the centuries before Saskatchewan was settled for farming. These sustained droughts have cumulative impacts that prevent the recovery of natural and social systems during intervening years with normal to above-average water supplies.

While terrestrial ecosystems and dryland farms receive their water by precipitation, irrigated farms, aquatic ecosystems, and most of our communities and population depend on water from rivers, streams and lakes. Under some climate change scenarios, with moderate degrees of climate change, annual streamflow may increase, as more water becomes available in winter and spring. This short-term increase in runoff is followed by a dramatic decline, as climate change progresses in the 21st century and winter snow cover becomes discontinuous. The other major change is the shift towards earlier runoff, leaving less surface water for mid-to-late summer in average years, and much less during the droughts that are expected with increased severity.

The projected climate changes will benefit some plants, animals and insects, and be to the detriment of others, often with economic consequences. Previously non-native plants and animals will appear on the landscape. Some native species will decline or disappear entirely. Other species already present will increase in numbers or geographic distribution, given adequate connectivity. New landscape ecosystems might evolve; for example, a drier climate in the southern Prairies could potentially support shortgrass prairie, which is currently found farther south. Change in terrestrial ecosystems will be most visible near sharp ecological gradients such as in the mountains, in island forests, and along the margins of the northern and western coniferous forests. Aquatic habitats will be stressed by the lesser amounts of surface water and the associated changes in water quality. The increased stress on aquatic ecosystems from warmer and drier conditions, and loss of wetlands, could place prairie aquatic species at risk and cause declines in migratory waterfowl populations. Changing ecosystems could make the habitats of disease carrying vectors more hospitable.

WE ARE LOSING THE ADVANTAGES OF A COLD WINTER

Much of the projected increase in temperature and precipitation will occur in winter and spring. There are several advantages of such changes, including reduced energy demand for heating and decreased mortality from extreme cold. However, there are advantages of cold winters: winter recreation, northern transportation over lake and river ice and frozen ground; fewer pests and diseases than in warmer climate; and the storage of water as ice and snow, historically the most abundant, reliable and predictable source of water supplies. Most of the water in prairie rivers, lakes, sloughs, and dugouts is from melting snow. Most of the impacts of climate change impacts on water resources will result from annual changes in the amount and timing of snow accumulation and melt, and the depth and duration of frozen ground.

INCREASED FARM AND FOREST PRODUCTIVITY DURING A LONGER WARMER GROWING SEASON WILL BE CONSTRAINED BY OTHER CLIMATE IMPACTS

With warming in winter and spring, the frost-free growing season is getting longer. More heat, higher concentrations of carbon dioxide (CO_2), and elevated water use efficiency at higher temperatures all favour diversification of prairie agriculture and crop, and grassland productivity. However, increased productivity will be limited by available soil moisture. While future crop yields will depend on many factors, climate change impact assessments tend to show increasing trends in the near-term until certain thresholds of climate change are reached. This upward trend is then followed by average decreases and interrupted by large losses from severe climatic events, such as droughts and excessive moisture, and the complex interactions of insects, diseases, and weeds. Extreme weather and climate are "wild cards." A trend of increasing frequency and severity of extreme events is fairly certain, but the detrimental effects are not considered well, or at all, in future estimates of agricultural production. As a result, predictions of major increases in crop and forage yield as the growing season get longer and warmer are probably much too simplistic and overstate the benefits of global warming for Western Canada.

Similarly, a potential increase in forest productivity likely will be limited or overwhelmed at many sites by moisture limitations and other constraints. Fires, insects and drought have major impacts on Saskatchewan's forests. Warmer drier conditions in the future will likely magnify these impacts. In particular, the southern margin of the boreal forest will become increasingly vulnerable to a range of climate change impacts and may eventually lose tree cover all together. On the positive side, there may be some locations where other conditions are not limiting, and a longer growing season and CO_2 fertilization may result in increased productivity.

THE MOST SERIOUS THREATS ARE PROJECTED WITH THE LEAST CERTAINTY

A common theme that emerged throughout the chapters of this book is the significant threat posed by the projected increase in climate variability and frequency of extreme events. Climatic extremes, and especially droughts, will limit the opportunities provided by a warmer climate and challenge our capacity to adapt to changing conditions. The most costly climate events in Canadian history have been prairie droughts. Flooding also is damaging, and so are the associated health impacts that include waterborne disease outbreaks, stress and anxiety. The historic recurrence of the social and economic impacts of drought suggest that future droughts of extreme severity or long duration will be the element of climate change and variability most likely to exceed the coping and adaptive capacities of prairie communities and industries. Water scarcity in some years will be a constraint for all sectors and communities and, ultimately, could limit the economic growth; including development related to oil sands and expanded irritation.

A distinct characteristic of prairie water levels is the wet and dry cycles at periodicities from a few years to multiple decades. In Western Canada, this variability in the hydroclimate has been linked to in sea surface temperature oscillations. The longer decadal to multi-decadal cycles present specific concerns, scientific and social. The length of the longer cycles can approximate or exceed the length of instrumental water and weather records, so that they either go undetected or are interpreted as linear trends spanning decades. What seems like conflicting trends from the observation or modelling of precipitation and water levels can rather represent a similar response of the hydrological system to climate forcing, but with different timing relative to the long wet and dry cycles. The socio-economic issue is the impacts and adaptation options for infrequent and prolonged drought; current drought mitigation strategies (e.g. water storage, insurance, relief) apply only to the relatively frequent droughts lasting one or two years. Sustained drought has cumulative impacts and prevents the recovery of natural and social systems during intervening years with normal to above-average water supplies. They also are more likely to exceed soil moisture thresholds beyond which communities and landscapes are much more vulnerable.

Unfortunately, global climate models (GCMs) simulate extreme events and the variability of hydroclimate with much less certainty than trends and variability in temperature variables. Nearly all climate assessments are based on climate change scenarios derived from GCMs. These scenarios give shifts in mean conditions between decades. The climate will actually change by fluctuating, from season to season and year to year, above and below these trends. Estimates of variability and changes in extreme values are available on a global scale, but there are few projections at the regional scale suitable for provincial vulnerability assessment. This critical gap in our knowledge of climate variability and change is problematic for evaluating impacts and developing appropriate adaptation strategies.

THE NET IMPACTS DEPEND ON DEGREES OF CLIMATE CHANGE AND ADAPTATION

The net impacts of climate change depend heavily on the degree of climate change, which will be determined in part by human actions, and the degree and effectiveness of adaptation that will limit exposure to future climate risks and provide new opportunities from more favourable conditions. In the past decade, much research has assessed the potential impacts of climate change on prairie resources and economic sectors. However, nearly all of these studies have assumed no adaptation, or made simple assumptions. This reflects a lack of understanding of adaptation processes and the difficulty of predicting changes in public policy and socio-economic factors that could favour or inhibit adaptation. There is a gap in our understanding of the extent to which existing policies might discourage or even prevent the use of adaptation options. There is also a need for determining the relative importance of adaptive responses versus other priorities for resource managers, and to develop approaches that incorporate climate change considerations into existing policy instruments.

Adaptation to climate change will involve not only the development of appropriate technologies and a more efficient use of existing resources, but also the need for new institutional arrangements in civil society, an area where social capital could be central for the development of a well-organized adaptive capacity. Planned adaptation to current and anticipated climate change has become an important consideration for sustainable economic development, requiring appropriate changes to management practices, public policy, municipal bylaws, and construction codes and design. Despite the relatively low population density of the Prairie Provinces, most of the ecosystems and water resources are managed. Increasingly, our communities, ecosystems, farms, forests, and water resources will be managed to prevent further global warming and to minimize the impacts of a changing climate.

ADAPTIVE CAPACITY IS GENERALLY HIGH, BUT UNEVENLY DISTRIBUTED

An evaluation of the conventional determinants of adaptive capacity (natural and human capital, infrastructure, technology, etc.) suggests that there is a relatively high level in the Prairie Provinces. A history of adaptation to a variable and harsh climate has built substantial adaptive capacity in the agriculture sector, which can now rely on various precedents for adapting to threats to productivity. Policies and management practices have been adjusted to address, for example, soil degradation, trade barriers, and changes in export markets and transportation subsidies. The history of prairie agriculture has been a continuous process of adaptation and drought proofing through innovation and improvements in water, soil, crop, and pasture management. More severe drought will test this accumulated adaptive capacity.

Moderate to high adaptive capacity in other sectors can be attributed to risk management strategies and adaptive management practices, although these mechanisms have generally not been tested with respect to climate change. Barriers to adaptation may include lack of financial capacity, lack of understanding of the implications of climate change among managers, and existing policies that may prevent the implementation of adaptation measures.

Adaptive capacity is uneven geographically and among segments of society by virtue of their demographic, health, regional, socio-economic, or cultural circumstances. Populations most vulnerable in the Prairie Provinces include: the elderly, children, those with underlying health problem, those with lower socio-economic status or are homeless, family farmers, and First Nations Peoples. The elderly, Aboriginal and immigrant populations are the fastest growing and also among the most vulnerable to health impacts. Economic vulnerability often precedes negative health outcomes associated with extreme weather.

The present uneven geographic distribution of people and resources, with population and wealth concentrated in Alberta, will likely be further amplified by changing climate. Economic and social stresses related to climate changes could encourage further migration from rural to urban communities and to regions with the most resources. A population shift from rural areas to large urban centres undermines the viability of rural communities, and may put addition social pressures on cities. Rural communities, especially isolated ones with limited economic diversity, are most at risk due to limited emergency response capacity and dependence on climate-sensitive economic sectors (agriculture, forestry). Rural Aboriginal communities will experience these same stresses, in addition to threats to subsistence-based livelihood resources.

Formal and informal institutions interact to either sustain or undermine capacities to deal with global challenges like climate change. Efforts to improve adaptive capacity must deal with the existing institutional factors. To the extent that governance institutions organize the relationships between the state and the civil society, they are fundamental in developing adaptive capacity. Social capital can be used to mobilize resources in order to ensure the well-being of persons, groups, and communities. It may be particularly important in dealing with the uncertainties and instabilities that climate change creates, complementing and even substituting efforts by governments. The few available studies show people with higher levels of social capital are, on average, more informed, more optimistic and more empowered when it comes to dealing with climate change and water quality issues.

ADJUSTMENT TO POLICY, MANAGEMENT PRACTICES AND DECISION-MAKING PROCESSES ARE REQUIRED

Growing demands on natural resources and ecological services, and the paradigm of sustainable economic development, have spawned policy and decision-making processes that are suitable for the planning of adaptation to climate change. Examples of relevant existing policy and management instruments include sustainable community initiatives, infrastructure renewal, environmental farm plans, watershed basin councils, and principles of adaptive forest management and integrated water resource management. Thus, there is an existing policy framework for an institutional adaptive response to climate change. This framework must be evaluated, however, in terms of how it supports adaptation or conversely fosters maladaptation by providing the wrong incentives or creating barriers to adaptation. Similarly, management practices and processes must be considered from the perspective of adaptation, to embed decision making about climate change in the planning and management process. Adaptation on the farm, in the forest, and in local communities is largely achieved by municipalities and individuals working collectively in social networks and as informal institutions (e.g. producer co-ops). Provincial governments play a key role in terms of facilitation and a policy framework that enables proactive and effective adaptation. With general depopulation of rural areas throughout the Prairies, strategies for sustainable urban growth and for sustaining rural economies need to include the evaluation of climate risks and opportunities relevant to different sectors of the population and regional economies. Rural economic development will be strongly influenced by the impacts of climate change on natural resources and especially water supplies.

Sustainable growth in agricultural productivity requires best management practices, with adaptive components to deal with climate change and other compounding effects. Appropriate integration of both adaptation and mitigation in agriculture is needed to ensure that they are coordinated and mutually supportive. Best management practices that enable coping with droughts and climate change include water well management, land management for soils at risk, cover crops, nutrient recovery from waste water, irrigation, enhancing biodiversity, grazing plans, and integrated pest management planning. Considering climate change in forest management will require providing information on impacts at a scale consistent with decision making. Consideration of new species, assisted migration of existing species and populations, and revised tenure agreements are examples of policy changes that could assist in more effective adaptation. Local autonomy and flexibility in decision making will become increasingly important in an environment in which conditions are changing rapidly, and where the past is no longer a guide to the future.

Prairie people and institutions have historically managed their natural resources to maintain relatively healthy aquatic and terrestrial ecosystems because there has been a relatively abundant supply of high quality water and ecological services. However, increasing demands for water and ecological services, and fluctuating water supplies in recent years, have stressed the need to make some major shifts in our approach to managing these renewable, but finite, resources. Uncertain water supplies could require major innovations in planning and managing how water is allocated, stored, used, and distributed. The preferred adaptation strategy for dealing with these uncertainties is integrated basin management plans with apportionment powers, enforceable land use controls, and agricultural management incentives. Under climate change, it will not be possible to maintain prairie ecosystems as they were, or as we know them now. The new climate-driven reality is that biodiversity managers need to think of themselves not as practitioners of preservation, but as "creation ecologists," since antecedent landscapes can no longer be effectively targeted. Passivity in the face of impacts may shrink our ecosystem options, particularly in prairie forests. However, active management entails some risk and expense. Whatever options we choose, the future ecosystems that result from climate change in Saskatchewan will be unprecedented. We have options, but the past is not one of them.

CONTRIBUTORS

ELAINE BARROW (PhD) is an Adjunct Professor at the University of Regina and works as a scientific consultant specializing in the construction of climate change scenarios. She provided the scenarios of climate change for Canada's National Assessment of Climate Change released by Natural Resources Canada in March 2008, and has also provided similar information for the governments of Alberta and Saskatchewan. As well as being the founding principal investigator of the Canadian Climate Impacts Scenarios Project (1999–2004), she was also a member of the IPCC's Task Group on Data and Scenario Support for Impact and Climate Analysis (TGICA) from 2004 until 2007.

DANNY BLAIR is a Professor of Geography at the University of Winnipeg. He is co-chair of Climate Change Connection, Manitoba's principal climate change outreach organization. His main research interests are climate variability and change in Canada's western interior, especially as recorded in the instrumental period and as represented in the synoptic climatology of the region. He was a contributing author of Canada's National Assessment of Climate Change released in 2008, and is a frequent presenter and participant in workshops about prairie climate change impacts and adaptation strategies.

JAMES M. BYRNE is Professor and Chair of Geography at the University of Lethbridge. He has served as the Project Leader of the Nat Christie Climate and Agriculture Research Program (1992–96); the National Theme Leader in Water Resources Management, and a founding member of the Canadian Water Network (2001–04); lead scientist and producer of the 2002 award-winning *Global Change* three-part television series, and the 2004 seven-part TV series *Water under Fire* broadcast on five Canadian television networks. He was the co-leader for the Microbial Ecology of the Oldman River Basin, Alberta Research Program. He has published more than 130 papers, reports and conference presentations, given over 40 invited keynote addresses to national and international science and applied science meetings, and has served as an expert witness on climate, water and related issues in the energy industry.

DARRELL R. CORKAL is a Senior Water Quality Engineer with Agri-Environment Services Branch (formerly known as Prairie Farm Rehabilitation Administration), Agriculture and Agri-Food Canada, Saskatoon. His main research interests are water quality and the impact of climate change on water resources, as these relate to the agricultural sector and rural populations.

DEBRA DAVIDSON is Associate Professor of Environmental Sociology, with a joint appointment between the Departments of Rural Economy and Renewable Resources at the University of Alberta. Over the past three years she has also served as the Director of the Environmental Research and Studies Centre, a non-profit, outreach organization the mandate of which is to breach the barriers between academic research in environmental sciences and civil society. Her primary areas of research and teaching include the social dimensions of global environmental change, political sociology of the environment and natural resources, governance, and environmental risk. She is also the co-editor of *Consuming Sustainability: Critical Social Analyses of Ecological Change* (Fernwood Publishing, 2005).

HARRY P. DIAZ is Professor of Sociology and Social Studies and Director of the Canadian Plains Research Center (CPRC) at the University of Regina. His fields of research include adaptation and vulnerability to climate change, water scarcities, and environmental governance in Canada and Latin America. He is currently leading a major research project on institutional adaptations to climate change focusing on dryland communities in the prairies of Canada and in Chile and participating in a similar international comparative study in Argentina, Bolivia, and Chile.

BRETT DOLTER is a writer, researcher and environmental educator. His main research interests are the motivations that lead people to take environmental action and how to encourage environmental action. Brett has worked in government, the non-profit sector, and as an instructor at the University of Regina. He also writes for media such as the *Sasquatch* magazine, the *Prairie Dog* magazine, and operates an environmental policy consulting firm called BD Green Solutions.

NORMAN HENDERSON is Director at the Prairie Adaptation Research Collaborative (PARC) at the University of Regina. His research interests include resource management and nature conservation. A particular focus is climate change impacts on the natural and semi-natural landscapes of the West and on the options we have to live with these impacts. He has published in *Ambio, Nature, Trends in Ecology and Evolution, Management Science, Plains Anthropologist, Great Plains Research,* and *Environmental and Resource Economics.*

TED HOGG is a Research Scientist with the climate change program with Natural Resources Canada, Canadian Forest Service in Edmonton. He has been involved in several large-scale research programs in the western Canadian boreal forest, including BOREAS, Fluxnet Canada, and the Canadian Carbon Program. His current research interest focuses on understanding the impacts of climate change and variation on the productivity, health and dieback of forests in the Prairie Provinces.

MARGOT HURLBERT is an Assistant Professor jointly appointed to the Department of Justice Studies and the Department of Sociology and Social Studies at the University of Regina. Prior to embarking on a full-time academic career, she practiced law for 19 years in a variety of diverse areas including human rights, family, agriculture, criminal and banking law, as well as corporate commercial, privacy, and legislative drafting and policy. She has authored numerous journal articles, book chapters and scholarly papers on a broad range of justice topics but more recently on the subjects of Aboriginal justice, water, and climate change adaptation. Her research interests focus on environment, climate change, water and Aboriginal peoples.

MARK JOHNSTON is Distinguished Scientist at the Saskatchewan Research Council and Adjunct Professor in the Soil Science Department at the University of Saskatchewan. His main research interests are the impacts of climate change on Canada's forests and how forest management can be adapted to climate change. He contributed a section on forests to Canada's National Assessment of Climate Change released in March 2008, and is currently helping lead a national study of the vulnerability of the Canadian forest sector to climate change under the auspices of the Canadian Council of Forest Ministers.

STEFAN KIENZLE is Associate Professor of Hydrology and GIS at the University of Lethbridge. His main research focus is watershed modeling, with an emphasis on simulating the impacts of climate change and land cover change on the hydrological cycle. He publishes widely in international journals on both basic and applied research, has co-authored chapters of *Canada's National Assessment of Climate Change* released in March 2008, and collaborates with a number of international research institutions.

JUSTINE KLAVER-KIBRIA has an MSc in Epidemiology. Her area of focus is prairie climate change and human health issues. She has several papers in the process of publication as well as having contributed to *Canada's National Assessment of Climate Change* released in March 2008, and an Alberta Economic Vulnerability Study. She is branching into other areas such as Environment-Health Consulting and she is a certified Building Biologist.

SUREN KULSHRESHTHA is a Professor at the Department of Bioresource Policy, Business and Economics at the University of Saskatchewan. His main interests include greenhouse gas emissions from agriculture, climate change impacts, drought impact assessment, irrigation economics and non-market valuation of ecosystem services.

RYAN MACDONALD is a doctoral student at the University of Lethbridge. He is particularly interested in how changes in hydrological regimes affect aquatic ecosystems, and how human populations can manage water resources to minimize their impacts on river systems. He has worked on a number of water-related research projects and has published internationally recognized work related to hydrological modeling and climate change.

JEREMY PITTMAN is a climate change adaptation specialist with the Saskatchewan Watershed Authority. His research interests include local knowledge and adaptive capacities of Indigenous and rural peoples in the context of climate change. He recently completed a Master's of Science in Geography from the University of Regina, where he was involved in many interdisciplinary research projects through the Prairie Adaptation Research Collaborative, including the Institutional Adaptations to Climate Change and Nikan Oti projects.

SUSAN PRADO is a Social Worker and Master's Candidate at the University of Regina. During her participation in the IACC project, she conducted the vulnerability assessment for the community of Taber, Alberta. Recently, she has been involved with work on literacy among immigrant families in Canada. Her professional interests include community development, social justice and research.

DAVID SAUCHYN is Professor of Geography and Senior Research Scientist at the Prairie Adaptation Research Collaborative at the University of Regina. His main research interests are the climate and hydrology of the past millennium in Canada's western interior and how knowledge of the past can inform scenarios of future climate and water supplies. He was a lead author of *Canada's National Assessment of Climate Change* released in March 2008 and has been an invited expert witness on climate change in the Canadian Senate and House of Commons, and at forums hosted by provincial premiers and environment ministers.

JEFF THORPE is a Principal Research Scientist at the Saskatchewan Research Council. His research focuses on application of ecological knowledge to sustainable management of the grasslands and forests of Saskatchewan. He has contributed to ecological land classification, rangeland inventory methods, predictive modeling of ecosites and plant communities, integrated resource management planning, conservation of rare plant and animal species, impacts of climate change on Saskatchewan's grasslands and forests, and adaptation to climate change in land management practices.

JOHANNA WANDEL is Assistant Professor in the Department of Geography and Environmental Management at the University of Waterloo. Her research interests are focused on human dimensions of environmental change, with particular emphasis on vulnerability of human communities in developed economies. She was a contributing author to the Third and Fourth Assessment Reports of the Intergovernmental Panel on Climate Change, and currently sits on the Science Steering Group of the International Study of Arctic Change.

ELAINE WHEATON is a Climate Scientist, Distinguished Scientist at the Saskatchewan Research Council (SRC) and Adjunct Professor at the University of Saskatchewan. Her main research interests are climate change, impacts, hazards, adaptations, and vulnerability. Her awards include the 2007 Nobel Peace Prize Certificate for substantial contributions to the work of the Intergovernmental Panel on Climate Change, 2007 Wolbeer Award for contributions to water resources, Distinguished Scientist appointment, and the YWCA Science and Technology award. She is widely published and is the author of the award-winning book, *But It's a Dry Cold! Weathering the Canadian Prairies.*

TIM WILLIAMSON is a senior forest economist with the Canadian Forest Service at the Northern Forestry Centre in Edmonton. His current research interests are in the areas of climate change vulnerability analysis, integrated assessment of climate change effects, and analysis of adaptive capacity.

VIRGINIA WITTROCK is a Research Scientist/Climatologist at the Saskatchewan Research Council. Her research interests have been in the areas of climate change, impacts, adaptations, hazards and vulnerability. She is on the Saskatchewan Board of Directors for the Canadian Water Resources Association and is Chair of the Saskatchewan Chapter of the Canadian Meteorological and Oceanographic Society. She won the Saskatoon YWCA Science and Technology award in 2004 and was appointed as one of Canada's "Leaders in Innovation" by the Partnership Group of Science and Engineering in 2006. She has over 100 publications in referred scientific journals, books, technical reports and conference proceedings.

INDEX

A

D

G

H

I

Q

R

U

V

W

Y

A NOTE ABOUT THE TYPE

The body of this book is set in *Arno*. Named after the river that runs through Florence, the centre of the Italian Renaissance, *Arno* draws on the warmth and readability of early humanist types of the 15th and 16th centuries. While inspired by the past, *Arno* is distinctly contemporary in both appearance and function. Designed by Robert Slimbach, Adobe principal designer, *Arno* is a meticulously crafted face in the tradition of early Venetian and Aldine book types. Embodying themes that Slimbach has explored in typefaces such as *Minion*® and *Brioso*™, *Arno* represents a distillation of his design ideals and a refinement of his craft.

The main accent font in this book is *Franchise Bold*. Designed by Derek Weathersbee, *Franchise Bold* is a contemporary display typeface meant to communicate messages quickly and with power.

Other accents are set in *Bodoni Antiqua*. In the late 1700s Italian master printer Giambattista Bodoni created the typeface that bears his name and has endured for over two centuries. *Bodoni Antiqua* was first introduced in 1930. H. Berthold then expanded the family, first in the 1970s to include condensed versions which were gaining popularity at that time, and again in the mid-1980s with the expert series. A popular, versatile family, *Bodoni Antiqua* is the corporate typeface of IBM.

ENVIRONMENTAL BENEFITS STATEMENT

Canadian Plains Research Center and CPRC Press saved the following resources by printing the pages of this book on chlorine free paper made with 100% post-consumer waste.

TREES	WATER	SOLID WASTE	GREENHOUSE GASES
5 FULLY GROWN	**2,488** GALLONS	**151** POUNDS	**517** POUNDS

Calculations based on research by Environmental Defense and the Paper Task Force. Manufactured at Friesens Corporation